Microwave
Circuit Analysis
and
Amplifier Design

Microwave Circuit Analysis and Amplifier Design

SAMUEL Y. LIAO

Professor of Electrical Engineering
California State University, Fresno

PRENTICE-HALL, INC., *Englewood Cliffs, New Jersey 07632*

Library of Congress Cataloging-in-Publication Data

Liao, Samuel Y.
   Microwave circuit analysis and amplifier design.

   Bibliography: p.
   Includes index.
   1. Microwave devices—Design and construction.
I. Title.
TK7876.L4769 1987   621.381'3   86-4942
ISBN 0-13-581786-2

The author dedicates this book
to the memory of
his brother Tsao Shiang, who
gave his life for his government,
and to his children

Editorial/production supervision: Madelaine Cooke
Cover design: 20/20 Services, Inc.
Manufacturing buyer: Gordon Osbourne

© 1987 by Prentice-Hall, Inc.
A Division of Simon & Schuster
Englewood Cliffs, New Jersey 07632

Printed in the United States of America

10 9 8 7 6 5 4 3 2

ISBN   0-13-581786-2   025

Prentice-Hall International (UK) Limited, *London*
Prentice-Hall of Australia Pty. Limited, *Sydney*
Prentice-Hall Canada Inc., *Toronto*
Prentice-Hall Hispanoamericana, S.A., *Mexico*
Prentice-Hall of India Private Limited, *New Dehli*
Prentice-Hall of Japan, Inc., *Tokyo*
Prentice-Hall of Southeast Asia Pte. Ltd., *Singapore*
Editora Prentice-Hall do Brasil, Ltda., *Rio de Janeiro*

Contents

7 LARGE-SIGNAL AND BROADBAND AMPLIFIER DESIGN   236

10 OPTICAL-FIBER WAVEGUIDES AND LIGHT MODULATOR DESIGN   *364*

Preface

This book is intended to serve as a text in a course on microwave circuit analysis and amplifier design at the senior or beginning graduate level in electrical engineering. Its primary purpose is to provide readers with design techniques for microwave amplifiers and oscillators. It is assumed that readers have had previous courses in solid-state electronics. Because the book is to a large extent self-contained, it can also be used as a reference book by electronics engineers working in the microwave area.

It has been observed for years that there is a learning gap between theoretical studies on microwave engineering in school and the practical design needed on the job in industry. This book is intended to narrow (but not to close) this gap by presenting design techniques for microwave amplifiers and oscillators, and related subjects.

This book does not include detailed discussion of microwave semiconductor devices but is, rather, a continuation of the author's *Microwave Solid-State Devices*. The text is divided into twelve chapters:

1. Chapter 1 is introductory.
2. Chapter 2 describes distributed lines, coaxial lines and impedance-matching techniques.
3. Chapter 3 studies S parameters and their application to amplifier design.
4. Chapter 4 deals with small-signal and narrowband amplifier design, including high-gain amplifier design, high-power amplifier design, low-noise amplifier design, and narrowband amplifier design.
5. Chapter 5 treats balanced amplifier design and power-combining techniques.

6. Chapter 6 discusses microwave striplines and stripline amplifier design.

7. Chapter 7 analyzes large-signal and broadband amplifier design, including high-power amplifier design, low-noise amplifier design, broadband amplifier design, and feedback techniques.

8. Chapter 8 investigates microwave waveguides and reflection amplifier design, including coaxial reflection amplifier design, reduced-height waveguide reflection amplifier design, circulator reflection amplifier design, and hybrid-coupler reflection amplifier design.

9. Chapter 9 discusses microwave oscillator circuits and oscillator design, including high-power oscillator design, broadband oscillator design, Gunn-diode oscillator design, and waveguide-cavity IMPATT oscillator design.

10. Chapter 10 describes optical-fiber waveguides and light modulator design, including step-index fibers, graded-index fibers, and Pockels-cell light modulator design.

11. Chapter 11 analyzes dielectric planar waveguides and film coating design.

12. Chapter 12 considers microwave measurements and evaluations, including the microwave amplifier test, the microwave oscillator test, measurement and microwave analysis, and the electromagnetic compatibility (EMC) test.

The arrangement of topics is flexible, and the instructor has a choice in the selection or order of the topics to suit either a one-semester or possibly a one-quarter course. Problems are included for each chapter to aid readers in further understanding the subjects discussed in the text. Instructors who have adopted the book for their courses may obtain a solutions manual from the publisher.

SAMUEL Y. LIAO

Chapter 1

Introduction

The purpose of this book is to present the analysis and design of microwave amplifier and oscillator circuits. Microwave techniques have been increasingly adopted in many electronic systems, such as space communications, radar systems, and missile electronic systems. As a result of the accelerating growth of microwave technology, research, design, and development in institutes and industries, students preparing for and electronics engineers working in the microwave field need to understand the analysis and design of microwave amplifiers and oscillators for the production of microwave electronic components and systems.

1-0 MICROWAVE FREQUENCIES

The term *microwave frequencies* traditionally refers to those frequencies from 1 to 300 GHz or wavelengths measured from 30 cm to 1 mm. However, microwaves really indicate wavelengths in the micron ranges—that is, microwave frequencies up to infrared and visible-light regions. In this book "microwave frequencies" means those from 1 to 10^6 GHz.

The microwave band designation that resulted from World War II radar security considerations was not officially recognized by any industrial, professional, or government organization until 1969. In August 1969 the U.S. Department of Defense, Office of Joint Chiefs of Staff, by a message to all services, directed the use of microwave frequency bands as listed in Table 1-0-1. On May 24, 1970, the Department of Defense adopted another band designation for microwave frequencies as shown in Table 1-0-2. In electronics industries and academic institutes, however, IEEE (Institute of Electrical and Electronics Engineers, Inc.) microwave frequency bands are commonly used, as shown in Table 1-0-3. These three band designations are given in Table 1-0-4 for comparison.

TABLE 1-0-1   OLD FREQUENCY BANDS

Designation	Frequency range (GHz)
P-band	0.225– 0.390
L-band	0.390– 1.550
S-band	1.550– 3.900
C-band	3.900– 6.200
X-band	6.200– 10.900
K-band	10.900– 36.000
Q-band	36.000– 46.000
V-band	46.000– 56.000
W-band	56.000–100.000

TABLE 1-0-2   NEW FREQUENCY BANDS

Designation	Frequency range (GHz)
A-band	0.100– 0.250
B-band	0.250– 0.500
C-band	0.500– 1.000
D-band	1.000– 2.000
E-band	2.000– 3.000
F-band	3.000– 4.000
G-band	4.000– 6.000
H-band	6.000– 8.000
I-band	8.000– 10.000
J-band	10.000– 20.000
K-band	20.000– 40.000
L-band	80.000– 60.000
M-band	60.000–100.000

1-1 MICROWAVE CIRCUITS

Prior to 1965, nearly all microwave equipment utilized waveguides, coaxial lines, or striplines. In the past two decades, the conventional technology of integrated circuits has been introduced in microwave frequencies. Today, microwave circuits can be classified into three categories according to circuit technology.

1. *Microwave discrete circuits (MDCs).* A microwave discrete circuit is made of separate elements connected together by conducting wires. The word *discrete* literally means separately distinct. Discrete circuits are still very useful in high-power microwave components and systems.

2. *Microwave monolithic integrated circuits (MMICs).* A microwave monolithic integrated circuit consists of a single-crystal chip of semiconductor on which all active and passive elements or components and their interconnections are formed. The word *monolithic* is derived from the Greek *monos,* meaning single, and *lithos,* meaning stone. Thus a monolithic integrated circuit is built on the surface

TABLE 1-0-3 IEEE FREQUENCY BANDS

Band number	Designation	Frequency		Wavelength		Applications
2	ELF (extreme low frequency)	30–300	Hz	10–1	Mm	
3	VF (voice frequency)	300–3000	Hz	1–0.1	Mm	
4	VLF (very low frequency)	3–30	kHz	100–10	km	Navigation, sonar
5	LF (low frequency)	30–300	kHz	10–1	km	Radio beacons, navigation
6	MF (medium frequency)	300–3000	kHz	1–0.1	km	AM broadcast, Coast Guard
7	HF (high frequency)	3–30	MHz	100–10	m	Telephone, telegraph
8	VHF (very high frequency)	30–300	MHz	10–1	m	TV, FM broadcast
9	UHF (ultrahigh frequency)	300–3000	MHz	100–10	cm	TV, satellite links
10	SHF (superhigh frequency)	3–30	GHz	10–1	cm	Radar, microwave links
11	EHF (extreme high frequency)	30–300	GHz	1–0.1	cm	Radar, experimental
12	Decimillimeter	300–3000	GHz	1–0.1	mm	
	P-band	0.23–1	GHz	130–30	cm	
	L-band	1–2	GHz	30–15	cm	
	S-band	2–4	GHz	15–7.5	cm	
	C-band	4–8	GHz	7.5–3.75	cm	
	X-band	8–12.5	GHz	3.75–2.4	cm	
	Ku-band	12.5–18	GHz	2.4–1.67	cm	
	K-band	18–26.5	GHz	1.67–1.13	cm	
	Ka-band	26.5–40	GHz	1.13–0.75	cm	
	Millimeter wave	40–300	GHz	7.5–1	mm	
	Submillimeter wave	300–3000	GHz	1–0.1	mm	

TABLE 1-0-4 COMPARISON OF IEEE BANDS, OLD BANDS, AND NEW BANDS

Frequency in GHz	0.1	0.15	0.2	0.3	0.4	0.5	0.6	0.75	1	1.5	2	3	4	5	6	7.5	10	15	20	30	40	50	60	75	100
Wavelength in cm	300	200	150	100	75	60	50	40	30	20	15	10	7.5	6	5	4	3	2	1.5	1	0.75	0.6	0.5	0.4	0.3

IEEE bands: VHF | UHF | L | S | C | X | Ku | K | Ka | MILLIMETER (Q V W)

Old bands: P | L | S | C | X | K | Q V | W

New bands: A | B | C | D | E | F | G | H | I | J | K | L | M

Wavelength in cm	300	200	150	100	75	60	50	40	30	20	15	10	7.5	6	5	4	3	2	1.5	1	0.75	0.6	0.5	0.4	0.3
Frequency in GHz	0.1	0.15	0.2	0.3	0.4	0.5	0.6	0.75	1	1.5	2	3	4	5	6	7.5	10	15	20	30	40	50	60	75	100

4

of a single semi-insulating substrate. Such circuits are produced by the processes of epitaxial growth, ion implantation, sputtering, evaporation, diffusion, and others. MMICs are very useful, for example, in satellite communication systems and airborne radar systems because a large number of identical circuits are required.

3. *Microwave integrated circuits (MICs).* Microwave integrated circuits are a combination of active and passive elements that are manufactured by successive diffusion processes on a semiconductor substrate in monolithic or hybrid form. However, MICs are quite different from MMICs. The MMICs contain very high packing densities, whereas the packing density of a typical MIC is quite low. An MIC whose elements are formed on a semi-insulating substrate such as GaAs is called a *film integrated circuit*. A microwave integrated circuit, which consists of a combination of two or more integrated circuit types, such as monolithic or film, together with discrete elements, is referred to as a *hybrid microwave integrated circuit*. MICs are very useful in low-power and low-packing-density microwave electronic systems such as digital circuits and military weapon systems.

1-1-1 Microwave Circuit Elements

There are two types of microwave circuit elements:

1. *Lumped-element circuits.* The term *lumped element* means true lumped—that is, no variation of L and C with frequencies, nor any variation of phase over the element. The lumped element can be much smaller than an equivalent distributed-line circuit at microwave frequencies.

2. *Distributed-line circuits.* The term *distributed* means that the line parameters R, L, G, and C are a function of the line length and that the values of L and C are varied with frequencies. For example, coaxial lines, microstrip lines, slotted lines, and coplanar waveguides are commonly used distributed circuits.

The choice of lumped or distributed elements for amplifier matching networks depends on the operating frequency. When the frequency is up to X-band, its wavelength is very short and a smaller lumped element exhibits a negligible phase shift. In monolithic microwave integrated circuits (MMICs), lumped resistors are very useful in thin-film resistive terminations for couplers, and lumped capacitors are absolutely essential for bias bypass applications. When the operating frequencies are over 20 GHz, distributed-element circuits are preferred.

1-1-2 Microwave Network Matching and Power Combining

When the source and load impedances do not match the input and output impedances of the active devices, matching networks must be designed to match the input and output ports. In general, the normal Smith chart is used for matching circuit design when the magnitude of the reflection coefficient is equal to or less than unity; the compressed Smith chart must be used for matching purposes when the magnitude of the reflection coefficient is greater than unity.

Microwave power combining and dividing techniques are sometimes very desirable in microwave electronic systems when a large amount of power cannot be supplied by a single power source or the input power is too far beyond the power capability of a single semiconductor device. In general, there are two categories of microwave power combiners and dividers: binary and nonbinary combining structures, which combine the output of N devices with multiple stages, and N-way combining structures, which combine the output of N devices in a single stage.

1-2 MICROWAVE AMPLIFIER AND OSCILLATOR DESIGN

Microwave amplifier and oscillator design is an important part of a microwave course in electrical engineering.

Microwave amplifiers.   Microwave amplifiers are being used increasingly in many microwave electronic systems, such as spaceborne communication systems and airborne radar systems. The design procedures for these microwave amplifiers vary from simple to very complicated. Table 1-2-1 tabulates the commonly used microwave amplifiers with design procedures described in the later chapters.

TABLE 1-2-1 COMMONLY USED MICROWAVE AMPLIFIERS

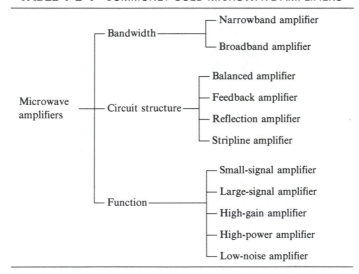

Microwave oscillators.   Theoretically, the design of a microwave oscillator is more difficult than that for a microwave amplifier. Several microwave oscillators discussed in the book are listed in Table 1-2-2.

TABLE 1-2-2 COMMONLY USED MICROWAVE OSCILLATORS

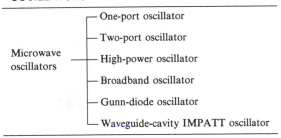

Microwave oscillators
- One-port oscillator
- Two-port oscillator
- High-power oscillator
- Broadband oscillator
- Gunn-diode oscillator
- Waveguide-cavity IMPATT oscillator

Chapter 2

Microwave Transmission Lines and Matching Techniques

2-0 INTRODUCTION

Conventional two-conductor transmission lines are commonly used to transmit microwave energy. If the line is properly matched to its characteristic impedance at each terminal, its efficiency can reach a maximum. If not properly matched, a matching technique must be utilized to reduce the standing-wave ratio for high efficiency. The purpose of this chapter is to describe microwave transmission lines and matching techniques. Several types of transmission lines are shown in Fig. 2-0-1.

Type 1: Two-wire line. This type of line consists of two parallel conducting wires separated by a uniform distance, as shown in Fig. 2-0-1(a). Examples are power lines and telephone lines. As the wires are spaced closer together, the proximity effect of one current on the other causes the current to be confined to the adjacent surfaces. The fields extend far beyond the wires and the radiation losses are excessive at higher frequencies. Hence two-wire lines are not often used at frequencies above a few hundred megahertz.

Type 2: Coaxial line. This line consists of an inner conductor and a concentric outer conducting sheath separated by a dielectric medium, as shown in Fig. 2-0-1(b). The electric and magnetic fields are confined entirely within the dielectric region. Its radiation loss is very low. Examples are telephone lines and TV cables. The coaxial lines are commonly used for precision measurements in the microwave laboratories.

Type 3: Striplines. This type of line consists of two parallel or coplanar strips separated by a dielectric slab of uniform thickness, as shown in Fig. 2-0-1(c).

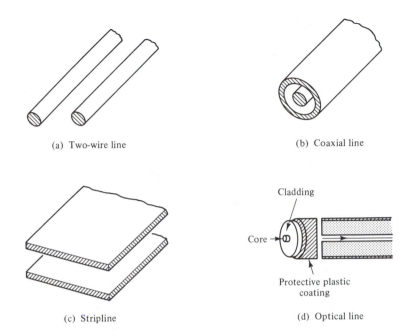

(a) Two-wire line   (b) Coaxial line

(c) Stripline   (d) Optical line

Figure 2-0-1   Common types of transmission lines.

Striplines include parallel striplines, microstrip lines, and coplanar striplines; and they are commonly used in microwave integrated circuits (MICs).

Type 4: Optical line. This line consists of an inner core of higher refractive index surrounded by an outer cladding of lower refractive index, as shown in Fig. 2-0-1(d). Its radiation loss is extremely low and its major applications are in computer links, industrial automation, medical instruments, telecommunications, and military command systems.

2-1 TRANSMISSION LINES

In ordinary circuit theory it is assumed that all impedance elements are lumped constants. This is not true for a long transmission line over a wide range of frequencies. Frequencies of operation are so high that inductances of short lengths of conductor and capacitances between short conductors and their surroundings cannot be neglected. These inductances and capacitances are distributed along the length of a conductor, and their effects combine at each point of the conductor. Since the wavelength is short in comparison to the physical length of the line, distributed parameters cannot be represented accurately by means of a lumped-parameter equivalent circuit. Hence microwave transmission lines can be analyzed in terms of voltage, current, and impedance only by the distributed-circuit theory. If the spacing between the lines is smaller than the wavelength of the transmitted signal, the transmission line needs to be analyzed as a waveguide.

2-1-1 Transmission-Line Equations and Solutions

Transmission-line equations. A transmission line may be analyzed by
either the solution of Maxwell's field equations or the methods of distributed-circuit
theory. The solution of Maxwell's equations involves three space variables in addition
to the time variable. However, the distributed-circuit method involves only one space
variable in addition to the time variable. In this section the latter method is used to
analyze a transmission line in terms of the voltage, current, impedance, and power
along the line.

In view of a uniformly distributed circuit, the schematic circuit of a conventional
two-conductor transmission line with constant parameters R, L, G, and C is shown
in Fig. 2-1-1. The parameters are expressed in their respective names per unit length,
and the propagation wave is assumed in the positive z direction.

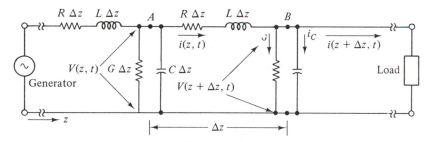

Figure 2-1-1   Elementary section of transmission line.

By Kirchhoff's voltage law, the summation of the voltage drops around the
central loop is given by

$$v(z, t) = i(z, t)R\ \Delta z + L\ \Delta z\ \frac{\partial i(z, t)}{\partial t} + v(z, t) + \frac{\partial v(z, t)}{\partial z}\ \Delta z \qquad (2\text{-}1\text{-}1)$$

By rearranging Eq. (2-1-1) and dividing it by Δz, then omitting the argument (z, t),
which is understood, the result becomes

$$-\frac{\partial v}{\partial z} = Ri + L\ \frac{\partial i}{\partial t} \qquad (2\text{-}1\text{-}2)$$

By Kirchhoff's current law, the summation of the currents at point B in Fig. 2-1-1
can be expressed as

$$i(z, t) = v(z + \Delta z, t)G\ \Delta z + C\ \Delta z\ \frac{\partial v(z + \Delta z, t)}{\partial t} + i(z + \Delta z, t)$$

$$= \left[v(z, t) + \frac{\partial v(z, t)}{\partial z}\ \Delta z \right] G\ \Delta z + C\ \Delta z\ \frac{\partial}{\partial t}\left[v(z, t) + \frac{\partial v(z, t)}{\partial z}\ \Delta z \right] \qquad (2\text{-}1\text{-}3)$$

$$+ \ i(z, t) + \frac{\partial i(z, t)}{\partial z}\ \Delta z$$

As Δz approaches zero, $\partial v / \partial t$ will also approach zero. By rearranging Eq. (2-1-3), dividing it by Δz, and omitting (z, t), the result yields the expression

$$-\frac{\partial i}{\partial z} = Gv + C\frac{\partial v}{\partial t} \qquad (2\text{-}1\text{-}4)$$

By differentiating Eq. (2-1-2) with respect to z and Eq. (2-1-4) with respect to t and combining the results, the final transmission-line equation in voltage form is found to be

$$\frac{\partial^2 v}{\partial z^2} = RGv + (RC + LG)\frac{\partial v}{\partial t} + LC\frac{\partial^2 v}{\partial t^2} \qquad (2\text{-}1\text{-}5)$$

Similarly, by differentiating Eq. (2-1-2) with respect to t and Eq. (2-1-4) with respect to z and combining the results, the final transmission-line equation in current form is found to be

$$\frac{\partial^2 i}{\partial z^2} = RGi + (RC + LG)\frac{\partial i}{\partial t} + LC\frac{\partial^2 i}{\partial t^2} \qquad (2\text{-}1\text{-}6)$$

All these transmission-line equations are applicable to the general transient solution. The voltage and current on the line are functions of both position z and time t. The instantaneous line voltage and current can be expressed as

$$v(z, t) = \text{Re } \mathbf{V}(z)e^{j\omega t} \qquad (2\text{-}1\text{-}7)$$

$$i(z, t) = \text{Re } \mathbf{I}(z)e^{j\omega t} \qquad (2\text{-}1\text{-}8)$$

where Re stands for "real part of." The factors $V(z)$ and $I(z)$ are complex quantities of the sinusoidal functions of position z on the line, and they are known as *phasors*. Phasors give the magnitudes and phases of the sinusoidal function at each position of z, and they can be expressed as

$$\mathbf{V}(z) = \mathbf{V}_+ e^{-\gamma z} + \mathbf{V}_- e^{\gamma z} \qquad (2\text{-}1\text{-}9)$$

$$\mathbf{I}(z) = \mathbf{I}_+ e^{-\gamma z} + \mathbf{I}_- e^{\gamma z} \qquad (2\text{-}1\text{-}10)$$

$$\gamma = \alpha + j\beta \text{ is the propagation constant} \qquad (2\text{-}1\text{-}11)$$

where $\mathbf{V}_+, \mathbf{I}_+$ = complex amplitudes in the positive z direction, and V_- and I_- complex amplitudes in the negative z direction

α = attenuation constant, nepers per unit length

β = phase constant, radians per unit length

By substituting $j\omega$ for $\partial / \partial t$ in Eqs. (2-1-2), (2-1-4), (2-1-5), and (2-1-6) and dividing each equation by $e^{j\omega t}$, the transmission-line equations in phasor form of the frequency domain become

$$\frac{d\mathbf{V}}{dz} = -\mathbf{Z}\mathbf{I} \qquad (2\text{-}1\text{-}12)$$

$$\frac{d\mathbf{I}}{dz} = -\mathbf{YV} \tag{2-1-13}$$

$$\frac{d^2\mathbf{V}}{dz^2} = \gamma^2\mathbf{V} \tag{2-1-14}$$

$$\frac{d^2\mathbf{I}}{dz^2} = \gamma^2\mathbf{I} \tag{2-1-15}$$

in which the following substitutions have been made:

$$\mathbf{Z} = R + j\omega L \qquad \text{ohms per unit length} \tag{2-1-16}$$

$$\mathbf{Y} = G + j\omega C \qquad \text{mhos per unit length} \tag{2-1-17}$$

$$\gamma = \sqrt{\mathbf{ZY}} = \alpha + j\beta, \qquad \text{propagation constant} \tag{2-1-18}$$

For a lossless line, $R = G = 0$, and the transmission-line equations are expressed as

$$\frac{d\mathbf{V}}{dz} = -j\omega L\mathbf{I} \tag{2-1-19}$$

$$\frac{d\mathbf{I}}{dz} = -j\omega C\mathbf{V} \tag{2-1-20}$$

$$\frac{d^2\mathbf{V}}{dz^2} = -\omega^2 LC\mathbf{V} \tag{2-1-21}$$

$$\frac{d^2\mathbf{I}}{dz^2} = -\omega^2 LC\mathbf{I} \tag{2-1-22}$$

It is interesting to note that Eqs. (2-1-14) and (2-1-15) for a transmission line are similar to equations of the electric and magnetic waves, respectively. The only difference is that the transmission-line equations are one dimensional.

Solutions of transmission-line equations.   The only possible solution for Eq. (2-1-14) is

$$\mathbf{V} = \mathbf{V}_+ e^{-\gamma z} + \mathbf{V}_- e^{\gamma z} = \mathbf{V}_+ e^{-\alpha z} e^{-j\beta z} + \mathbf{V}_- e^{\alpha z} e^{j\beta z} \tag{2-1-23}$$

The factors $\mathbf{V}_+$ and $\mathbf{V}_-$ represent complex quantities. The term involving $e^{-j\beta z}$ shows a wave traveling in the positive z direction, and the term with the factor $e^{j\beta z}$ is a wave going in the negative z direction. The quantity βz is called the *electrical length* of the line and is measured in radians.

Similarly, the only possible solution for Eq. (2-1-15) is

$$\mathbf{I} = \mathbf{Y}_0(\mathbf{V}_+ e^{-\gamma z} - \mathbf{V}_- e^{\gamma z})$$
$$= \mathbf{Y}_0(\mathbf{V}_+ e^{-\alpha z} e^{-j\beta z} - \mathbf{V}_- e^{\alpha z} e^{j\beta z}) \tag{2-1-24}$$

In Eq. (2-1-24), the characteristic impedance of the line is defined as

$$\mathbf{Z}_0 = \frac{1}{\mathbf{Y}_0} \equiv \sqrt{\frac{\mathbf{Z}}{\mathbf{Y}}} = \sqrt{\frac{R + j\omega L}{G + j\omega C}} = R_0 \pm jX_0 \qquad (2\text{-}1\text{-}25)$$

The magnitudes of the voltage and current waves on the line are shown in Fig. 2-1-2.

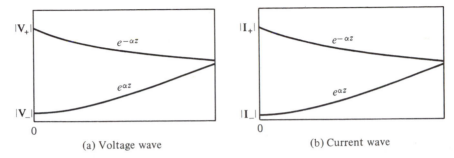

| (a) Voltage wave | (b) Current wave |

Figure 2-1-2   Magnitude of voltage and current traveling waves.

At microwave frequencies, it can be seen that

$$R \ll \omega L \qquad \text{and} \qquad G \ll \omega C \qquad (2\text{-}1\text{-}26)$$

By using the binomial expansion, the propagation constant can be expressed as

$$\gamma = \sqrt{(R + j\omega L)(G + j\omega C)}$$

$$= \sqrt{(j\omega)^2 LC} \sqrt{\left(1 + \frac{R}{j\omega L}\right)\left(1 + \frac{G}{j\omega C}\right)}$$

$$\simeq j\omega \sqrt{LC}\left[\left(1 + \frac{1}{2}\frac{R}{j\omega L}\right)\left(1 + \frac{1}{2}\frac{G}{j\omega C}\right)\right] \qquad (2\text{-}1\text{-}27)$$

$$\simeq j\omega \sqrt{LC}\left[1 + \frac{1}{2}\left(\frac{R}{j\omega L} + \frac{G}{j\omega C}\right)\right]$$

$$= \frac{1}{2}\left(R\sqrt{\frac{C}{L}} + G\sqrt{\frac{L}{C}}\right) + j\omega \sqrt{LC}$$

Therefore, the attenuation and phasor constants are, respectively, given by

$$\alpha = \frac{1}{2}\left(R\sqrt{\frac{C}{L}} + G\sqrt{\frac{L}{C}}\right) \qquad (2\text{-}1\text{-}28)$$

$$\beta = \omega \sqrt{LC} \qquad (2\text{-}1\text{-}29)$$

Similarly, the characteristic impedance is found to be

$$\mathbf{Z}_0 = \sqrt{\frac{R + j\omega L}{G + j\omega C}}$$

$$= \sqrt{\frac{L}{C}}\left(1 + \frac{R}{j\omega L}\right)^{1/2}\left(1 + \frac{G}{j\omega C}\right)^{-1/2}$$

$$\simeq \sqrt{\frac{L}{C}}\left(1 + \frac{1}{2}\frac{R}{j\omega L}\right)\left(1 - \frac{1}{2}\frac{G}{j\omega C}\right) \qquad (2\text{-}1\text{-}30)$$

$$\simeq \sqrt{\frac{L}{C}}\left[1 + \frac{1}{2}\left(\frac{R}{j\omega L} - \frac{G}{j\omega C}\right)\right]$$

$$\simeq \sqrt{\frac{L}{C}}$$

The phase velocity is found from Eq. (2-1-29) to be

$$v_p = \frac{\omega}{\beta} = \frac{1}{\sqrt{LC}} \qquad (2\text{-}1\text{-}31)$$

The product of LC is independent of the size and separation of the conductors and depends only on the permeability μ and permittivity ϵ of the insulating medium. If a lossless transmission line used for microwave frequencies has an air dielectric and contains no ferromagnetic materials, free-space parameters can be assumed. Thus the numerical value of $1/\sqrt{LC}$ for air-insulated conductors is approximately equal to the velocity of light. That is,

$$v_p = \frac{1}{\sqrt{LC}} = \frac{1}{\sqrt{\mu_0\epsilon_0}} = c = 3 \times 10^8 \text{ m/s} \qquad (2\text{-}1\text{-}32)$$

When the dielectric of a lossy microwave transmission line is not air, the phase velocity is smaller than the velocity of light and is given by

$$v_\epsilon = \frac{1}{\sqrt{\mu\epsilon}} = \frac{c}{\sqrt{\mu_r\epsilon_r}} \qquad (2\text{-}1\text{-}33)$$

In general, the relative phase-velocity factor may be defined as

$$\text{velocity factor} = \frac{\text{actual phase velocity}}{\text{velocity of light}}$$

$$v_r = \frac{v_\epsilon}{c} = \frac{1}{\sqrt{\mu_r\epsilon_r}} \qquad (2\text{-}1\text{-}34)$$

A low-loss transmission line filled with dielectric medium, such as a coaxial line with solid dielectric between conductors, has a velocity factor on the order of about 0.65.

Example 2-1-1:   Line Characteristic Impedance and Propagation Constant

A transmission line has the following parameters:

$$R = 2\ \Omega/m \qquad G = 0.5\ \text{mmho}/m \qquad f = 1\ \text{GHz}$$
$$L = 8\ \text{nH}/m \qquad C = 0.23\ \text{pF}$$

Calculate (a) the characteristic impedance, and (b) the propagation constant.

Solution   (a) From Eq. (2-1-25) the line characteristic impedance is

$$Z_0 = \sqrt{\frac{R + j\omega L}{G + j\omega C}} = \sqrt{\frac{2 + j2\pi \times 10^9 \times 8 \times 10^{-9}}{0.5 \times 10^{-3} + j2\pi \times 10^9 \times 0.23 \times 10^{-12}}}$$

$$= \sqrt{\frac{50.31\ \underline{/87.72°}}{15.29 \times 10^{-4}\ \underline{/70.91°}}} = 181.39\ \underline{/8.40°} = 179.44 + j26.50$$

(b) From Eq. (3-1-18) the propagation constant

$$\Gamma = \sqrt{(R + j\omega L)(G + j\omega C)} = \sqrt{(50.31\ \underline{/87.72°})(15.29 \times 10^{-4}\ \underline{/70.91°})}$$

$$= \sqrt{769.24 \times 10^{-4}\ \underline{/158.63°}}$$

$$= 0.2774\ \underline{/79.31°} = 0.051 + j0.273$$

2-1-2 Reflection Coefficient and Transmission Coefficient

Reflection coefficient.   In the analysis of the solutions of transmission-line equations in Section 2-1-1, the traveling wave along the line contains two components: one traveling in the positive z direction, and the other traveling in the negative z direction. However, if the load impedance is equal to the line characteristic impedance, the reflected traveling wave does not exist.

Figure 2-1-3 shows a transmission line terminated in an impedance Z_ℓ. It is usually more convenient to start solving the transmission-line problem from the receiving end rather than from the sending end, since the voltage-to-current relationship at the load point is fixed by the load impedance. The incident voltage and current waves traveling along the transmission line are given by

$$\mathbf{V} = \mathbf{V}_+ e^{-\gamma z} + \mathbf{V}_- e^{+\gamma z} \tag{2-1-35}$$

$$\mathbf{I} = \mathbf{I}_+ e^{-\gamma z} + \mathbf{I}_- e^{+\gamma z} \tag{2-1-36}$$

in which the current wave can be expressed in terms of the voltage by

$$\mathbf{I} = \frac{\mathbf{V}_+}{\mathbf{Z}_0} e^{-\gamma z} - \frac{\mathbf{V}_-}{\mathbf{Z}_0} e^{\gamma z} \tag{2-1-37}$$

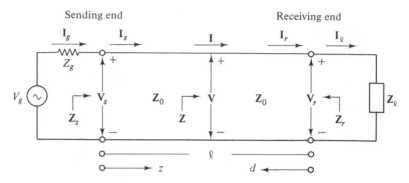

Figure 2-1-3 Transmission line terminated in a load impedance.

If the line has a length of ℓ, the voltage and current at the receiving end become

$$\mathbf{V} = \mathbf{V}_+ e^{-\gamma\ell} + \mathbf{V}_- e^{\gamma\ell} \qquad (2\text{-}1\text{-}38)$$

$$\mathbf{I} = \frac{1}{\mathbf{Z}_0}(\mathbf{V}_+ e^{-\gamma\ell} - \mathbf{V}_- e^{\gamma\ell}) \qquad (2\text{-}1\text{-}39)$$

The ratio of the voltage to the current at the receiving end is the load impedance; that is,

$$\mathbf{Z}_\ell = \frac{\mathbf{V}_\ell}{\mathbf{I}_\ell} = \mathbf{Z}_0 \frac{\mathbf{V}_+ e^{-\gamma\ell} + \mathbf{V}_- e^{\gamma\ell}}{\mathbf{V}_+ e^{-\gamma\ell} - \mathbf{V}_- e^{\gamma\ell}} \qquad (2\text{-}1\text{-}40)$$

The reflection coefficient, which is designated by Γ (gamma), is defined as

$$\text{reflection coefficient} \equiv \frac{\text{reflected voltage or current}}{\text{incident voltage or current}} \qquad (2\text{-}1\text{-}41)$$

$$\Gamma \equiv \frac{\mathbf{V}_{\text{ref}}}{\mathbf{V}_{\text{inc}}} = -\frac{\mathbf{I}_{\text{ref}}}{\mathbf{I}_{\text{inc}}}$$

If Eq. (2-1-40) is solved for the ratio of the reflected voltage at the receiving end, which is $V_- e^{\gamma\ell}$, to the incident voltage at the receiving end, which is $V_+ e^{-\gamma\ell}$, the result is the reflection coefficient at the receiving end.

$$\Gamma_\ell = \frac{\mathbf{V}_- e^{\gamma\ell}}{\mathbf{V}_+ e^{-\gamma\ell}} = \frac{\mathbf{Z}_\ell - \mathbf{Z}_0}{\mathbf{Z}_\ell + \mathbf{Z}_0} \qquad (2\text{-}1\text{-}42)$$

If the load impedance and/or the characteristic impedance are complex quantities, which is usually true, the reflection coefficient is usually a complex quantity which can be expressed as

$$\Gamma_\ell = |\Gamma_\ell| e^{j\theta_\ell} \qquad (2\text{-}1\text{-}43)$$

where

$|\Gamma_\ell|$ = magnitude and equal to or less than unity i.e., $|\Gamma_\ell| \cdot \leq 1$

θ_ℓ = phase angle between the incident and reflected voltages at the receiving end; it is usually called the phase angle of the reflection coefficient

   The general solution of the reflection coefficient at any point on the line, then, corresponds to the incident and reflected waves at that point, each attenuated in the direction of its own progress along the line. The generalized reflection coefficient is defined as

$$\Gamma \equiv \frac{V_- e^{\gamma z}}{V_+ e^{-\gamma z}} \tag{2-1-44}$$

From Fig. 2-1-3, let $z = \ell - d$; then the reflection coefficient at some point located a distance d from the receiving end is

$$\Gamma_d = \frac{V_- e^{\gamma(\ell-d)}}{V_+ e^{-\gamma(\ell-d)}} = \frac{V_- e^{\gamma \ell}}{V_+ e^{-\gamma \ell}} e^{-2\gamma d} = \Gamma_\ell e^{2\gamma d} \tag{2-1-45}$$

Then the reflection coefficient at that point can be expressed in terms of the reflection coefficient at the receiving end as

$$\Gamma_d = \Gamma_\ell e^{-2\alpha d} e^{-j2\beta d} = |\Gamma_\ell| e^{-2\alpha d} e^{j(\theta_\ell - 2\beta d)} \tag{2-1-46}$$

This is a very useful equation for determining the reflection coefficient at any point along the line. For a lossy line, both the magnitude and phase of the reflection coefficient are changing in an inward-spiral way, as shown in Fig. 2-1-4.

   For a lossless line, $\alpha = 0$, the magnitude of the reflection coefficient remains constant; only the phase of Γ is changing circularly toward the generator, with an angle of $-2\beta d$, as shown in Fig. 2-1-5. It is evident that Γ_ℓ will be zero and that there will be no reflection from the receiving end when the terminating impedance is equal to the characteristic impedance of the line. Thus a terminating inpedance that differs from the characteristic impedance will create a reflected wave traveling toward the source from the termination. The reflection wave, upon reaching the sending end, will itself be reflected if the source impedance is different from the line impedance at the sending end.

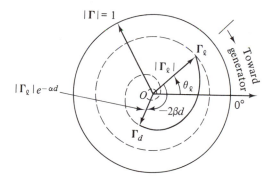

Figure 2-1-4 Reflection coefficient for lossy line.

Transmission coefficient.   A transmission line terminated in its characteristic impedance Z_0 is called a *properly terminated line* or *nonresonant line*. Otherwise, it is called an *improperly terminated line* or *resonant line*. As described in the preceding

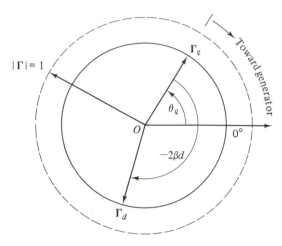

Figure 2-1-5 Reflection coefficient for lossless line.

section, there is a reflection coefficient Γ at any point along an improperly terminated line. From the principle of conservation of energy, the incident power minus the reflected power must be equal to the power transmitted to the load. This can be expressed as

$$1 - |\mathbf{\Gamma}_\ell|^2 = \frac{\mathbf{Z}_0}{\mathbf{Z}_\ell} |\mathbf{T}|^2 \tag{2-1-47}$$

Equation (2-1-47) will be verified later. The letter T represents the transmission coefficient, which is defined as

$$T \equiv \frac{\text{transmitted voltage or current}}{\text{incident voltage or current}} = \frac{\mathbf{V}_{tr}}{\mathbf{V}_{inc}} = \frac{\mathbf{I}_{tr}}{\mathbf{I}_{inc}} \tag{2-1-48}$$

Figure 2-1-6 shows the transmission of power along a transmission line. In the figure, P_{inc} is the incident power, P_{ref} is the reflected power, and P_{tr} is the transmitted power. Let the traveling waves at the receiving end be

$$\mathbf{V}_+ e^{-\gamma\ell} + \mathbf{V}_- e^{\gamma\ell} = \mathbf{V}_{tr} e^{-\gamma\ell} \tag{2-1-49}$$

$$\frac{\mathbf{V}_+}{\mathbf{Z}_0} e^{-\gamma\ell} - \frac{\mathbf{V}_-}{\mathbf{Z}_0} e^{\gamma\ell} = \frac{\mathbf{V}_{tr}}{\mathbf{Z}_\ell} e^{-\gamma\ell} \tag{2-1-50}$$

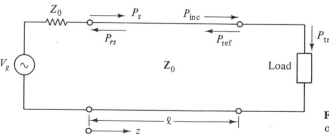

Figure 2-1-6 Power transmission on a line.

Multiplication of Eq. (2-1-50) by Z_l and substitution of the result in Eq. (2-1-49) yields

$$\Gamma_\ell = \frac{V_-e^{\gamma\ell}}{V_+e^{-\gamma\ell}} = \frac{Z_\ell - Z_0}{Z_\ell + Z_0} \tag{2-1-51}$$

which, in turn, upon substitution back into Eq. (2-1-49) results in

$$T = \frac{V_{tr}}{V_+} = \frac{2Z_\ell}{Z_\ell + Z_0} \tag{2-1-52}$$

The average power carried by the two waves in the incident and reflected sides is

$$\langle P_{inr} \rangle = \langle P_{inc} \rangle - \langle P_{ref} \rangle = \frac{|V_+e^{-\alpha\ell}|^2}{2Z_0} - \frac{|V_-e^{\alpha\ell}|^2}{2Z_0} \tag{2-1-53}$$

The average power carried to the load by the transmitted wave is

$$\langle P_{tr} \rangle = \frac{|V_{tr}e^{-\alpha\ell}|^2}{2Z_\ell} \tag{2-1-54}$$

By setting $\langle P_{inr} \rangle = \langle P_{tr} \rangle$, and with the aid of Eqs. (2-1-51) and (2-1-52), the result is

$$|T|^2 = \frac{Z_\ell}{Z_0}(1 - |\Gamma_\ell|^2) \tag{2-1-55}$$

This relation verifies the previous statement that the transmitted power is equal to the algebraic sum of the incident power and reflected power.

Example 2-1-2:   Reflection Coefficient and Transmission Coefficient

A certain transmission line has a characteristic impedance of $75 + j0.01 \ \Omega$ and is terminated in a load impedance of $70 + j50 \ \Omega$. Compute (a) the reflection coefficient, and (b) the transmission coefficient. Verify (c) that the relationship is as shown in Eq. (2-1-55), and (d) that the transmision coefficient is equal to the algebraic sum of 1 plus the reflection coefficient.

Solution   (a) From Eq. (2-1-51) the reflection coefficient is

$$\Gamma = \frac{Z_\ell - Z_0}{Z_\ell + Z_0} = \frac{70 + j50 - (75 + j0.01)}{70 + j50 + (75 + j0.01)}$$

$$= \frac{50.24 \ \underline{/95.71°}}{153.38 \ \underline{/19.03°}} = 0.33 \ \underline{/76.68°} = 0.08 + j0.32$$

(b) From Eq. (2-1-52) the transmission coefficient is

$$T = \frac{2Z_\ell}{Z_\ell + Z_0} = \frac{2(70 + j50)}{70 + j50 + (75 + j0.01)}$$

$$= \frac{172.05 \ \underline{/35.54°}}{153.38 \ \underline{/19.03°}} = 1.12 \ \underline{/16.51°} = 1.08 + j0.32$$

(c) $\mathbf{T}^2 = (1.12 \,\underline{/16.51°})^2 = 1.25 \,\underline{/33.02°}$

$$\frac{\mathbf{Z}_\ell}{\mathbf{Z}_0}(1 - \mathbf{\Gamma}^2) = \frac{70 + j50}{75 + j0.01}\left[1 - (0.33 \,\underline{/76.68°})^2\right]$$

$$= \frac{86 \,\underline{/35.54°}}{75 \,\underline{/0°}} \times 1.10\underline{/-2.6°} = 1.25 \,\underline{/33°}$$

Equation (2-1-55) is verified.

(d) The transmission coefficient is

$$\mathbf{T} = 1.08 + j0.32 = 1 + 0.08 + j0.32 = 1 + \mathbf{\Gamma}$$

2-1-3 Line Impedance and Admittance

Line impedance. The line impedance of a transmission line is the complex ratio of the voltage phasor at any point to the current phasor at that point. It is defined as

$$\mathbf{Z} \equiv \frac{\mathbf{V}(z)}{\mathbf{I}(z)} \tag{2-1-56}$$

Figure 2-1-7 shows the equivalent circuit of a transmission line.

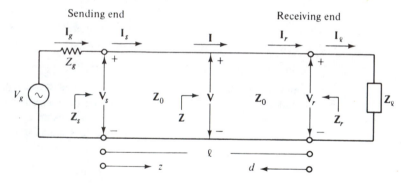

Figure 2-1-7 Diagram of a transmission line, showing notation.

In general, the voltage or current along a line is the sum of the respective incident wave and reflected wave, that is,

$$\mathbf{V} = \mathbf{V}_{\text{inc}} + \mathbf{V}_{\text{ref}} = \mathbf{V}_+e^{-\gamma z} + \mathbf{V}_-e^{\gamma z} \tag{2-1-57}$$

$$\mathbf{I} = \mathbf{I}_{\text{inc}} + \mathbf{I}_{\text{ref}} = \mathbf{Y}_0(\mathbf{V}_+e^{-\gamma z} - \mathbf{V}_-e^{\gamma z}) \tag{2-1-58}$$

At the sending end, $z = 0$; then Eqs. (2-1-57) and (2-1-58) become, respectively,

$$\mathbf{I}_s\mathbf{Z}_s = \mathbf{V}_+ + \mathbf{V}_- \tag{2-1-59}$$

$$\mathbf{I}_s\mathbf{Z}_0 = \mathbf{V}_+ - \mathbf{V}_- \tag{2-1-60}$$

Solving these two equations for V_+ and V_- yields

$$V_+ = \frac{I_s}{2} (Z_s + Z_0) \tag{2-1-61}$$

$$V_- = \frac{I_s}{2} (Z_s - Z_0) \tag{2-1-62}$$

Substitution of V_+ and V_- in Eqs. (2-1-57) and (2-1-58) gives us

$$V = \frac{I_s}{2} [(Z_s + Z_0)e^{-\gamma z} + (Z_s - Z_0)e^{\gamma z}] \tag{2-1-63}$$

$$I = \frac{I_s}{2Z_0} [(Z_s + Z_0)e^{-\gamma z} - (Z_s - Z_0)e^{\gamma z}] \tag{2-1-64}$$

Then the line impedance at any point z from the sending end in terms of Z_s and Z_0 is expressed as

$$Z = Z_0 \frac{(Z_s + Z_0)e^{-\gamma z} + (Z_s - Z_0)e^{\gamma z}}{(Z_s + Z_0)e^{-\gamma z} - (Z_s - Z_0)e^{\gamma z}} \tag{2-1-65}$$

At $z = \ell$, the line impedance at the receiving end in terms of Z_s and Z_0 is given by

$$Z_r = Z_0 \frac{(Z_s + Z_0)e^{-\gamma \ell} + (Z_s - Z_0)e^{\gamma \ell}}{(Z_s + Z_0)e^{-\gamma \ell} - (Z_s - Z_0)e^{\gamma \ell}} \tag{2-1-66}$$

Alternatively, the line impedance can be expressed in terms of Z_ℓ and Z_0. At $z = \ell$, $V_r = I_\ell Z_\ell$; then

$$I_\ell Z_\ell = V_+ e^{-\gamma \ell} + V_- e^{\gamma \ell} \tag{2-1-67}$$

$$I_\ell Z_0 = V_+ e^{-\gamma \ell} - V_- e^{\gamma \ell} \tag{2-1-68}$$

Solving these two equations for V_+ and V_-, we obtain

$$V_+ = \frac{I_\ell}{2} (Z_\ell + Z_0)e^{\gamma \ell} \tag{2-1-69}$$

$$V_- = \frac{I_\ell}{2} (Z_\ell - Z_0)e^{-\gamma \ell} \tag{2-1-70}$$

Substitution of the results above in Eqs. (2-1-57) and (2-1-58), and letting $z = l - d$, yields

$$V = \frac{I_\ell}{2} [(Z_\ell + Z_0)e^{\gamma d} + (Z_\ell - Z_0)e^{-\gamma d}] \tag{2-1-71}$$

$$I = \frac{I_\ell}{2Z_0} [(Z_\ell + Z_0)e^{\gamma d} - (Z_\ell - Z_0)e^{-\gamma d}] \tag{2-1-72}$$

Then the line impedance at any point from the receiving end of terms of Z_s and Z_0 is expressed as

$$\mathbf{Z} = \mathbf{Z}_0 \frac{(\mathbf{Z}_\ell + \mathbf{Z}_0)e^{\gamma d} + (\mathbf{Z}_\ell - \mathbf{Z}_0)e^{-\gamma d}}{(\mathbf{Z}_\ell + \mathbf{Z}_0)e^{\gamma d} - (\mathbf{Z}_\ell - \mathbf{Z}_0)e^{-\gamma d}} \tag{2-1-73}$$

The line impedance at the sending end can also be found from Eq. (2-1-73) by setting $d = l$, and it is

$$\mathbf{Z}_s = \mathbf{Z}_0 \frac{(\mathbf{Z}_\ell + \mathbf{Z}_0)e^{\gamma \ell} + (\mathbf{Z}_\ell - \mathbf{Z}_0)e^{-\gamma \ell}}{(\mathbf{Z}_\ell + \mathbf{Z}_0)e^{\gamma \ell} - (\mathbf{Z}_\ell - \mathbf{Z}_0)e^{-\gamma \ell}} \tag{2-1-74}$$

It is a tedious work to solve Eq. (2-1-55), (2-1-56), (2-1-73), or (2-1-74) for the line impedance. These equations can be simplified by replacing the exponential factors with either hyperbolic functions or circular functions. The hyperbolic functions are obtained from

$$e^{\pm \gamma z} = \cosh(\gamma z) \pm \sinh(\gamma z) \tag{2-1-75}$$

Substitution of the hyperbolic functions in Eq. (2-1-65) yields the line impedance at any point from the sending end in terms of the hyperbolic functions as

$$\mathbf{Z} = \mathbf{Z}_0 \frac{\mathbf{Z}_s \cosh(\gamma z) - \mathbf{Z}_0 \sinh(\gamma z)}{\mathbf{Z}_0 \cosh(\gamma z) - \mathbf{Z}_s \sinh(\gamma z)} = \mathbf{Z}_0 \frac{\mathbf{Z}_s - \mathbf{Z}_0 \tanh(\gamma z)}{\mathbf{Z}_0 - \mathbf{Z}_s \tanh(\gamma z)} \tag{2-1-76}$$

Similarly, substitution of the hyperbolic functions in Eq. (2-1-73) yields the line impedance from the receiving end in terms of the hyperbolic function as

$$\mathbf{Z} = \mathbf{Z}_0 \frac{\mathbf{Z}_\ell \cosh(\gamma d) + \mathbf{Z}_0 \sinh(\gamma d)}{\mathbf{Z}_0 \cosh(\gamma d) + \mathbf{Z}_\ell \sinh(\gamma d)} = \mathbf{Z}_0 \frac{\mathbf{Z}_\ell + \mathbf{Z}_0 \tanh(\gamma d)}{\mathbf{Z}_0 + \mathbf{Z}_\ell \tanh(\gamma d)} \tag{2-1-77}$$

For a lossless line, $\gamma = j\beta$, and employing the following relationships between hyperbolic and circular functions,

$$\sinh(j\beta z) = j \sin(\beta z) \tag{2-1-78}$$

$$\cosh(j\beta z) = \cos(\beta z) \tag{2-1-79}$$

the impedance of a lossless transmission line ($\mathbf{Z}_0 = R_0$) can be expressed in terms of the circular functions as

$$\mathbf{Z} = R_0 \frac{\mathbf{Z}_s \cos(\beta z) - jR_0 \sin(\beta z)}{R_0 \cos(\beta z) - j\mathbf{Z}_s \sin(\beta z)} = R_0 \frac{\mathbf{Z}_s - jR_0 \tan(\beta z)}{R_0 - j\mathbf{Z}_s \tan(\beta z)} \tag{2-1-80}$$

and

$$\mathbf{Z} = R_0 \frac{\mathbf{Z}_l \cos(\beta d) + jR_0 \sin(\beta d)}{R_0 \cos(\beta d) + j\mathbf{Z}_\ell \sin(\beta d)} = R_0 \frac{\mathbf{Z}_\ell + jR_0 \tan(\beta d)}{R_0 + j\mathbf{Z}_\ell \tan(\beta d)} \tag{2-1-81}$$

Impedance in terms of reflection coefficient.   Rearrangement of Eq. (2-1-73) gives the line impedance looking at it from the receiving end as

$$\mathbf{Z} = \mathbf{Z}_0 \frac{1 + \Gamma_\ell e^{-2\gamma d}}{1 - \Gamma_\ell e^{-2\gamma d}} \tag{2-1-82}$$

in which the following substitution is made:

$$\Gamma_\ell = \frac{\mathbf{Z}_\ell - \mathbf{Z}_0}{\mathbf{Z}_\ell + \mathbf{Z}_0} \tag{2-1-83}$$

From Eq. (2-1-46), the reflection coefficient at a distance d from the receiving end is given by

$$\Gamma = \Gamma_\ell e^{-2\gamma d} = |\Gamma_\ell| e^{-2\alpha d} e^{j(\theta_\ell - 2\beta d)} \tag{2-1-84}$$

Then the simple equation for the impedance of a line at a distance d from the load is expressed by

$$\mathbf{Z} = \mathbf{Z}_0 \frac{1 + \Gamma}{1 - \Gamma} \tag{2-1-85}$$

The reflected coefficient is usually a complex quantity and can be expressed as

$$\Gamma = |\Gamma| e^{j\phi} \tag{2-1-86}$$

where $|\Gamma| = |\Gamma_\ell| e^{-2\alpha d}$
$\phi = \theta_\ell - 2\beta d$.

The impedance variation along a lossless line ($\mathbf{Z}_0 = R_0$) can be found as follows:

$$\mathbf{Z}(d) = \mathbf{Z}_0 \frac{1 + |\Gamma| e^{j\phi}}{1 - |\Gamma| e^{j\phi}} = R_0 \frac{1 + |\Gamma|(\cos\phi + j\sin\phi)}{1 - |\Gamma|(\cos\phi + j\sin\phi)}$$
$$= R(d) + jX(d) = |\mathbf{Z}(d)| e^{j\theta_d} \tag{1-87}$$

where

$$|\mathbf{Z}(d)| = R_0 \sqrt{\frac{1 + 2|\Gamma|\cos\phi + |\Gamma|^2}{1 - 2|\Gamma|\cos\phi + |\Gamma|^2}} \tag{2-1-88}$$

$$R(d) = R_0 \frac{1 - |\Gamma|^2}{1 - 2|\Gamma|\cos\phi + |\Gamma|^2} \tag{2-1-89}$$

$$X(d) = R_0 \frac{2|\Gamma|\sin\phi}{1 - 2|\Gamma|\cos\phi + |\Gamma|^2} \tag{2-1-90}$$

$$\theta_d = \arctan\left(\frac{X}{R}\right) = \arctan\left(\frac{2|\Gamma|\sin\phi}{1 - |\Gamma|^2}\right) \tag{2-1-91}$$

Since $\phi = \theta_\ell - 2\beta d$, then $\phi = \theta_\ell - 2\pi$ if $\beta d = \pi$. However, $\cos(\theta_\ell - 2\pi) = \cos\theta_\ell$ and $\sin(\theta_\ell - 2\pi) = \sin\theta_\ell$; then

$$\mathbf{Z}(d) = \mathbf{Z}\left(d + \frac{\pi}{\beta}\right) = \mathbf{Z}\left(d + \frac{\lambda}{2}\right) \tag{2-1-92}$$

It is concluded that the impedance along a lossless line will be repeated for every interval at a half-wavelength distance.

Determination of characteristic impedance.   A common procedure for determining the characteristic impedance and propagation constant of a given transmission line is to take two measurements.

1. Measure the sending-end impedance with the receiving end short-circuited, and record the result:

$$\mathbf{Z}_{sc} = \mathbf{Z}_0 \tanh(\gamma l) \tag{2-1-93}$$

2. Measure the sending-end impedance with the receiving end open-circuited, and record the result:

$$\mathbf{Z}_{0c} = \mathbf{Z}_0 \coth(\gamma l) \tag{2-1-94}$$

Then the characteristic impedance of the measured transmission line is given by

$$\mathbf{Z}_0 = \sqrt{\mathbf{Z}_{sc}\mathbf{Z}_{0c}} \tag{2-1-95}$$

and the propagation constant of the line can be computed from

$$\gamma = \alpha + j\beta = \frac{1}{\ell}\operatorname{arctanh}\sqrt{\frac{\mathbf{Z}_{sc}}{\mathbf{Z}_{0c}}} \tag{2-1-96}$$

In addition, the computation method as described in Eqs. (2-1-18) and (2-1-25) and the matching method are also commonly used.

Normalized impedance.   The normalized impedance of a transmission line is defined as

$$\mathbf{z} = \frac{\mathbf{Z}}{\mathbf{Z}_0} = \frac{1 + \Gamma}{1 - \Gamma} = r + jx \tag{2-1-97}$$

It should be noted that lowercase letters are commonly used for normalized quantities in describing the distributed transmission-line circuits.

An examination of Eqs. (2-1-93), through (2-1-95) shows that the normalized impedance for a lossless line has the following significant features.

1. The maximum normalized impedance is

$$z_{max} = \frac{Z_{max}}{R_0} = \frac{|V_{max}|}{Z_0|I_{min}|} = \frac{1 + |\Gamma|}{1 - |\Gamma|} = \rho \qquad (2\text{-}1\text{-}98)$$

z_{max} is a positive real value and is equal to the standing-wave ratio ρ at the location of any maximum voltage on the line.

2. The minimum normalized impedance is

$$z_{min} = \frac{Z_{min}}{R_0} = \frac{|V_{min}|}{Z_0|I_{max}|} = \frac{1 - |\Gamma|}{1 + |\Gamma|} = \frac{1}{\rho} \qquad (2\text{-}1\text{-}99)$$

z_{min} is also a positive real number, but equals the reciprocal of the standing-wave ratio at the location of any minimum voltage on the line.

3. z_{max} or z_{min} is repeated for every interval of a half-wavelength distance along the line.

$$z_{max}(z) = z_{max}\left(z \pm \frac{\lambda}{2}\right) \qquad (2\text{-}1\text{-}100)$$

$$z_{min}(z) = z_{min}\left(z + \frac{\lambda}{2}\right) \qquad (2\text{-}1\text{-}101)$$

4. Since V_{max} and V_{min} are separated by a quarter-wavelength, then z_{max} is equal to the reciprocal of z_{min} for every $\lambda/4$ separation.

$$z_{max}\left(z \pm \frac{\lambda}{4}\right) = \frac{1}{z_{min}(z)} \qquad (2\text{-}1\text{-}102)$$

Admittance.   When a transmission line is branched, it is better to solve the line equations for the line voltage, current, and transmitted power in terms of admittance instead of impedance. The characteristic admittance and the generalized admittance are defined as

$$\mathbf{Y}_0 = \frac{1}{\mathbf{Z}_0} = G_0 \pm jB_0 \qquad (2\text{-}1\text{-}103)$$

$$\mathbf{Y} = \frac{1}{\mathbf{Z}} = G \pm jB \qquad (2\text{-}1\text{-}104)$$

Then normalized admittance can be expressed as

$$y = \frac{\mathbf{Y}}{\mathbf{Y}_0} = \frac{\mathbf{Z}_0}{\mathbf{Z}} = \frac{1}{z} = g \pm jb \qquad (2\text{-}1\text{-}105)$$

Example 2-1-3:   Line Impedance

A lossless line has a characteristic impedance of 50 Ω and is terminated in a load resistance of 75 Ω. the line is energized by a generator which has an output impedance of 50 Ω

and an open-circuit output voltage of 30 V (rms). The line is assumed to be 2.25 wavelengths long. Determine (a) the input impedance, (b) the magnitude of the instantaneous load voltage, and, (c) the instantaneous power delivered to the load.

Solution   (a) From Eq. (2-1-101) the line that is 2.25 wavelengths long looks like a quarter-wave line. Then

$$\beta d = \frac{2\pi}{\lambda} \frac{\lambda}{4} = \frac{\pi}{2}$$

From Eq. (2-1-81) the input impedance is

$$Z_{\text{in}} = \frac{R_0^2}{R_\ell} = \frac{(50)^5}{75} = 33.33 \ \Omega$$

(b) The reflection coefficient is

$$\Gamma_\ell = \frac{R_\ell - R_0}{R_\ell + R_0} = \frac{75 - 50}{75 + 50} = 0.20$$

Then the instantaneous voltage at the load is

$$V_\ell = V_+ e^{-\gamma \ell} \ (1 + \Gamma_\ell) = 30(1 + 0.20) = 36 \text{ V}$$

(c) The instantaneous power delivered to the load is

$$P_\ell = \frac{(36)^2}{75} = 17.28 \text{ W}$$

2-2 STANDING WAVE AND STANDING-WAVE RATIO

2-2-1 Standing Wave

The general solutions of the transmission-line equation consist of two waves traveling in opposite directions with unequal amplitude, as shown in Eqs. (2-1-23) and (2-1-24). Equation (2-1-23) can be written as

$$\mathbf{V} = \mathbf{V}_+ e^{-az} e^{-j\beta z} + \mathbf{V}_- e^{az} e^{j\beta z}$$

$$= \mathbf{V}_+ e^{-az}[\cos (\beta z) - j \sin (\beta z)] + \mathbf{V}_- e^{az}[\cos (\beta z) + j \sin (\beta z)] \quad (2\text{-}2\text{-}1)$$

$$= (\mathbf{V}_+ e^{-az} + \mathbf{V}_- e^{az}) \cos (\beta z) - j(\mathbf{V}_+ e^{-az} - \mathbf{V}_- e^{az}) \sin (\beta z)$$

With no loss in generality, it can be assumed that $\mathbf{V}_+ e^{-az}$ and $\mathbf{V}_- e^{az}$ are real; then the voltage-wave equation can be expressed as

$$\mathbf{V}_s = \mathbf{V}_0 e^{-j\phi} \quad (2\text{-}2\text{-}2)$$

This is called the *equation of the voltage standing wave*, where

$$\mathbf{V}_0 = (\mathbf{V}_+ e^{-az} + \mathbf{V}_- e^{az})^2 \cos^2 (\beta z) + (\mathbf{V}_+ e^{-az} - \mathbf{V}_- e^{-az})^2 \sin^2 (\beta z) \qquad (2\text{-}2\text{-}3)$$

which is called the *standing-wave pattern of the voltage wave* or the *amplitude of the standing wave*, and

$$\phi = \arctan \left[\frac{\mathbf{V}_+ e^{-az} - \mathbf{V}_- e^{az}}{\mathbf{V}_+ e^{-az} + \mathbf{V}_- e^{az}} \tan (\beta z) \right] \qquad (2\text{-}2\text{-}4)$$

which is called the *phase pattern of the standing wave*.

The maximum and minimum values of Eq. (2-2-3) can be found as usual by differentiating the equation with respect to βz and equating the resultant to zero. By doing so and substituting the proper values of βz in the equation:

1. The maximum amplitude is

$$\mathbf{V}_{max} = \mathbf{V}_+ e^{-az} + \mathbf{V}_- e^{az} = \mathbf{V}_+ e^{-az}(1 + |\mathbf{\Gamma}|) \qquad (2\text{-}2\text{-}5)$$

and this occurs at $\beta z = n\pi, n = 0, \pm 1, \pm 2,....$

2. The minimum amplitude is

$$\mathbf{V}_{min} = \mathbf{V}_+ e^{-az} - \mathbf{V}_- e^{az} = \mathbf{V}_+ e^{-az}(1 - |\mathbf{\Gamma}|) \qquad (2\text{-}2\text{-}6)$$

and this occurs at $\beta z = (2n - 1)\pi/2, n = 0, \pm 1, \pm 2,....$

3. The distance between any two sucessive maxima or minima is one-half wavelength since

$$\beta z = n\pi, \quad z = \frac{n\pi}{\beta} = \frac{n\pi}{2\pi/\lambda} = n\frac{\lambda}{2}, \qquad n = 0, \pm 1, \pm 2,... \qquad (2\text{-}2\text{-}7)$$

Then $z_1 = \lambda/2$.

It is evident that there are no zeros in the minimum. Similarly,

$$\mathbf{I}_{max} = \mathbf{I}_+ e^{-az} + \mathbf{I}_- e^{az} = \mathbf{I}_+ e^{-az}(1 + |\mathbf{\Gamma}|) \qquad (2\text{-}2\text{-}8)$$

$$\mathbf{I}_{min} = \mathbf{I}_+ e^{-az} - \mathbf{I}_- e^{az} = \mathbf{I}_+ e^{-az}(1 - |\mathbf{\Gamma}|) \qquad (2\text{-}2\text{-}9)$$

The standing-wave patterns of two oppositely traveling waves with unequal amplitude in lossy or lossless line are shown in Fig. 2-2-1 and 2-2-2, respectively.

Further study of Eq. (2-2-3) reveals that:

1. When $\mathbf{V}_+ \neq 0$ and $\mathbf{V}_- = 0$, the standing-wave pattern becomes

$$\mathbf{V}_0 = \mathbf{V}_+ e^{-az} \qquad (2\text{-}2\text{-}10)$$

2. When $\mathbf{V}_+ = 0$ and $\mathbf{V}_- \neq 0$, the standing-wave pattern becomes

$$\mathbf{V}_0 = \mathbf{V}_- e^{az} \qquad (2\text{-}2\text{-}11)$$

3. When the positive wave and the negative wave have equal amplitudes (i.e.,

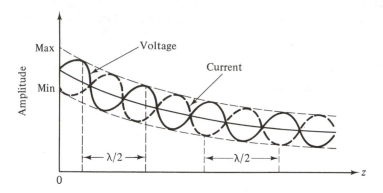

Figure 2-2-1   Standing-wave pattern in a lossy line.

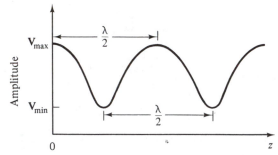

Figure 2-2-2   Voltage standing-wave pattern in a lossless line.

$|\mathbf{V}_+e^{-\alpha z}| = |\mathbf{V}_-e^{\alpha z}|$) or the magnitude of the reflection coefficient is unity, the standing-wave pattern with a zero phase is given by

$$\mathbf{V}_s = 2\mathbf{V}_+e^{-\alpha z}\cos{(\beta z)} \qquad (2\text{-}2\text{-}12)$$

which is called a *pure standing wave*. Similarly, the equation of a pure standing wave for the current is

$$\mathbf{I}_s = -j2\mathbf{Y}_0\mathbf{V}_+e^{-\alpha z}\sin{(\beta z)} \qquad (2\text{-}2\text{-}13)$$

Equations (2-2-12) and (2-2-13) show that the voltage and current standing waves are 90° out of phase along the line. The points of zero current are called the *current nodes*. The voltage nodes and current notes are interlaced a quarter-wavelength apart.

The voltage and current may be expressed as real functions of time and space. These are

$$v_s(z, t) = \text{Re}\,[\mathbf{V}_s(z)e^{j\omega t}] = 2\mathbf{V}_+e^{-\alpha z}\cos{(\beta z)}\cos{(\omega t)} \qquad (2\text{-}2\text{-}14)$$

$$i_s(z, t) = \text{Re}\,[\mathbf{I}_s(z)e^{j\omega t}] = 2\mathbf{Y}_0\mathbf{V}_+e^{-\alpha z}\sin{(\beta z)}\sin{(\omega t)} \qquad (2\text{-}2\text{-}15)$$

The amplitudes of Eqs. (2-2-14) and (2-2-15) vary sinusoidally with time, the voltage being a maximum at *time instant* when the current is zero, and vice versa. Figure 2-

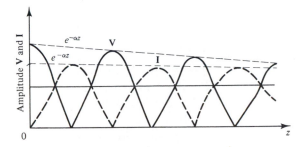

Figure 2-2-3 Pure standing waves of voltage and current.

2-3 shows the pure standing-wave patterns of the phasor of Eqs. (2-2-12) and (2-2-13) for an open-terminal line.

2-2-2 Standing-Wave Ratio

Standing waves result from the simultaneous presence of waves traveling in opposite directions on a transmission line. The ratio of the maximum of the standing-wave pattern to the minimum is defined as the standing-wave ratio, designated by ρ. That is,

$$\text{standing-wave ratio} \equiv \frac{\text{maximum voltage or current}}{\text{minimum voltage or current}} \qquad (2\text{-}2\text{-}16)$$

$$\rho \equiv \frac{|\mathbf{V}_{max}|}{|\mathbf{V}_{min}|} = \frac{|\mathbf{I}_{max}|}{|\mathbf{I}_{min}|}$$

The standing-wave ratio is caused by the fact that the two traveling-wave components of Eq. (2-2-1) add in phase at some points and subtract at the other. The distance between two successive maxima or minima is $\lambda/2$. The standing-wave ratio of a pure traveling wave is unity and that of a pure standing wave is infinite. It should be noted that since the voltage and current standing-wave ratios are identical, there are no distinctions made between VSWR and ISWR.

When the standing-wave ratio is unity, there is no reflected wave, and the line is called a *flat line*. The standing-wave ratio cannot be defined on a lossy line because the standing-wave pattern changes markedly from one position to another. On a low-loss line, the ratio remains fairly constant and may be defined for some region. For a lossless line, the ratio stays the same throughout the line.

Since the reflected wave is defined as the product of an incident wave and its reflection coefficient, the standing-wave ratio ρ is related to the reflection coefficient Γ by

$$\rho = \frac{1 + |\mathbf{\Gamma}|}{1 - |\mathbf{\Gamma}|} \qquad (2\text{-}2\text{-}17)$$

and vice versa:

$$|\Gamma| = \frac{\rho - 1}{\rho + 1} \tag{2-2-18}$$

This relation is very useful for determining the reflection coefficient from the standing-wave ratio, which is usually determined from the Smith chart. The curve in Fig. 2-2-4 shows the relationship between the reflection coefficient $|\Gamma|$ and the standing-wave ratio ρ. In view of Eq. (2-2-17), since $|\Gamma| \leq 1$, the standing-wave ratio is a positive real number, and it is never less than unity, $\rho \geq 1$. In view of Eq. (2-2-18), the magnitude of the reflected coefficient is equal to or less than unity.

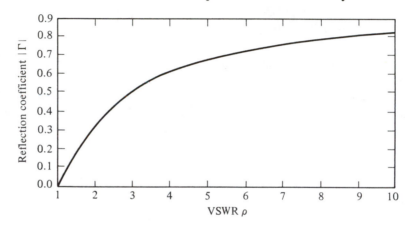

Figure 2-2-4   SWR versus reflection coefficient.

Example 2-2-1:   Standing-Wave Ratio

A transmission line has a characteristic impedance of $50 + j0.01 \ \Omega$ and is terminated in a load impedance of $73 - j42.5 \ \Omega$. Calculate (a) the reflection coefficient, and (b) the standing-wave ratio.

Solution   (a) From Eq. (2-1-42) the reflection coefficient is

$$\Gamma = \frac{\mathbf{Z}_\ell - \mathbf{Z}_0}{\mathbf{Z}_\ell + \mathbf{Z}_0} = \frac{73 - j42.5 - (50 + j0.01)}{73 - j42.5 + (50 + j0.01)}$$

$$= 0.377 \ \underline{/-42.7°}$$

(b) From Eq. (2-2-17) the standing-wave ratio is

$$\rho = \frac{1 + |\Gamma|}{1 - |\Gamma|} = \frac{1 + 0.377}{1 - 0.377} = 2.21$$

2-3 COAXIAL LINES AND IMPEDANCE TRANSFORMATION

A coaxial line consists of a round inner conductor and a concentric outer conductor separated by a dielectric medium. The lines can be classified into four types according to the structures of their outer conductor, as shown in Fig. 2-3-1. The four types of coaxial lines are:

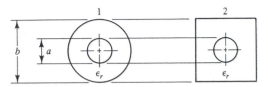

(a) Round coaxial line and square coaxial line

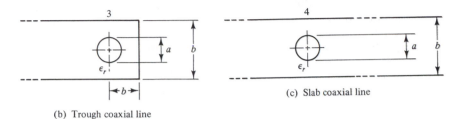

(b) Trough coaxial line

(c) Slab coaxial line

Figure 2-3-1 Coaxial-line configurations.

1. Round coaxial line
2. Square coaxial line
3. Trough line
4. Slab line

When coaxial lines are operating in their *transverse electromagnetic* (**TEM**) *dominant mode*, their characteristic impedance is independent of frequency. Consequently, they are widely used in microwave systems. In the dominant TEM mode, the inductance and capacitance of a coaxial line are given by

$$L = Z_0\sqrt{\mu\epsilon} \qquad \text{H/m} \qquad\qquad (2\text{-}3\text{-}1)$$

and

$$C = \frac{1}{Z_0}\sqrt{\mu\epsilon} \qquad \text{F/m} \qquad\qquad (2\text{-}3\text{-}2)$$

For a uniform and lossless line the characteristic impedance is expressed by

$$Z_0 = \sqrt{\frac{L}{C}} \qquad \text{ohms} \qquad\qquad (2\text{-}3\text{-}3)$$

2-3-1 Coaxial Lines

Round coaxial line.   A round coaxial line consists of a round inner conductor and a round concentric outer conductor separated by a dielectric medium such as polyethylene ($\epsilon_r = 2.25$), as shown in Fig. 2-3-2. The electric and magnetic fields are entirely confined within the dielectric region. Its radiation loss is very low and they

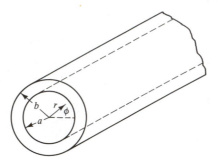

Figure 2-3-2 Section of a coaxial line.

are commonly used for precision measurements in microwave laboratories. The inductance and capacitance are given by

$$L = \frac{\mu_0 \mu_r}{2\pi} \ell n\left(\frac{b}{a}\right) \qquad \text{H/m} \tag{2-3-4}$$

and

$$C = \frac{2\pi\epsilon_0\epsilon_r}{\ell n(b/a)} \qquad \text{F/m} \tag{2-3-5}$$

Its characteristic impedance is expressed by

$$Z_0 = \frac{\eta_0}{2\pi\sqrt{\epsilon_r}} \ell n\left(\frac{b}{a}\right) \qquad \text{ohms} \tag{2-3-6}$$

or

$$Z_0 = \frac{138}{\sqrt{\epsilon_r}} \ell og\left(\frac{b}{a}\right) \qquad \text{ohms} \tag{2-3-7}$$

where η_0 = 377 Ω for free-space intrinsic impedance
ϵ_r = relative dielectric constant of the medium
μ_r = 1 is assumed for relative permeability
a = inner radius
b = outer radius

Figure 2-3-3 shows graphically the characteristic impedances of coaxial lines as a function of the cross-sectional ratio b/a.

The flexible round coaxial lines are available in different types. Their diameters vary from 0.635 mm (0.25 in.) to about 2.54 cm (1 in.), depending on the power requirement. In some coaxial lines, the inner conductor is stranded or a solid wire, but the outer conductor is a single braid or double. The dielectric material used in these round coaxial lines is polyethylene, which has a low loss at microwave frequencies. Particularly for RG series, the dielectric is either solid polyethylene or foam type. The loss per unit length for a foam polyethylene is even appreciably less than the equivalent solid polyethylene.

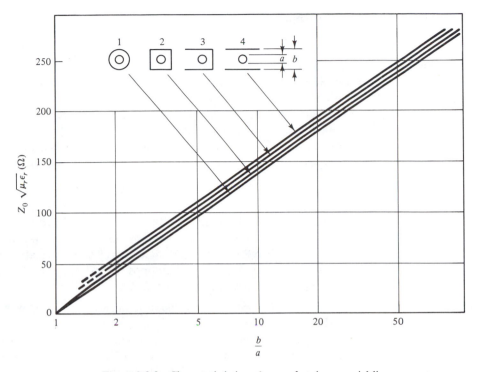

Figure 2-3-3   Characteristic impedances of various coaxial lines.

Square coaxial line.   A square coaxial line consists of a round inner conductor and a square concentric outer conductor separated by a dielectric medium as shown in Fig. 2-3-1(a). Its characteristic impedance is approximately given by

$$Z_0 = \frac{138}{\sqrt{\epsilon_r}} \ell og \left(1.0787 \frac{b}{a}\right) \quad \text{ohms} \qquad (2\text{-}3\text{-}8)$$

Trough line.   A trough line consists of a round inner conductor and a trough outer conductor as shown in Fig. 2-3-1(b). Its characteristic impedance is approximated by

$$Z_0 = \frac{138}{\sqrt{\epsilon_r}} \ell og \left(1.17 \frac{b}{a}\right) \quad \text{ohms} \qquad (2\text{-}3\text{-}9)$$

Slab line.   A slab line consists of a round inner conductor and an outer slab conductor as shown in Fig. 2-3-1(c). Its characteristic impedance is approximately expressed as

$$Z_0 = \frac{138}{\sqrt{\epsilon_r}} \ell og \left(\frac{4b}{\pi a}\right) \quad \text{ohms} \qquad (2\text{-}3\text{-}10)$$

Equation (2-3-10) gives a value 1% high at 50 Ω and better accuracy above 50 Ω.

2-3-2 Coaxial Connectors

For high-frequency operation the average circumference of a coaxial cable must be limited to about one wavelength in order to reduce multimodal propagation for elimination of erratic reflection coefficients, power losses, and signal distortion. The early standardization of coaxial connectors to interconnect electronic or electric system equipment during World War II was mandatory for microwave operation and a low reflection coefficient [or low voltage-standing-wave ratio (VSWR)]. Ever since, many modifications and new designs for microwave connectors have been proposed and developed. Seven types of microwave coaxial connectors are listed below, six of which are shown in Figure 2-3-4.

APC-3.5. The APC-3.5 (Amphenol Precision Connector—3.5 mm) was originally developed by Hewlett-Packard, but it is now manufactured by Amphenol. The connector provides repeatable connections and has a very low voltage-standing-wave ratio (VSWR). Either the male (plug) or the female (jack) end of this 50-Ω connector can mate with the opposite type of an SMA connector. The APC-3.5 connector can work up to 34 GHz.

APC-7. The APC-7 (Amphenol Precision Connector—7 mm) was also developed by Hewlett-Packard in the mid-1960s, but it has lately been improved and is now manufactured by Amphenol. The connector provides a coupling mechanism without male or female distinction and is the most repeatable connecting device. Its VSWR is extremely low, in the range 1.02 to 18 GHz. Maury Microwave also has a MPC series available.

BNC. The BNC (Bayonet Navy Connector) connector was originally designed for military system applications during World War II. The connector operates very well up to about 4 GHz; beyond that it intends to radiate electromagnetic energy. The BNC can accept flexible cables with diameter of up to 6.35 mm (0.25 in.) and characteristic impedance of 50 or 75 Ω. It is now the most commonly used connector under 1 GHz.

SMA. The SMA (Sub-Miniature A) connector was originally designed by Bendix Scintilla Corporation, but it has been manufactured by Omni-Spectra Inc. as the OSM connector and by many other electronic companies. The main application of SMA connectors is on components for microwave systems. The connector is seldom used above 24 GHz because of higher-order modes.

SMC. The SMC (Sub-Miniature C) is a 50-Ω connector and is smaller than SMA. The connector is manufactured by Sealectro Corporation and can accept flexible cables with a diameter of up to 3.17 mm (0.125 in.) for a frequency range of up to 7 GHz.

TNC. The TNC (Threaded Navy Connector) is merely threaded BNC. The function of the thread is to stop radiation at higher frequencies so that the connector can work at frequencies up to 12 GHz.

New coupling nut

APC-3.5 male

BNC female

Old coupling nut

(a) APC-3.5

(b) APC-7

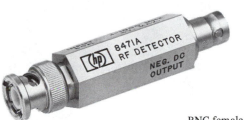

BNC female

SMA female

BNC male

SMA male

(c) BNC

(d) SMA

APC-3.5
male

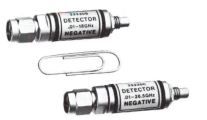

Type N female

SMC male
(plug)

Type N male

(e) SMC

(f) Type N

Figure 2-3-4   Microwave coaxial connectors.

35

Type N.   The Type N (Navy) was originally designed for military systems during World War II and is the most popular measurement connector for the frequency range 1 to 18 GHz. It is a 50-Ω connector and its VSWR is extremely low, within 1.02.

In microwave measurements the test fixtures are usually connected to the test equipment, such as the HP automatic network analyzer (ANA8409B) through microwave connectors and/or adapters. A problem often occurs at the transition section because the connection is not matched well enough. For example, the small round pin of an APC-7 connector has a very small contact area onto the flat surface of a 50-Ω microstrip line. Another problem is that it is very difficult to position the center pin of an APC-7 connector onto the microstrip line without an unforeseeable gap. To improve the transition section and minimize its VSWR (voltage-standing-wave ratio), it is recommended that the square center pin of an APC-7 connector be used and that the contact position of the center pin be stabilized with Teflon fiber glass.

2-3-3 Impedance Transformers

A section of coaxial line can be used to transform a 50-Ω resistor to any resistance. Three special cases are described below.

1. *Quarter-wave-line transformer.* The input impedance of a lossless coaxial line is given by

$$Z_{in} = Z_0 \frac{Z_\ell + jZ_0 \tan(\beta\ell)}{Z_0 + jZ_\ell \tan(\beta\ell)} \qquad (2\text{-}3\text{-}11)$$

where Z_0 = character
$\quad\ Z_\ell$ = load impedance
$\quad\ \beta\ell$ = electrical length, degrees or radians

For a quarter-wave line, $\ell = \lambda/4$; then $\beta\ell = \pi/2 = 90°$. Equation (2-3-11) becomes

$$Z_0 = \sqrt{Z_{in}Z_\ell} \qquad (2\text{-}3\text{-}12)$$

Equation (2-3-11) indicates that a quarter-wavelength of coaxial line can transform a certain value of input impedance to any load impedance value.

2. *Open-circuited termination.* If $Z_\ell = \infty$, Eq. (2-3-11) becomes

$$Z_{in} = -jZ_0 \cot(\beta\ell) \qquad (2\text{-}3\text{-}13)$$

3. *Short-circuited termination.* If $Z_\ell = 0$, Eq. (2-3-11) becomes

$$Z_{in} = +jZ_0 \tan(\beta\ell) \qquad (2\text{-}3\text{-}14)$$

If the section length of a coaxial line is chosen properly, a section of coaxial line can be used as either an open-circuited or a short-circuited stub for impedance-matching purposes.

Example 2-3-1:   Impedance Transformation

The receiving antenna has an impedance of 76 Ω, but the TV cable connecting to the TV set has a characteristic impedance of 300 Ω. It is necessary to design an impedance transformer to match the antenna impedance to the TV cable. (a) What type of impedance transformer should be used? (b) What is the characteristic impedance of the transformer?

Solution   (a) A quarter-wave line can serve the purpose.

(b) From Eq. (2-3-12) the characteristic impedance of the quarter-wave-line transformer is

$$Z_0 = \sqrt{75 \times 300} = 150 \ \Omega$$

2-4 SMITH CHART AND COMPRESSED SMITH CHART

Many of the computations required to solve transmission-line problems involve the use of rather complicated equations. The solution of such problems is tedious and difficult because they necessitate accurate manipulation of numerous equations. To simplify solution of such problems, a graphic method is needed to arrive at a quick answer. A number of impedance charts have been developed to facilitate the graphical solution of transmission-line problems. Basically, all of the charts are derived from the fundamental relationships expressed in the transmission equations. The most popular chart is that developed by Phillip H. Smith [1].

The Smith chart can be used in solving transmission-line problems with a magnitude of reflection coefficient equal to or less than unity, which means that the line resistance must be positive. When the magnitude of the reflection coefficient is greater than unity or the line resistance is negative, the compressed Smith chart must be used. This section deals with both the Smith chart and the compressed Smith chart.

2-4-1 Smith Chart

Basically, the Smith chart consists of a plot of the normalized impedance or admittance with the angle and magnitude of a generalized complex reflection coefficient in a unity circle. The chart is applicable to the analysis of a lossless line as well as a lossy line. By simple rotation on the chart, the effect of the position on the line may be determined. To see how a Smith chart is developed, consider the equation of reflection coefficient at the load [Eq. (2-1-42)],

$$\Gamma_\ell = \frac{\mathbf{Z}_\ell - \mathbf{Z}_0}{\mathbf{Z}_\ell + \mathbf{Z}_0} = |\Gamma_\ell| e^{j\theta_\ell} = \Gamma_r + j\Gamma_i \qquad (2\text{-}4\text{-}1)$$

Since $|\Gamma_\ell| \leq 1$, the plot of Γ_ℓ must lie on or within the unity circle. The reflection coefficient at any other location along a line is expressed in terms of Γ_ℓ as [Eq. (2-1-46)].

$$\Gamma_d = |\Gamma_\ell| e^{-2\alpha d} e^{j(\theta_\ell - 2\beta d)} = |\Gamma_d| e^{j(\theta_\ell - 2\beta d)} \qquad (2\text{-}4\text{-}2)$$

which is also within or on the unity circle. Figure 2-4-1 shows circles for a constant reflection coefficient Γ and constant electrical-length radials βd.

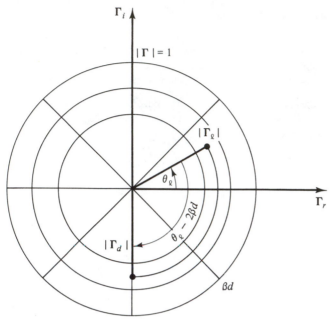

Figure 2-4-1 Constant Γ circles and electrical-length radials βd.

As described earlier, the normalized impedance along a line is given by

$$\cdot\mathbf{z} = \frac{\mathbf{Z}}{\mathbf{Z}_0} = \frac{1 + \Gamma_\ell e^{-2\gamma d}}{1 - \Gamma_\ell e^{-2\gamma d}} \tag{2-4-3}$$

With no loss in generality, it is assumed that $d = 0$; then

$$\mathbf{z} = \frac{1 + \Gamma_\ell}{1 - \Gamma_\ell} = \frac{\mathbf{Z}_\ell}{\mathbf{Z}_0} = \frac{R_\ell + jX_\ell}{\mathbf{Z}_0} = r + jx \tag{2-4-4}$$

and

$$\Gamma_\ell = \frac{\mathbf{z} - 1}{\mathbf{z} + 1} = \Gamma_r + j\Gamma_i \tag{2-4-5}$$

Substitution of Eq. (2-4-5) into Eq. (2-4-4) yields

$$r = \frac{1 - \Gamma_r^2 - \Gamma_i^2}{(1 - \Gamma_r)^2 + \Gamma_i^2} \tag{2-4-6}$$

and

$$x = \frac{2\Gamma_i}{(1 - \Gamma_r)^2 + \Gamma_i^2} \tag{2-4-7}$$

Equations (2-4-6) and (2-4-7) can be rearranged as

$$\left(\Gamma_r - \frac{r}{1 + r}\right)^2 + \Gamma_i^2 = \left(\frac{1}{1 + r}\right)^2 \tag{2-4-8}$$

and

$$(\Gamma_r - 1)^2 + \left(\Gamma_i - \frac{1}{x}\right)^2 = \left(\frac{1}{x}\right)^2 \tag{2-4-9}$$

Equation (2-4-8) represents a family of circles, where each circle has a constant resistance r. The radius of any circle is $1/(1 + r)$, and the center of any circle is $r/(1 + r)$ along the real axis in the unity circle, where r varies from 0 to ∞. All constant-resistance circles are plotted in Figure 2-4-2 according to Eq. (2-4-8).

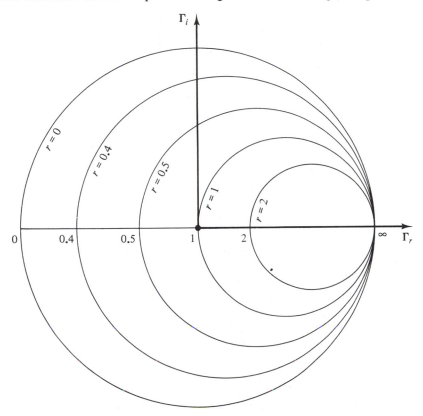

Figure 2-4-2   Constant resistance r circles.

Equation (2-4-9) also describes a family of circles, but each of these circles specifies a constant reactance x. The radius of any circle is $1/x$, and the center of any circle is at

$$\Gamma_r = 1, \qquad \Gamma_i = \frac{1}{x}, \qquad \text{where } -\infty \leq x \leq \infty$$

All constant-reactance circles are plotted in Figure 2-4-3 according to Eq. (2-4-9).

There are relative distance scales in wavelength along the circumference of the Smith chart. Also, there is a phase scale specifying the angle of the reflection coefficient.

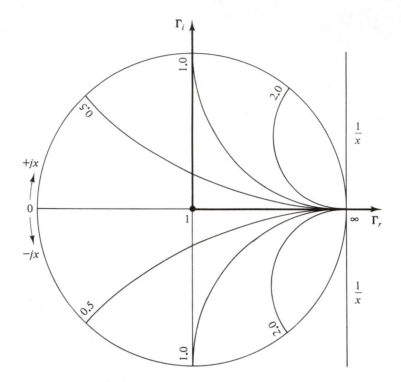

Figure 2-4-3   Constant reactance x circles.

When a normalized impedance $\mathbf{z}$ is located on the chart, the normalized impedance of any other location along the line can be found by the use of [Eq. (2-4-3)].

$$\mathbf{z} = \frac{1 + \Gamma_\ell e^{-2\gamma d}}{1 - \Gamma_\ell e^{-2\gamma d}} \tag{2-4-10}$$

where

$$\Gamma_\ell e^{-2\gamma d} = |\Gamma_\ell| e^{-2\alpha d} e^{j(\theta_\ell - 2\beta d)}$$

The Smith chart may also be used for normalized admittance. This is evident
since

$$\mathbf{Y} = \frac{1}{\mathbf{Z}_0} = G_0 + jB_0 \quad \text{and} \quad \mathbf{Y} = \frac{1}{\mathbf{Z}} = G + jB \tag{2-4-11}$$

Then the normalized admittance is

$$y = \frac{\mathbf{Y}}{\mathbf{Y}_0} = \frac{\mathbf{Z}_0}{\mathbf{Z}} = \frac{1}{\mathbf{z}} = g + jb \tag{2-4-12}$$

Figure 2-4-4 shows a Smith chart that superimposes Figs. 2-4-2 and 2-4-3 into
one chart.

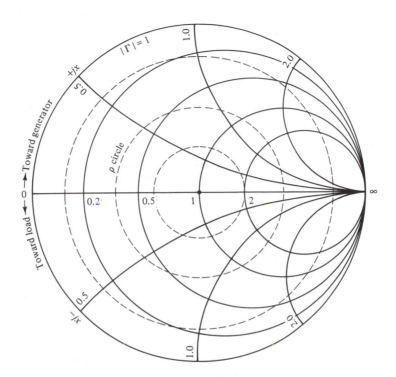

Figure 2-4-4 Smith chart.

Figure 2-4-5 shows a complete Smith chart. It is interesting to note that the chart was set up with the reflection coefficient as the radial coordinate and that the circles concentric with the center of the unity circle are circles of constant reflection coefficient. Since the standing-wave ratio is determined only by the magnitude of the reflection coefficient, these circles are also contours of constant standing-wave ratio. Since the standing-wave ratio is never less than unity, the scale for the standing-wave ratio varies from unity to infinity on the real axis. It should be noted that the distances are given in wave-lengths toward the generator and also toward the load, so that it is easy to determine which direction to advance as the position on the line is changed.

The characteristics of the Smith chart are summarized as follows.

1. The constant r and the constant x loci form two families of orthogonal circles in the chart.
2. The constant r and constant x circles all pass through the point ($\Gamma_r = 1$, $\Gamma_i = 0$).
3. The upper half of the diagram represents $+jx$.
4. The lower half of the diagram represents $-jx$.
5. For admittance the constant r circles become constant g circles, and the constant x circles become constant susceptance b circles.
6. The distance once around the Smith chart is one-half wavelength ($\lambda/2$).

IMPEDANCE OR ADMITTANCE COORDINATES

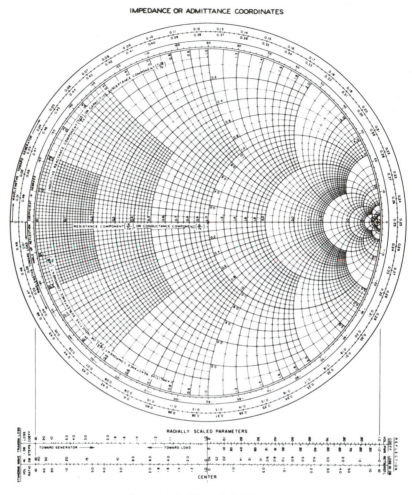

RADIALLY SCALED PARAMETERS

Figure 2-4-5 Normal Smith chart.

7. At a point of $z_{min} = 1/\rho$, there is V_{min} on the line.

8. At a point of $z_{max} = \rho$, there is V_{max} on the line.

9. The horizontal radius to the right of the chart center corresponds to V_{max}, I_{min}, z_{max}, and ρ (SWR).

10. The horizontal radius to the left of the chart center corresponds to V_{min}, I_{max}, z_{min}, and $1/\rho$.

11. Since the normalized admittance **y** is a reciprocal of the normalized impedance z, the corresponding quantities in the admittance chart are 180° out of phase with those in the impedance chart.

12. The normalized impedance or admittance is repeated for every half-wavelength of distance.

The magnitude of the reflection coefficient is related to the standing-wave ratio by the following expression:

$$|\Gamma| = \frac{\rho - 1}{\rho + 1} \tag{2-4-13}$$

A Smith chart or slotted line may be used to measure a standing-wave pattern directly from which the magnitude of the reflection coefficient, reflected power, transmitted power, and the load impedance may be calculated. Typical values are shown in the nomograph of Fig. 2-4-6 or at the bottom of the Smith chart. The use of the Smith chart is illustrated in the following examples.

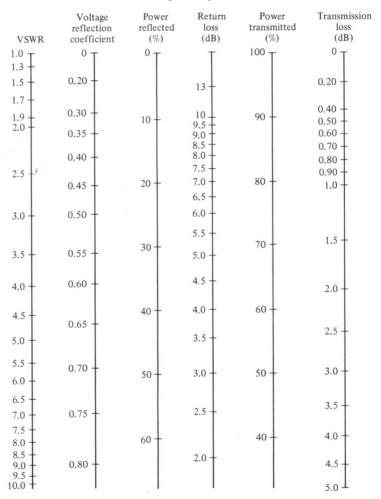

Figure 2-4-6 Nomograph of transmission line.

Example 2-4-1: Location Determination of Voltage Maximum and Minimum from Load

Consider the circuit shown in Fig. 2-4-7. Given the normalized load impedance $z_\ell = 1$

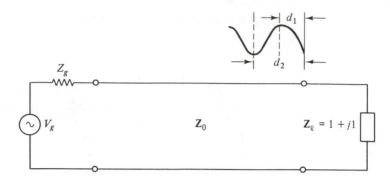

Figure 2-4-7 Diagram for Example 2-4-1.

$+ j1 \; \Omega$ and the operating wavelength $\lambda = 5$ cm, determine the first V_{max}, the first V_{min} from the load, and the VSWR ρ.

Solution

1. Enter $z_\ell = 1 + j1$ on the chart (Fig. 2-4-8).
2. Read 0.162λ on the distance scale by drawing a dashed straight line from the center of the chart through the load point and intersecting the distance scale.
3. Move a distance from the point at 0.162λ toward the generator and stop at the voltage maximum on the right-hand real axis at 0.25λ. Then

$$d_1(V_{max}) = (0.25 - 0.162)\lambda = (0.088)(5) = 0.44 \text{ cm}$$

4. Similarly, move a distance from the point of 0.162λ toward the generator and stop

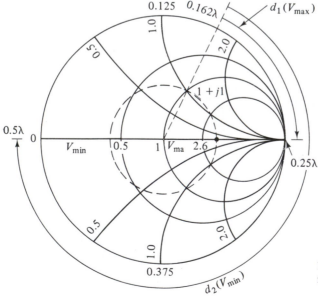

Figure 2-4-8 Graphical solution for Example 2-4-1.

at the voltage minimum on the left-hand real axis at 0.5λ. Then

$$d_2(V_{min}) = (0.5 - 0.162)\lambda = (0.338)(5) = 1.69 \text{ cm}$$

5. Make a standing-wave circle with the center at (1, 0) and a radius of $\sqrt{1 + 1}$. The intersection of the circle and the right portion of the real axis is the SWR. That is, $\rho = 2.6$.

Example 2-4-2:   Line-Impedance Determination with Resistive Load

Consider the circuit shown in Fig. 2-4-9. Given the characteristic impedance of the line $Z_0 = 50 \, \Omega$ and the terminating impedance $Z_\ell = 100 \, \Omega$, what impedance is seen at the points 2.15λ and 3.75λ away from the termination?

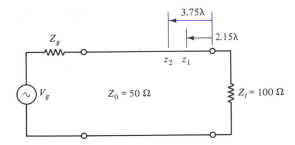

Figure 2-4-9   Diagram for Example 2-4-2.

Solution

1. Compute the normalized load impedance $z_\ell = 2$ and enter it on the chart at point z_ℓ (Fig. 2-4-10).

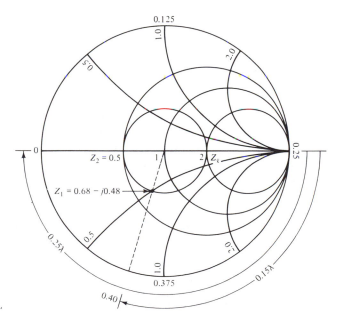

Figure 2-4-10   Graphical solution for Example 2-4-2.

2. Plot the VSWR circle ($\rho = 2$).

3. Move a distance of 2.5λ and 3.75λ from the point of 0.25 along the distance scale toward the generator and stop at the locations of 0.40 and 0.50. Note that the distance once around the chart is 0.5λ.

4. The intercepting points of the SWR circle ($\rho = 2$) and the lines linking the center to the stopping points of 0.40 and 0.50 are

$$z_1 = 0.68 - j0.48$$

$$z_2 = 0.5$$

5. Convert the normalized values to the line impedances.

$$\mathbf{Z}_1 = (0.68 - j0.48)(50) = 34 - j24 \ \Omega$$

$$\mathbf{Z}_2 = (0.5)(50) = 25 \ \Omega$$

Example 2-4-3:   Line-Impedance Determination with a Complex Load Admittance

Consider the circuit shown in Fig. 2-4-11. Given the characteristic impedance $Z_0 = 50$ Ω and the load admittance $\mathbf{Y}_\ell = 0.005 - j0.005$ Ω, determine the impedance at the point 3.15λ away from the load and the SWR ρ at that point.

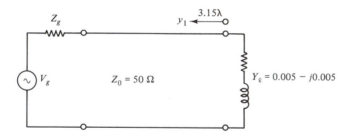

Figure 2-4-11   Diagram for Example 2-4-3.

Solution

1. Calculate the normalized admittance

$$y_\ell = \frac{\mathbf{Y}_\ell}{\mathbf{Y}_0} = \mathbf{Y}_\ell \mathbf{Z}_0 = (0.005 - j0.005)(50) = 0.25 - j0.25$$

2. Enter y_ℓ on the chart (Fig. 2-4-12).

3. Plot a SWR ρ circle and read

$$\rho = 4.2$$

4. Move a distance of 3.15λ from the load y_ℓ at 0.458 on the distance scale toward the generator and stop at 0.108. Remember that 3λ does not count because the admittance will repeat in that interval.

5. Read the normalized admittance on the ρ circle.

$$y_1 = 0.38 + j0.74$$

6. Read the normalized impedance on the ρ circle 180° out of phase with y_1.

$$z_1 = 0.55 - j1.10$$

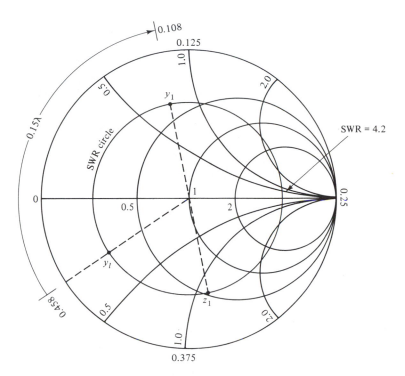

Figure 2-4-12   Graphical solution for Example 2-4-7.

7. The impedance at that point is

$$\mathbf{Z}_1 = (0.55 - j1.10)(50) = 27.5 - j55 \; \Omega$$

Example 2-4-4:   Impedance Determination with Short-Circuit Minima Shift

The location of a minimum is usually specified instead of a maximum because it can be determined more accurately than that of the maximum. Consider the circuit shown in Fig. 2-4-13, where the characteristic impedance of the line $Z_0 = 50 \; \Omega$, and the SWR ρ

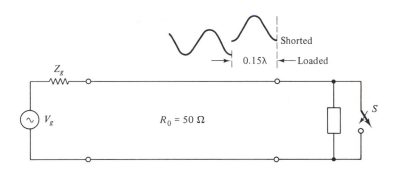

Figure 2-4-13   Diagram for Example 2-4-4.

= 2 when the line is loaded. When the load is shorted, the minima shift 0.15λ toward the load. Determine the load impedance.

Solution

1. When the line is shorted, the first voltage minimum occurs at the place of the load.
2. When the line is loaded, the first voltage minimum shifts 0.15λ from the load. The distance between two successive minima is one-half wavelength.
3. Plot a SWR circle for ρ = 2 (Fig. 2-4-14).
4. Move a distance of 0.15λ from the minimum point along the distance scale toward the load and stop at 0.15λ.
5. Draw a line from this point to the center of the chart.

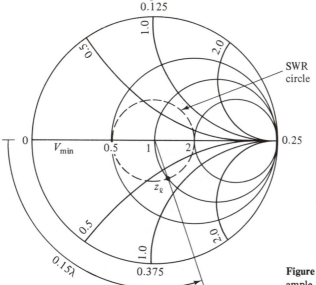

Figure 2-4-14   Graphical solution for Example 2-4-4.

6. The intersection between the line and the SWR circle is

$$\mathbf{z}_\ell = 1 - j0.65$$

7. The load impedance is

$$\mathbf{Z}_\ell = (1 - j0.65)(50) = 50 - j32.5 \ \Omega$$

Example 2-4-5:   Determination of Γ, ρ, Return Loss, and Transmission Loss on a Lossy Line

The voltage-reflection coefficient Γ is shown at the upper right-hand scale in the bottom of the chart, and the voltage-standing-wave ratio is shown at the lower left-hand scale. The return loss is shown on the lower right-hand scale, indicating the reverse power in decibels traveling in a line from a mismatched load below the reference point incident

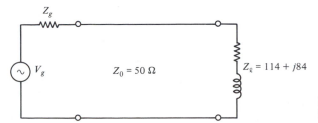

Figure 2-4-15 Diagram for Example 2-4-5.

on that mismatched load. The transmission loss is shown on the upper left-hand scale, indicating the power loss in a lossy line.

A coaxial line (Fig. 2-4-15) has the following characteristics:

$$Z_0 = 50 \; \Omega$$

$$Z_\ell = 114 + j84 \; \Omega$$

$$\alpha = 0.207 \; \text{Np/wavelength}$$

$$\beta = 6.3 \; \text{rad/wavelength}$$

$$f = 1 \; \text{GHz}$$

$$\ell = 1\text{m}$$

$$\epsilon_r = 2.25$$

Determine (a) the reflection coefficient Γ at the load and at a distance of 0.1 m from the load, (b) the standing-wave ratio ρ at the load and the point of 0.1 m away from the load, (c) the return loss and transmitted power, and (d) the transmission loss coefficients.

Solution (a) Find Γ.

1. Enter the normalized load $z_\ell = 2.28 + j1.68$ on the chart (Fig. 2-4-16).
2. Draw the SWR circle and read

$$\rho = 3.7$$

3. Measure the distance from the center to the point z_ℓ and the angle of z_ℓ.

$$\Gamma_\ell = |\Gamma_\ell| e^{j\theta_\ell} = 0.565 e^{j26°}$$

4. Read the power reflection coefficient on the lower right-hand scale.

$$|\Gamma|^2 = 0.32$$

5. Change the unit of 0.1 m in wavelength.

$$\ell = \frac{0.1}{\lambda} = \frac{0.1}{(3 \times 10^8)/10^9} = 0.333\lambda$$

6. The effective reflection coefficient at the point 0.1 m away from the load is

$$|\Gamma| = |\Gamma_\ell| e^{-2\alpha\ell} = 0.565 e^{-(2)(0.207)(0.333)} = 0.492$$

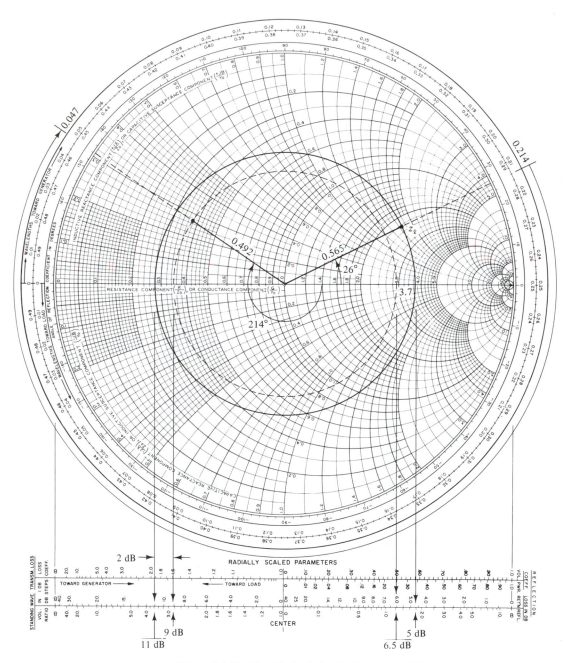

Figure 2-4-16   Graphical solution for Example 2-4-5.

7. Plot a circle with a radius of 0.492 centered at (1, 0). This circle is the new SWR circle, and the new SWR is 2.9.

8. Move a distance of 0.333λ from 0.214 on the distance scale toward the generator and stop at 0.047.

9. Draw a line from the point of 0.047 through the new SWR circle to the center at (1, 0) and read the value of the intersection for the normalized impedance.

$$\mathbf{z} = 0.37 + j0.27$$

10. The new reflection coefficient is

$$\Gamma = 0.492e^{j(\theta_\ell - 2\beta\ell)} = 0.942e^{j(26 - 240)} = 0.492e^{-j214°}$$

11. The power reflection coefficient is

$$|\Gamma|^2 = 0.24$$

(b) Find ρ.

1. The SWR at the load is

$$\rho_\ell = 3.7$$

2. The SWR at the point 0.1 m away from the load is

$$\rho = 2.9$$

3. The SWR in decibels are measured, respectively, on the standing-wave scale at the lower left-hand edge of the chart.

$$\rho_\ell = 11 \text{ dB}$$

$$\rho = 9 \text{ dB}$$

It should be noted that due to the loss on the line, the standing-wave ratio is decreased in a spiral way from $\rho_\ell = 3.7$ to $\rho = 2.9$, as indicated on the chart. This is because the power being transmitted toward the load on a lossy line and the power being reflected back toward the generator suffer an attenuation in both directions.

4. The attenuation loss is 2 dB, as shown on the chart.

(c) Find the return loss and transmitted power.

1. Draw a line perpendicular to the return-loss scale at the lower right-hand edge of the chart from the points where the SWR circle intersects the real axis on the right, and read

$$\text{Return loss (at load)} = 5.0 \text{ dB}$$

$$\text{Return loss (at 0.1 m from load)} = 6.5 \text{ dB}$$

These values mean that the power reflected from the points is 5 dB (or 6.5 dB) below the power incident on that point.

2. The transmitted power (reflection loss) is read on the lower side of the same scale as used in step 1.

$$\text{transmitted power (at load)} = 2 \text{ dB}$$
$$\text{transmitted power (at 1 m away)} = 1 \text{ dB}$$

These values indicate that the power absorbed by the load is about 2 dB below the power incident on the load, and the power transmitted at the point of 0.1 m away from the load is about 1 dB below.

(d) Find the transmission loss coefficient. Draw a line perpendicular to the transmission-loss scale at the upper left-hand scale of the bottom of the Smith chart from the points at which the SWR circles intersect the real axis on the left, and read

$$\text{transmission-loss coefficient (at load)} \doteq 2.0$$
$$\text{transmission-loss coefficient (at 0.1 m away)} \doteq 1.6$$

2-4-2 Compressed Smith Chart

The Smith chart is a plot of reflection coefficient with a magnitude equal to or less than unity. It is called the normal Smith chart. On the contrary, however, when the magnitude of reflection coefficient is greater than unity, the load resistance is negative and plot is known as the *compressed Smith chart*, shown in Fig. 2-4-17.

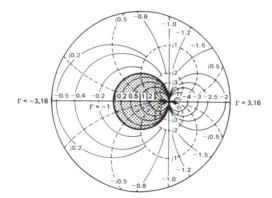

Figure 2-4-17   Compressed Smith chart.

It can be seen that the compressed Smith chart includes the normal Smith chart plus an additional negative-impedance portion. The compressed Smith chart is for the magnitude of reflection coefficient equal to or less than -3.16 (i.e., 10 dB of return gain). The chart is useful for plotting the variations of Γ_{in} and Γ_{out} for oscillator design. The impedance and admittance properties of the normal Smith chart are retained for the compressed chart. The following example illustrates the use of the compressed chart.

Example 2-4-6:   Determination of the Load Impedance Using the Compressed Chart

A certain reflection-type amplifier has the load reflection coefficient

$$\Gamma_\ell = 1.50 \; \underline{/155°}$$

Determine the normalized load impedance and admittance by using the compressed Smith chart.

Solution

1. Enter $\Gamma_\ell = 1.50 \underline{/155°}$ on the compressed chart as shown at point A in Fig. 2-4-18. Read $z_\ell = -0.2 + j0.2$ at point A.

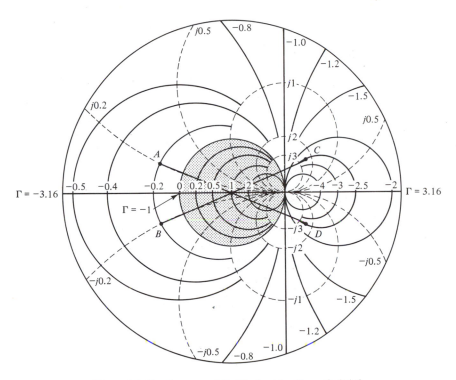

Figure 2-4-18   Compressed Smith chart for Example 2-4-6.

2. Read the conjugate normalized impedance as

$$z_\ell^* = -0.2 - j0.2 \text{ at point } B$$

3. Read the normalized admittance as

$$y_\ell = -2.5 - j2.5 \text{ at point } D$$

4. Read the conjugate normalized admittance as

$$y_\ell^* = -2.5 + j2.5 \text{ at point } C$$

An alternative method of determining negative resistance for $|\Gamma| > 1$ is to plot the value of $1/\Gamma^*$ on the normal Smith chart and read the values of the resistance circles as being negative and the reactance circles as marked.

Example 2-4-7: Determination of Negative Resistance from the Normal Smith Chart

The reflection coefficient of a load is given by

$$\Gamma_\ell = 1.50 \ \underline{/155°}$$

Determine the normalized load impedance by using the normal Smith chart.

Solution

1. $1/\Gamma_\ell^* = 1/1.50 \ \underline{/-155°} = 0.667 \ \underline{/155°}$.
2. Plot the values of $1/\Gamma_\ell^*$ on the normal Smith chart as shown in Fig. 2-4-19.
3. Read $z_\ell = -0.2 + j0.2$.

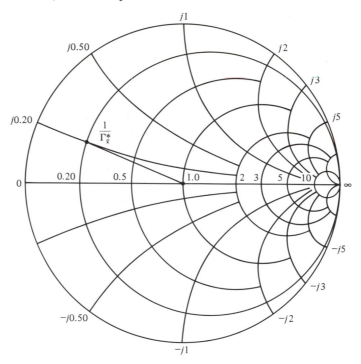

Figure 2-4-19   Graphical solution for Example 2-4-7.

2-5 IMPEDANCE-MATCHING TECHNIQUES

Impedance matching is very desirable at the microwave frequency range. Standing waves lead to increased losses and frequently cause the source transmitter to malfunction. A line terminated in its characteristic impedance has a standing-wave ratio of unity and transmits a given power without reflection. Also, the transmission efficiency is optimum when there is no power reflected.

"Matching" a transmission line has a special meaning, not the one used in circuit theory to indicate equal resistance seen looking in both directions from a given terminal pair for maximum power transfer. In circuit theory, maximum power transfer requires the load impedance to be equal to the complex conjugate of the generator. This condition is sometimes referred to as a *conjugate match*. In transmission-line problems, *matching* means simply terminating the line in its characteristic impedance. A section of transmission line, together with a short-, open-, or single- or double-circuited shunt stub, can be used to transform a 50-Ω characteristic impedance to any value of impedance. Figure 2-5-1 shows a matched transmission-line system.

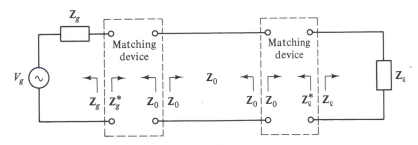

Figure 2-5-1   Matched transmission-line system.

2-5-1 Series and Shunt Element Matching

Since the matching problems involve parallel connections on the transmission line, it is necessary to work out the problems with admittances rather than impedances. The Smith chart itself can be used as a computer to convert the normalized impedance to admittance by rotation of 180°. The following examples illustrate series and shunt element matching.

Example 2-5-1:   Load with a Series Reactance

A section of 50-Ω line is terminated in a normalized load of $1 + j1$ Ω and its input normalized impedance is $1 - j1$ Ω. Determine the series element to match the two ports by using the Smith chart.

Solution

1. Plot $\mathbf{z}_\ell = 1 + j1$ and $\mathbf{z}_{\text{in}} = 1 - j1$ at points A and B on the Smith chart shown in Fig. 2-5-2(a).

2. Read the series element C as a capacitive reactance.

$$jx_c = -j2.0$$

3. The equivalent matching network is shown in Fig. 2-5-2(b).

Example 2-5-2:   Load Admittance with a Shunt Matching Element

A section of 50-Ω line has the following load and input admittances:

$$\mathbf{y}_\ell = 0.5 + j2.0$$

$$\mathbf{y}_{\text{in}} = 0.5 - j2.0$$

Determine the shunt element L to match the two ports by using the Smith chart

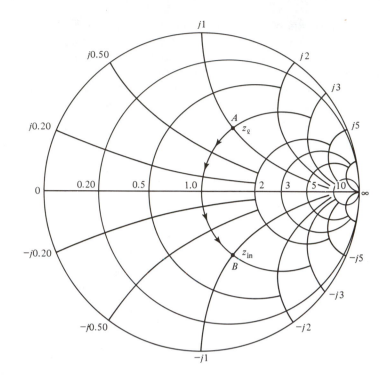

(a) Graphical solution

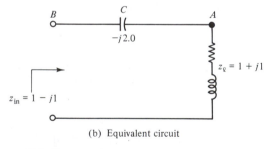

(b) Equivalent circuit

Figure 2-5-2 Graphical solution for Example 2-5-1.

Solution

1. Plot y_ℓ and y_{in} at points A and B on the Smith chart as shown in Fig. 2-5-3(a).
2. Read the shunt element L as an inductive susceptance.

$$y_L = -j4.0$$

This means that $jx_L = j0.25$.

3. The equivalent matching network is given by Fig. 2-5-3(b).

Design Example 2-5-3:   Matching a Load to a 50-Ω Line

A microwave device has an output impedance

$$\mathbf{Z}_{out} = 15 + j15 \ \Omega$$

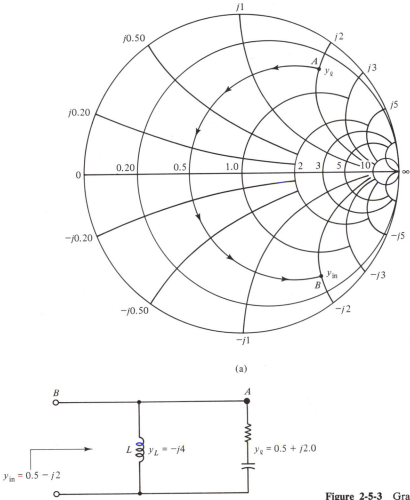

(a)

B ○ ———————————●——————————— ● A

$y_{in} = 0.5 - j2$

L ⎰ $y_L = -j4$ $y_\varrho = 0.5 + j2.0$

(b)

Figure 2-5-3 Graphical solution for Example 2-5-2.

It is necessary to design a matching network to transform the device output impedance to a 50-Ω line. Determine the matching network by using the Smith chart.

Solution

1. Plot the normalized device output impedance at point A on the Smith chart as shown in Fig. 2-5-4(a).

$$\mathbf{z}_{out} = 0.3 + j0.3 \text{ at point } A$$

2. Make a unity constant-conductance circle. This circle intersects the 0.3 constant-resistance circle at point B. Read

$$\mathbf{z}_B = 0.3 + j0.45$$

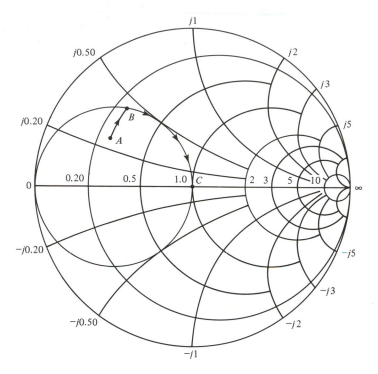

(a) Graphical solution

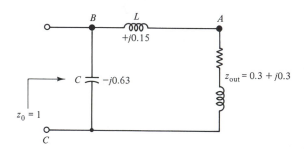

(b) Equivalent circuit

Figure 2-5-4 Graphical solution for Example 2-5-3.

Then the series element L is

$$\mathbf{z}_{\text{series}} = +j0.15$$

3. Point B corresponds to the point at the admittance chart at

$$\mathbf{y}_B = 1 - j1.60$$

The value of the shunt element C from point B to the center C is

$$\mathbf{y}_c = +j1.60$$

That means that the shunt element is a capacitor, as shown in Fig. 2-5-4(b).

$$jx_c = -j0.63$$

Design Example 2-5-4:   Matching the Device Impedance to a 50-Ω Load

A certain microwave device has the following impedance parameters:

$$\mathbf{Z}_{out} = 100 - j100 \qquad \mathbf{z}_{out} = 2 - j2$$

$$\mathbf{Y}_{out} = 0.005 + j0.005 \qquad \mathbf{y}_{out} = 0.25 + j0.25$$

Design a network to match the device admittance to the 50-Ω load by using the Smith chart.

Solution

1. Plot the device impedance $\mathbf{z}_{out}$ at point D on the Smith chart as shown in Fig. 2-5-5(a).

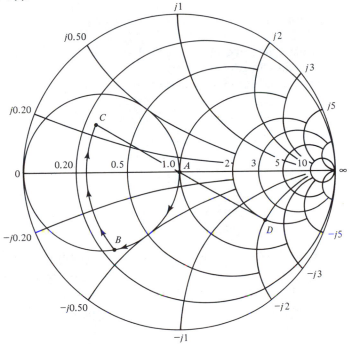

(a) Graphical solution

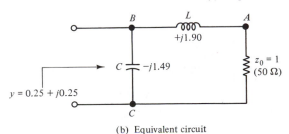

(b) Equivalent circuit

Figure 2-5-5   Graphical solution for Example 2-5-4.

2. Read the device admittance at point C.

$$\mathbf{y}_{out} = 0.25 + j0.25$$

3. Draw a unity constant-conductance circle that intersects the constant 0.25-resistance circle at point B Read

$$\mathbf{z}_{series} = +j1.90 \qquad \text{from the impedance chart}$$

$$\mathbf{y}_B = 0.25 - j0.42 \qquad \text{from the admittance chart}$$

4. The value from point B to point C is

$$\mathbf{y}_{shunt} = +j0.67$$

$$\mathbf{z}_{shunt} = -j1.49$$

5. The equivalent matching network is shown in Fig. 2-5-5(b).

2-5-2 Open- and Short-Stub Matching

The open-circuited and short-circuited shunt stubs are often used in microwave integrated circuits for admittance matching as shown in Fig. 2-5-6. The following example will illustrate the use of striplines in matching networks.

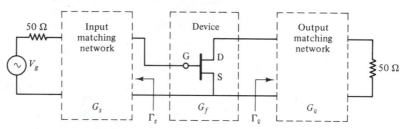

Figure 2-5-6   Matching network for an amplifier.

Design Example 2-5-5:   Open-Circuited and Short-Circuited Stub Matching

A GaAs MESFET amplifier has the following source and load reflection coefficients measured at 9 GHz with a 50-Ω line:

$$\Gamma_s = 0.606 \; \underline{/155°}$$

$$\Gamma_\ell = 0.707 \; \underline{/120°}$$

Design the input and output matching networks by using open-circuited and short-circuited shunt stubs.

Solution

1. Enter the source reflection coefficient Γ_s at point A on the Smith chart as shown in Fig. 2-5-7. Read the normalized source impedance and admittance.

$$\mathbf{z}_s = 0.22 + j0.215$$

$$\mathbf{y}_s = 2.38 - j2.20$$

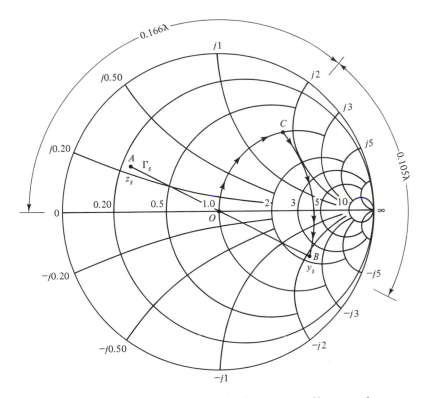

Figure 2-5-7   Graphical solution for the source matching network.

2. Draw an arc with $O-B$ as the radius and read the value at point C.

$$\mathbf{y}_c = 1 + j1.70$$

3. From point O at the origin to point C a shunt open-circuited stub with a length of 0.166λ is needed.

$$\ell_{shunt} = 0.166\lambda \qquad \text{for } + j1.70 \text{ at point } C$$

4. From point C to point B a series line of 50 Ω with a length of 0.105λ is needed.

$$\ell_{series} = 0.105\lambda \qquad \text{for transformation from point } C \text{ to point } B$$

5. Enter Γ_ℓ at point A on the Smith chart as shown in Fig. 2-5-8. Read the normalized load impedance and admittance as

$$\mathbf{z}_\ell = 0.22 + j0.56$$
$$\mathbf{y}_\ell = 0.65 - j1.55$$

6. From point O at the origin to point C a shunt short-circuited stub with a length of 0.074λ is needed.

$$\ell_{shunt} = 0.074\lambda \qquad \text{for } -j2.0 \text{ at point } C$$

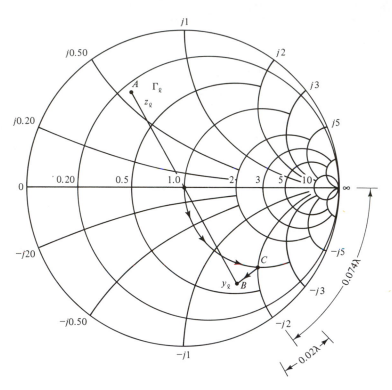

Figure 2-5-8 Graphical solution for the load matching network.

7. From point C to point B a series line of 50 Ω with a length of 0.02λ is needed.

$$\ell_{\text{series}} = 0.02\lambda \qquad \text{for transformation from point } C \text{ to point } B$$

8. The complete matching equivalent networks are shown in Fig. 2-5-9.

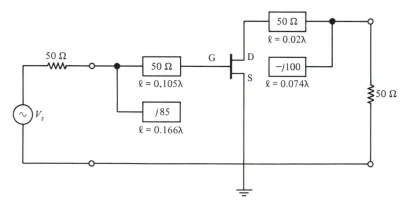

Figure 2-5-9 Matching equivalent networks.

2-5-3 Single- and Double-Stub Matching

Sometimes matching can be accomplished with a single-lumped susceptance. It is assumed that the load is an antenna which, because of design considerations, can be matched only to a normalized impedance of $0.4 - j0.4$ with a SWR $\rho = 2.9$. At some point on the line the conductance is equal to the characteristic admittance of the line. To match the line, [i.e., the characteristic admittance must be unity ($g = 1$) at that point], it is necessary to insert a shunt susceptance to cancel out the susceptive part of the admittance at the point. In general, the admittance at that point is

$$\mathbf{Y} = \mathbf{Y}_0 \pm jB \qquad (2\text{-}5\text{-}1)$$

and the normalized admittance for matching purposes is

$$y = 1 \pm jb \qquad (2\text{-}5\text{-}2)$$

$$\mathbf{Y}_s = \mp jb\mathbf{Y}_0 = \mp jB_s \qquad (2\text{-}5\text{-}3)$$

Example 2-5-6:   Double-Susceptance Matching

A transmission line (Fig. 2-5-10) has the following parameters:

$$\mathbf{Z}_\ell = 62.5 + j62.5 \; \Omega$$

$$\mathbf{Z}_0 = 50 \; \Omega$$

$$\rho = 2.9$$

$$f = 1 \text{ GHz}$$

Determine the value of a lumped inductance or capacitance which can match the transmission line, and the distance from the load to the point where the tuner will be placed.

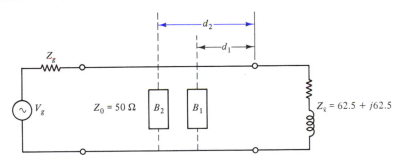

Figure 2-5-10   Double-susceptance matching for Example 2-5-6.

Solution

1. Compute the normalized load admittance and enter it on the chart (Fig. 2-5-11).

$$y_\ell = \frac{\mathbf{Z}_0}{\mathbf{Z}_\ell} = \frac{50}{62.5 + j62.5} = 0.4 - j0.4$$

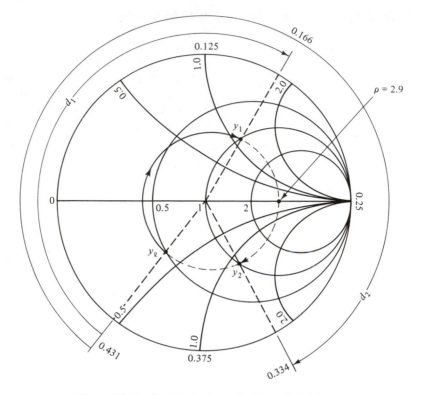

Figure 2-5-11 Graphical solution for Example 2-5-6.

2. Plot a SWR ρ circle.

3. The normalized line admittances to be matched are

$$\mathbf{y}_1 = 1 + j1.15$$
$$\mathbf{y}_2 = 1 - j1.15$$

4. The susceptance B must be inductive.

$$\mathbf{Y}_1 = -jb_1 \qquad \mathbf{Y}_0 = -j(1.15)\left(\frac{1}{50}\right) = -jB_1$$

$$B_1 = -0.023 = -\frac{1}{\omega L}$$

$$L = \frac{1}{2\pi(10^9)(0.023)} = 0.0071 \ \mu\text{H}$$

5. The distance for the first tuner from the load is

$$d_1 = [0.166 + (0.50 - 0.431)] \lambda = (0.231)(30) = 6.93 \text{ cm}$$

6. The susceptance B_2 must be capacitive.

$$\mathbf{Y}_2 = +jB_2 = +jb_2\,\mathbf{Y}_0 = +j(1.15)\left(\frac{1}{50}\right) = j0.023$$

$$B_2 = 0.023 = \omega C$$

$$C = \frac{B_2}{\omega} = \frac{0.023}{2\pi 10^9} = 3.68 \text{ pF}$$

7. The distance between the second tuner and the load is
   $d_2 = [0.334 + (0.50 - 0.431)]\lambda = (0.403)(30) = 12.09$ cm.

Single-stub matching.   Although single-lumped inductors or capacitors can match the transmission line, it is more common to use the susceptive properties of short-circuited sections of transmission lines. Short-circuited sections are preferable to open-circuited sections because a good short circuit is easier to obtain than a good open circuit.

For a lossless line with $\mathbf{Y}_g = \mathbf{Y}_0$, maximum power transfer requires $\mathbf{Y}_{11} = \mathbf{Y}_0$, where $\mathbf{Y}_{11}$ is the total admittance of the line and stub looking to the right at the point 1–1. The stub must be located at that point on the line where the real part of the admittance looking toward the load is $\mathbf{Y}_0$. In a normalized unit, $\mathbf{y}_{11}$ must be in the form

$$\mathbf{y}_{11} = \mathbf{y}_d \pm \mathbf{y}_s = 1 \tag{2-5-4}$$

if the stub has the same characteristic impedance as that of the line, or, otherwise,

$$\mathbf{Y}_{11} = \mathbf{Y}_d \pm \mathbf{Y}_s = \mathbf{Y}_0 \tag{2-5-5}$$

The stub length is then adjusted so that its susceptance just cancels out the susceptance of the line at the junction.

Example 2-5-7:   Single-Stub Matching

A lossless line of characteristic impedance $Z_0 = 50\ \Omega$ is to be matched to a load $\mathbf{Z}_\ell = 50/[2 + j(2 + \sqrt{3})]\ \Omega$ by means of a lossless short-circuited stub (Fig. 2-5-12). The characteristic impedance of the stub is $100\ \Omega$. Find the stub position (closest to the load) and length so that match is obtained.

Solution

1. Compute the normalized load admittance and enter it on the Smith chart (Fig. 2-5-13).

$$\mathbf{y}_\ell = \frac{1}{\mathbf{z}_\ell} = \frac{\mathbf{Z}_0}{\mathbf{Z}_\ell} = 2 + j(2 + \sqrt{3}) = 2 + j3.732$$

2. Draw a SWR circle in the clockwise direction and intersect the unity circle at the point $\mathbf{y}_d$.

$$\mathbf{y}_d = 1 - j2.6$$

Note that there are an infinite number of values of $\mathbf{y}_d(s)$. Take the one that permits

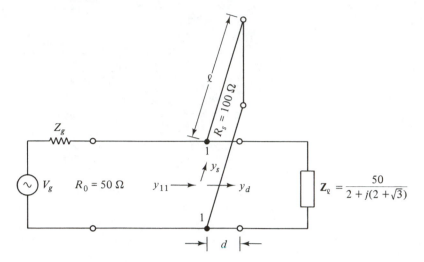

Figure 2-5-12   Single-stub matching for Example 2-5-7.

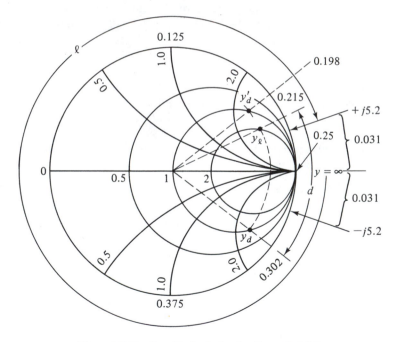

Figure 2-5-13   Graphical solution for Example 2-5-7.

the stub to be attached as closely as possible to the load so that the stub-line losses are minimized.

3. Since characteristic impedance of the stub is different from that of the line, the

condition for impedance matching at the junction requires

$$\mathbf{Y}_{11} = \mathbf{Y}_d + \mathbf{Y}_s$$

where $\mathbf{Y}_s$ is the susceptance the stub will contribute.

It is very clear that the stub and the portion of the line from the load to the junction are in parallel as seen by the main line extending to the generator. It is necessary to convert the admittances to normalized values for matching on the Smith chart. Then, Eq. (2-5-5) becomes

$$y_{11}Y_0 = y_d Y_0 + y_s Y_{0s}$$

$$y_s = (y_{11} - y_d)\frac{Y_0}{Y_{0s}} = [1 - (1 - j2.6)]\frac{100}{50} = +j5.20$$

4. The distance between the load and the stub position can be calculated from the distance scale as

$$d = (0.302 - 0.215)\lambda = 0.087\lambda$$

5. Since the stub contributes a susceptance of $+j5.20$, enter $+j5.20$ on the chart and determine the required distance ℓ from the short-circuited end ($z = 0$, $y = \infty$), which corresponds to the right side of the real axis on the chart, by transversing the chart toward the generator until the point of $+j5.20$ is reached. Then

$$\ell = (0.50 - 0.031)\lambda = 0.469\lambda$$

When a line is matched at the junction, there will be no standing wave in the line from the stub to the generator.

6. If an inductive stub is required,

$$\mathbf{y}'_d = 1 + j2.6$$

and the susceptance of the stub will be

$$\mathbf{y}'_s = -j5.2$$

7. The position of the stub from the load is

$$d' = [0.50 - (0.215 - 0.198)]\lambda = 0.483\lambda$$

and the length of the short-circuited stub is

$$\ell' = 0.031\lambda$$

As described previously, for maximum transmission of energy in the line, it is equally important and necessary to match the internal impedance of the generator as well as the load impedance to the line.

Double-stub matching.   Since single-stub matching is sometimes impractical because the stub cannot be placed physically in the ideal location, double-stub matching is needed. Double-stub devices consist of two short-circuited stubs connected in parallel with a fixed length between them. The length of the fixed section is usually 1/8, 3/8, or 5/8 of a wavelength. The stub that is nearest to the load is used to adjust the susceptance and is located at a fixed wavelength from the constant-conductance unity circle ($g = 1$) on an appropriate constant standing-wave-ratio circle. Then the admittance of the line at the second stub as shown in Fig. 2-5-14 is

$$\mathbf{y}_{22} = \mathbf{y}_{d2} \pm \mathbf{y}_{s2} = 1 \tag{2-5-6}$$

$$\mathbf{Y}_{22} = \mathbf{Y}_{d2} \pm \mathbf{Y}_{s2} = \mathbf{Y}_0 \tag{2-5-7}$$

With the positions and lengths of the stubs chosen properly, there will be no standing wave on the line to the left of the second stub measured from the load.

Example 2-5-8:   Double-Stub Matching

For the circuit shown in Fig. 2-5-14, the terminating impedance $\mathbf{Z}_\ell$ is $100 + j100$ Ω, and the characteristic impedance $\mathbf{Z}_0$ of the line and stub is 50 Ω. The first stub is placed at 0.40λ away from the load. The spacing between the two stubs is $\frac{3}{8}\lambda$. Determine the length of the short-circuited stubs when the matching is achieved. What terminations are forbidden for matching the line by the double-stub device?

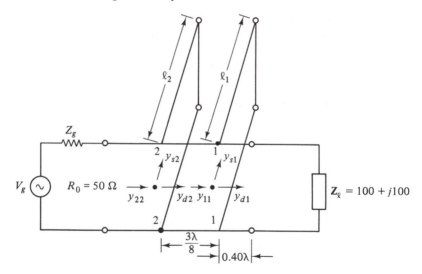

Figure 2-5-14   Double-stub matching for Example 2-5-8.

Solution

1. Compute the normalized load impedance z_ℓ and enter it on the chart (Fig. 2-5-15).

$$\mathbf{z}_\ell = \frac{100 + j100}{50} = 2 + j2$$

2. Plot a SWR ρ circle and read the normalized load admittance 180° out of phase with z_ℓ on the SWR circle.

$$\mathbf{y}_\ell = 0.25 - j0.25$$

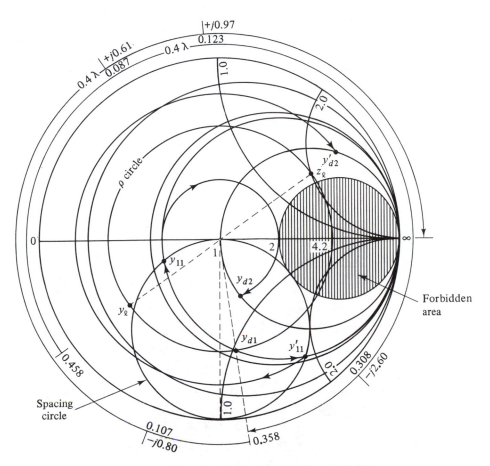

Figure 2-5-15 Graphical solution for Example 2-5-8.

3. Draw the spacing circle of $\frac{3}{8}\lambda$ by rotating the constant-conductance unity circle ($g = 1$) through a distance of $2\beta d = 2\beta \frac{3}{8}\lambda = \frac{3}{2}\pi$ toward the load. $\mathbf{y}_{11}$ must be on this spacing circle since $\mathbf{y}_{d2}$ will be on the $g = 1$ circle ($\mathbf{y}_{11}$ and $\mathbf{y}_{d2}$ are $\frac{3}{8}\lambda$ apart).

4. Move $\mathbf{y}_l$ for a distance of 0.40λ from 0.458 to 0.358 along the SWR ρ circle toward the generator and read $\mathbf{y}_{d1}$ on the chart.

$$\mathbf{y}_{d1} = 0.55 - j1.08$$

5. There are two possible solutions for $\mathbf{y}_{11}$. These can be found by carrying $\mathbf{y}_{d1}$ along the constant conductance ($g = 0.55$) circle that intersects the spacing circle at two points

$$\mathbf{y}_{11} = 0.55 - j0.11$$
$$\mathbf{y}'_{11} = 0.55 - j1.88$$

6. At the junction 1–1,

$$\mathbf{y}_{11} = \mathbf{y}_{d1} + \mathbf{y}_{s1}$$

Then $\mathbf{y}_{s1} = \mathbf{y}_{11} - \mathbf{y}_{d1} = (0.55 - j0.11) - (0.55 - j1.08) = +j0.97$
Similarly,

$$\mathbf{y}'_{s1} = -j0.80$$

7. The lengths of stub 1 are found as

$$\ell = (0.25 + 0.123)\lambda = 0.373\lambda$$
$$\ell'_1 = (0.25 - 0.107)\lambda = 0.143\lambda$$

8. The $\frac{3}{8}\lambda$-section of line transforms $\mathbf{y}_{11}$ to $\mathbf{y}_{d2}$, and $\mathbf{y}'_{11}$ to $\mathbf{y}'_{d2}$ along their constant-standing-wave circles, respectively. That is,

$$\mathbf{y}_{d2} = 1 - j0.61$$
$$\mathbf{y}'_{d2} = 1 + j2.60$$

9. Then stub 2 must contribute

$$\mathbf{y}_{s2} = +j0.61$$
$$\mathbf{y}'_{s2} = -j2.60$$

10. The lengths of stub 2 are found as

$$\ell_2 = (0.25 + 0.087)\lambda = 0.337\lambda$$
$$\ell'_2 = (0.308 - 0.25)\lambda = 0.058\lambda$$

11. It can be seen from Fig. 2-5-16 that a normalized admittance y_ℓ located inside the

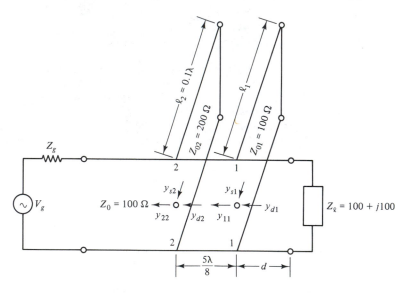

Figure 2-5-16 Double-stub matching for Example 2-5-9.

hatched area cannot be brought to lie on the locus of y_{11} or y'_{11} for a possible match by the parallel connection of any short-circuited stub, because the spacing circle and $g = 2$ circle are mutually tangent. Hence the area of $g = 2$ circle is called the *forbidden region* of the normalized load admittance for possible match.

Normally, the solution of a double-stub-matching problem can be worked out backward from the load toward the generator since the load is known and the distance of the first stub away from the load can be chosen arbitrarily. However, there are quite a few varieties of practical matching problems in which, for instance, some stubs have different Z_0 from that of the line, the length of a stub may be fixed, and so on. So it is difficult to describe a definite procedure for solving double-matching problems. The following example demonstrates a good variation of the matching method.

Example 2-5-9: Double-Stub Matching

The characteristic impedance Z_0 of a line is 100 Ω, and the load impedance Z_ℓ is 100 $- j100$ Ω. Stub 1 is located a distance d from the load and has a characteristic impedance of 100 Ω. Stub 2 is located $\frac{5}{8}\lambda$ from stub 1, and it has a Z_0 of 200 Ω and a length of 0.1λ. Both stubs are lossless and terminated in short circuits. Find the distance d and the length ℓ_1 of stub 1, then, with the line and load properly matched, determine the SWR ρ on the section of line between the stubs.

Solution

1. Compute the normalized load impedance and enter it on the chart.

$$z_\ell = \frac{Z_\ell}{Z_0} = \frac{100 + j100}{100} = 1 + j1$$

2. Draw a SWR ρ circle and read the normalized load admittance.

$$y_\ell = 0.5 - j05$$

3. Draw the spacing circle of $\frac{5}{8}\lambda$ as shown in Fig. 2-5-17.
4. Since the length of stub 2 is given as $\ell_2 = 0.1\lambda$ for a short-circuited termination, translate a distance of (0.1λ) from 0.25λ $(y = \infty)$ to 0.35λ at $-j1.38$. This means that stub 2 must contribute

$$y_{s2} = -j1.38$$

5. At junction 2, stub 2 has a Z_0 different from that of the main line; then

$$Y_{22} = Y_{d2} + Y_{s2} = Y_0$$

$$y_{d2} = 1 - y_{s2} \frac{Z_0}{Z_{02}} = 1 - (-j1.38)\left(\frac{100}{200}\right) = 1 + j0.69$$

6. Draw a SWR ρ circle through point y_{d2} and the circle intersects the spacing circle at

$$y_{11} = 0.52 + j0.125$$

Note: Another point $y'_{11} = 1.80 + j0.40$ can also be used to match the admittance.

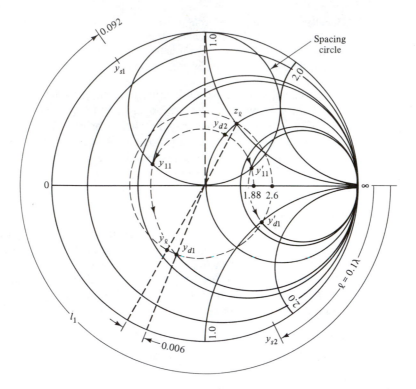

Figure 2-5-17　Graphical solution for Example 2-5-9.

7. Transform $\mathbf{y}_{11}$ along the constant-conductance circle ($g = 0.52$) to y_{d1} on the SWR circle ($\rho = 2.6$) and read

$$\mathbf{y}_{d1} = 0.52 - j0.53$$

Note: Another point $\mathbf{y}'_{d1} = 1.80 - j1.00$ can also be used for matching purposes.

8. Since stub 1 has the same Z_0 as that of the main line, then

$$y_{11} = y_{d1} + y_{s1}$$

$$y_{s1} = y_{11} - y_{d1} = (0.52 + j0.125) - (0.52 - j0.53)$$

$$= j0.655$$

Hence stub 1 contributes $+j0.655$.

9. The length of stub 1 is found as

$$\ell_1 = (0.25 + 0.092)\lambda = 0.342\ \lambda$$

10. The distance d of stub 1 away from the load is found as

$$d = (0.50 - 0.006)\lambda = 0.494\ \lambda$$

11. The SWR ρ in the section of line between the stubs is about 2.

REFERENCES

1. Smith, Philip H., Transmission line calculator. *Electronics*, Vol. 12, 1939, pp. 29–31; An improved transmission line calculator. *Electronics*, Vol. 17, 1944, pp. 130–133 and 318–325; Smith charts—Their development and use (a series published at intervals by the Key Electric Co.; No. 1 is dated March 1962, and No. 9 is dated December 1966).

PROBLEMS

2-1 Transmission Lines

2-1-1. An open-wire line has the following parameters:

$$R = 5 \; \Omega/m \qquad\qquad G = 6.20 \times 10^{-3} \; mho/m$$
$$L = 5.36 \times 10^{-8} \; H/m \qquad C = 2.07 \times 10^{-11} \; F/m$$
$$f = 0.1 \; GHz$$

Calculate:

(a) The characteristic impedance of the line in both rectangular and polar form.

(b) The propagation constant of the line.

(c) The normalized impedance of a load $100 + j100 \; \Omega$.

(d) The reflection coefficient at the load.

(e) The sending-end impedance if the line is assumed to be a quarter-wavelength long.

2-1-2. A lossless line has the following parameters:

Characteristic impedance	Z_0	$= 75 \; \Omega$
Load impedance	Z_ℓ	$= 300 \; \Omega$
Sending-end voltage	V_s	$= 20$ V (rms)
Source output impedance	Z_s	$= 75 \; \Omega$
Line length	ℓ	$= 2.25 \; \lambda$

(a) Find the sending-end impedance.

(b) Determine the magnitude of the receiving-end voltage.

(c) Calculate the receiving-end power at the load.

2-1-3. A lossless transmission line has a characteristic impedance of $100 \; \Omega$ and is branched out into two lines at a point A. Each of the branches has a same characteristic impedance of the main line. A pulse wave of 10 V (rms) is propagated along the main line and strikes the point A where the line is branched.

(a) Determine the magnitude of the pulse voltage that is reflected from the branched point A.

(b) Find the magnitude of the pulse voltage that is transmitted from the branched point A.

2-2 Standing-Wave Ratios

2-2-1. A lossless coaxial line has a characteristic impedance of $50 \; \Omega$ and is terminated in a load of $100 \; \Omega$. The magnitude of a voltage wave incident to the line is 20 V(rms).

Determine:
(a) The VSWR on the line.
(b) The maximum voltage V_{max} and minimum voltage V_{min} on the line.
(c) The maximum current I_{max} and minimum current I_{min} on the line.
(d) The power transmitted by the line.

2-2-2. A coaxial line has the following parameters:

Characteristic impedance	$\mathbf{Z}_0 = 49.4 + j7.6 \ \Omega$
Load impedance	$\mathbf{Z}_\ell = 100 + j100 \ \Omega$
Attenuation constant	$\alpha = 0.207$ Np/m
Phase constant	$\beta = 0.69$ rad/m
Line length	$\ell = 1$ m
Frequency	$f = 1$ GHz

(a) Determine the reflection coefficients at the load and at the point of 1 m away from the load.
(b) Find the standing-wave ratios at the load and at the point of 1 m away from the load.

2-3 Coaxial Lines

2-3-1. A coaxial line has the following parameters:

$R = 6 \ \Omega$/m		$f = 1$ GHz	
$L = 5.2 \times 10^{-8}$ H/m		$\mathbf{Z}_\ell = 100 + j100 \ \Omega$	
$G = 6 \times 10^{-3}$ mho/m			
$C = 2.136 \times 10^{-10}$ F/m			

Compute:
(a) The characteristic impedance and the propagation constant of the line.
(b) The reflection and transmission coefficients at the load.

2-3-2. A lossless coaxial line has a characteristic impedance of 75 Ω and is terminated in a load of $100 + j100 \ \Omega$. The line is 2 wavelengths long. Find the sending-end impedance of the line.

2-3-3. A coaxial line has the following parameters:

Characteristic impedance	$Z_0 = 50 \ \Omega$
Relative dielectric constant of polyethylene	$\epsilon_r = 2.25$
Attenuation constant at 0.2 GHz	$\alpha = 0.061$ Np/m
Line length	$\ell = 2$ m
Load impedance (open circuit)	$Z_\ell = \infty$
Sending-end voltage	$V_s = 50$ V (rms)

(a) Determine the sending-end impedance.
(b) Find the magnitude of the receiving-end voltage.

2-4 Smith Chart and Compressed Smith Chart

2-4-1. In Section 2-4-1 it is stated that the maximum normalized resistance along a transmission line is numerically equal to the voltage-standing-wave ratio (VSWR) along the line and that the minimum normalized resistance along the line is numerically equal to the reciprocal of the VSWR along the line. Verify the two statements by using equations.

2-4-2. The distance around the Smith chart once is equal to one-half wavelength of the transmission line. Prove the statement mathematically.

2-4-3. The compressed Smith chart is used to determine the negative resistance. A load reflection coefficient is measured to be $\Gamma_\ell = 1.50 \;\underline{/125°}$. Determine:
  (a) The normalized load impedance.
  (b) The normalized load admittance.
  (c) The normalized conjugate load impedance.
  (d) The normalized conjugate load admittance.

2-4-4. The negative resistance can be found by plotting the reciprocal of the conjugate reflection coefficient on a normal Smith chart and reading the values of the resistance circles as being negative and the reactance circles as marked.
  (a) Verify the statement mathematically.
  (b) If the load reflection coefficient is $\Gamma_\ell = 1.50 \;\underline{/125°}$, find the normalized load impedance and admittance.

2-5 Impedance Matching

2-5-1. A microwave device has an output impedance of $10 + j10$. Design a matching network to match the device impedance to a 50-Ω line.

2-5-2. A linear subsystem has a load impedance of $50\ \Omega$ at one end and an active device impedance of $25 - j100\ \Omega$ at the other. Design a matching network to match the device impedance to the load impedance and draw the equivalent circuit.

2-5-3. A lossless transmission line has a characteristic impedance of $100\ \Omega$ and is loaded by $100 + j100\ \Omega$. A single shorted stub with same characteristic impedance is inserted $\lambda/4$ from the load to match the line. The load current is measured to be 2 A. The length of the stub is $\lambda/8$.
  (a) Determine the magnitude and the phase of the voltage across the stub location.
  (b) Find the magnitude and the phase of the current flowing through the end of the stub.

2-5-4. A double-stub matching line is shown in Fig. P2-5-4. The characteristic resistances of

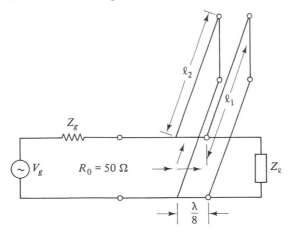

Figure P2-5-4

the line and the stubs are 50 Ω, respectively. The spacing between the two stubs is $\lambda/8$. The load is $50 + j50$. One stub is located at the load. Determine:

(a) The reactances contributed by the stubs.

(b) The lengths of the two shorted double-stub tuners. (*Note*: There are two sets of solutions.)

2-5-5. A double-stub tuner is to match a 60-Ω line to a load of 28 Ω. The distance from the load to the first stub is $\lambda/6$. The spacing between stubs is $3\lambda/8$. Determine the proper lengths for the two shorted stubs. (Express as fractions of the wavelength. There are two possible sets of solutions.)

2-5-6. The normalized load at the end of a lossless transmission line is $z_\ell = 1 + j1$. The guide wavelength is 5 cm.

(a) Find the distances from the load where the impedances are real. Since there is an infinite number, find the first two closest to the load. Use a Smith chart for solution.

(b) What are the values of these impedances?

(c) What is the voltage-standing-wave ratio (VSWR)?

2-5-7. The characteristic impedance Z_0 of a lossless transmission line is 50 Ω and the load is $60 - j80$ Ω. A double-stub matching device is designed to match the load to the line. One stub is at the load and the second one is $3\lambda/8$ away from the first one.

(a) Find the lengths of shorted stubs when a match is achieved.

(b) Locate and crosshatch the "forbidden region" of the normalized load for impossible match.

2-5-8 A lossless transmission line has a characteristic impedance Z_0 of 100 Ω and is loaded by an unknown impedance. Its voltage-standing-wave ratio is 4 and the first voltage maximum is $\lambda/8$ from the load.

(a) Find the load impedance.

(b) To match the load to the line, a quarter section of different line with a characteristic impedance $Z_{01} < Z_\rho$ is to be inserted somewhere between (in cascade with) the load and the original line. Determine the minimum distance between the load and matching section, and the characteristic impedance Z_{01} in terms of Z_0.

2-5-9. A lossless transmission line has a characteristic impedance of 100 Ω and is loaded by an unknown impedance. The standing-wave ratio along the line is 2. The adjacent minima are located at $z = -10$ and -35 cm from the load where $z = 0$. Determine the load impedance.

2-5-10. A matched transmission line is shown in Fig. P2-5-10.

(a) Find ℓ_1 and ℓ_2 that provide a proper match.

(b) With the line and load properly matched, determine the VSWR on the section of line between the stubs.

2-5-11. A lossless transmission line to be matched with double-stub tuners is shown in Fig. P2-5-11. The normalized load admittance is $y_\ell = 0.55 + j0.27$ and the characteristic impedance Z_0 of the line is 50 Ω. The signal frequency is 1 GHz.

(a) Determine ℓ_1 and ℓ_2 in centimeters.

(b) Find the susceptances y_{s1} and y_{s2} contributed by the stubs.

(c) Determine the SWR between the tuners.

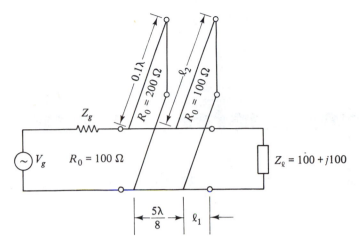

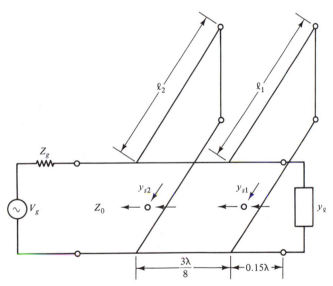

Figure P2-5-10

Figure P2-5-11

Chapter 3

S-Parameter Theory
and Applications

3-0 INTRODUCTION

In the design of a microwave module or subsystem it is often desirable to interconnect many active and passive elements together. The commonly used microwave solid-state devices are in the form of two ports. In this chapter we are concerned with the S-parameter theory, which will be used in later chapters for designing microwave matching networks.

From network theory, a two-port device as shown in Fig. 3-0-1 can be described by a number of parameter sets, such as the H, Y, and Z parameters.

H parameters:

$$V_1 = h_{11}I_1 + h_{12}V_2 \qquad (3\text{-}0\text{-}1)$$

$$I_2 = h_{21}I_1 + h_{22}V_2 \qquad (3\text{-}0\text{-}2)$$

Y parameters:

$$I_1 = y_{11}V_1 + Y_{12}V_2 \qquad (3\text{-}0\text{-}3)$$

$$I_2 = y_{21}V_1 + y_{22}V_2 \qquad (3\text{-}0\text{-}4)$$

Z parameters:

$$V_1 = z_{11}I_1 + z_{12}I_2 \qquad (3\text{-}0\text{-}5)$$

$$V_2 = z_{21}I_1 + z_{22}I_2 \qquad (3\text{-}0\text{-}6)$$

All of these network parameters relate total voltages and total currents at each of the

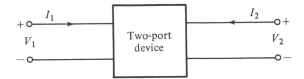

Figure 3-0-1 Two-port network.

two ports. For instance,

$$h_{11} = \frac{V_1}{I_1}\bigg|_{V_2=0} \qquad \text{short circuit} \qquad (3\text{-}0\text{-}7)$$

$$h_{12} = \frac{V_1}{V_2}\bigg|_{I_1=0} \qquad \text{open circuit} \qquad (3\text{-}0\text{-}8)$$

However, if the frequencies are in the microwave range, the H, Y, and Z parameters cannot be measured. This is because:

1. Equipment is not readiy available to measure total voltage and total current at the ports of the network.
2. Short and open circuits are difficult to achieve over a broad band of frequencies.
3. Active devices, such as power transistors and tunnel diodes, very often will not be short- or open-circuit stable.

Consequently, some new method of characterization is to be found to overcome these problems. The logical variables to use at the microwave frequencies are traveling waves rather than total voltages and total currents. These are the *S parameters*, which are expressed as

$$b_1 = S_{11}a_1 + S_{12}a_2 \qquad (3\text{-}0\text{-}9)$$

$$b_2 = S_{21}a_1 + S_{22}a_2 \qquad (3\text{-}0\text{-}10)$$

Figure 3-0-2 shows the S parameters of a two-port network.

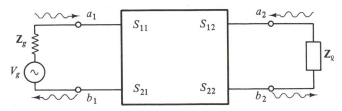

Figure 3-0-2 Two-port network for S parameters.

3-1 S-PARAMETER MATRIX

This section is devoted to presenting the S-parameter matrix and its properties in general which are applicable to any n-port devices in the frequencies of the microwave range. A microwave junction may have n ports, each of which is a lossless uniform transmission line, as shown in Fig. 3-1-1, where a_j is the incident traveling wave coming toward the junction, and b_i is the reflected traveling wave coming outward from the junction. From transmission-line theory, the incident and reflected waves are related by

$$b_i = \sum_{j}^{n} S_{ij}a_j \qquad \text{for } i = 1, 2, 3, \ldots, n \qquad (3\text{-}1\text{-}1)$$

where $S_{ij} = \Gamma_{ij}$ is the reflection coefficient of the ith port if $i = j$ with all other ports matched

$S_{ij} = T_{ij}$ is the forward transmission coefficient of the jth port if $i > j$ with all other ports matched

$S_{ij} = T_{ij}$ is the reverse transmission coefficient of the jth port if $i < j$ with all other ports matched

In general, Eq. (3-1-1) can be written as

$$b_1 = S_{11}a_1 + S_{12}a_2 + S_{13}a_3 + \cdots + S_{1n}a_n$$

$$b_2 = S_{21}a_1 + S_{22}a_2 + S_{23}a_3 + \cdots + S_{2n}a_n \qquad (3\text{-}1\text{-}2)$$

$$\cdots\cdots\cdots\cdots\cdots\cdots\cdots\cdots\cdots\cdots\cdots\cdots$$

$$b_n = S_{n1}a_1 + S_{n2}a_2 + S_{n3}a_3 + \cdots + S_{nn}a_n$$

In matrix notations, boldface Roman letters are used to represent matrix quantities and Eq. (3-1-2) is then expressed as

$$\mathbf{b} = \mathbf{Sa} \qquad (3\text{-}1\text{-}3)$$

where both $\mathbf{b}$ and $\mathbf{a}$ are column matrices, which are usually written as

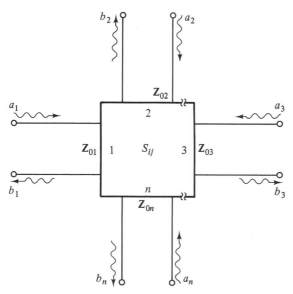

Figure 3-1-1. A microwave junction with n-ports.

$$\mathbf{b} = \begin{bmatrix} b_1 \\ b_2 \\ \cdot \\ \cdot \\ \cdot \\ b_n \end{bmatrix} \quad \text{and} \quad \mathbf{a} = \begin{bmatrix} a_1 \\ a_2 \\ \cdot \\ \cdot \\ \cdot \\ a_n \end{bmatrix} \tag{3-1-4}$$

The $n \times n$ matrix $\mathbf{S}$ is called the *scattering matrix,* which is

$$\mathbf{S} = \begin{bmatrix} S_{11} & S_{12} & S_{13} \cdots S_{1n} \\ S_{21} & S_{22} & S_{23} \cdots S_{2n} \\ \cdot \\ \cdot \\ \cdot \\ S_{n1} & S_{n2} & S_{n3} \cdots S_{nn} \end{bmatrix} \tag{3-1-5}$$

The coefficients S_{11}, S_{12}, $\cdots$, S_{nn} are called the *scattering parameters (S* parameters) or *scattering coefficients.* As a corollary to the S parameters, n-port voltages are linearly related to n-port currents by the impedance matrix of the junction. That is,

$$V_i = \sum_{j}^{n} Z_{ij} I_j \qquad \text{for } i = 1, 2, 3, \cdots, n \tag{3-1-6}$$

In matrix form, Eq. (3-1-6) can be expressed as

$$\begin{bmatrix} V_1 \\ V_2 \\ \cdot \\ \cdot \\ \cdot \\ V_n \end{bmatrix} = \begin{bmatrix} Z_{11} & Z_{12} & \cdots & Z_{1n} \\ Z_{21} & Z_{22} & \cdots & Z_{2n} \\ \cdots\cdots\cdots\cdots\cdots \\ Z_{n1} & Z_{n2} & \cdots & Z_{nn} \end{bmatrix} \begin{bmatrix} I_1 \\ I_2 \\ \cdot \\ \cdot \\ \cdot \\ I_n \end{bmatrix} \tag{3-1-7}$$

Symbolically,

$$\mathbf{V} = \mathbf{ZI} \tag{3-1-8}$$

3-2 PROPERTIES OF S PARAMETERS

Several properties of S parameters are described below.

3-2-1 Symmetry Property

The symmetry property states that if a microwave junction satisfies a reciprocity condition or if there are not any active solid-state devices at the junction, the junction is a linear passive circuit, and the S parameters are equal to their corresponding transposes. That is,

$$\mathbf{S} = \tilde{\mathbf{S}} \tag{3-2-1}$$

where $\tilde{\mathbf{S}} = S_{ji} = \mathbf{S} = S_{ij}$. $\tilde{\mathbf{S}}$ is the transpose of matrix $\mathbf{S}$.

The steady-state total voltage and current at the kth port are

$$V_k = V_k^+ + V_k^- \tag{3-2-2}$$

$$I_k = \frac{V_k^+}{Z_{0k}} - \frac{V_k^-}{Z_{0k}} \tag{3-2-3}$$

Therefore, the incident and reflected voltages at the kth port are

$$V_k^+ = \frac{1}{2}(V_k + Z_{0k}I_k) \tag{3-2-4}$$

$$V_k^- = \frac{1}{2}(V_k - Z_{0k}I_k) \tag{3-2-5}$$

The average incident power (complex) of the kth port is

$$\frac{1}{2}V_kI_k^* = \frac{|V_k^+|^2}{2Z_{0k}^*} \tag{3-2-6}$$

The normalized incident and reflected voltages at the kth port can be defined as

$$a_k = \frac{V_k^+}{\sqrt{Z_{0k}}} = \frac{1}{2}\left(\frac{V_k}{\sqrt{Z_{0k}}} + \sqrt{Z_{0k}}\,I_k\right) \tag{3-2-7}$$

$$b_k = \frac{V_k^-}{\sqrt{Z_{0k}}} = \frac{1}{2}\left(\frac{V_k}{\sqrt{Z_{0k}}} - \sqrt{Z_{0k}}\,I_k\right) \tag{3-2-8}$$

If the characteristic impedance is also normalized so that $\sqrt{Z_{0k}} = 1$

$$V_k = a_k + b_k \tag{3-2-9}$$

$$I_k = a_k - b_k \tag{3-2-10}$$

$$a_k = \frac{1}{2}(V_k + I_k) \tag{3-2-11}$$

$$b_k = \frac{1}{2}(V_k - I_k) \tag{3-2-12}$$

Since, from Eq. (3-1-6)

$$V_k = \sum_j^n Z_{kj}I_j \qquad \text{for } k = 1, 2, 3, \ldots, n \tag{3-2-13}$$

it follows that

$$a_k = \frac{1}{2} \sum_j^n (Z_{kj} + \delta_{kj}) I_k \qquad (3\text{-}2\text{-}14)$$

$$b_k = \frac{1}{2} \sum_j^n (Z_{kj} - \delta_{kj}) I_k \qquad (3\text{-}2\text{-}15)$$

where δ_{kj} is called the *Kronecker delta,* which is defined as

$$\delta_{kj} = \begin{cases} 1 & \text{if } k = j \\ 0 & \text{if } k \not\approx j \end{cases} \qquad \begin{matrix} (3\text{-}2\text{-}16) \\ (3\text{-}2\text{-}17) \end{matrix}$$

In matrix notation, Eqs. (3-2-14) and (3-2-15) can be written as

$$\mathbf{a} = \frac{1}{2}(\mathbf{Z} + [\mathbf{I}])\mathbf{I} \qquad (3\text{-}2\text{-}18)$$

$$\mathbf{b} = \frac{1}{2}(\mathbf{Z} - [\mathbf{I}])\mathbf{I} \qquad (3\text{-}2\text{-}19)$$

where $\mathbf{a}$ and $\mathbf{b}$ are column matrices and $[\mathbf{I}]$ is the identity matrix. Since the impedance matrix $\mathbf{Z}$ and the identity matrix $[\mathbf{I}]$ are square matrices $(n \times n)$, the matrix $(\mathbf{Z} - [\mathbf{I}])$ is surely $n \times n$ and may have an inverse. Thus

$$\mathbf{I} = 2(\mathbf{Z} + [\mathbf{I}])^{-1}\mathbf{a} \qquad (3\text{-}2\text{-}20)$$

Therefore, Eq. (3-2-19) becomes

$$\mathbf{b} = (\mathbf{Z} - [\mathbf{I}])(\mathbf{Z} + [\mathbf{I}])^{-1}\mathbf{a} \qquad (3\text{-}2\text{-}21)$$

In comparing Eq. (3-2-21) with Eq. (3-1-3), the matrix $\mathbf{S}$ can be written as

$$\mathbf{S} = (\mathbf{Z} - [\mathbf{I}])(\mathbf{Z} + [\mathbf{I}])^{-1} \qquad (3\text{-}2\text{-}22)$$

Let the matrices $\mathbf{P}$ and $\mathbf{Q}$ be so defined that

$$\mathbf{P} = \mathbf{Z} - [\mathbf{I}] \qquad (3\text{-}2\text{-}23)$$

$$\mathbf{Q} = \mathbf{Z} + [\mathbf{I}] \qquad (3\text{-}2\text{-}24)$$

Since the impedance matrix $\mathbf{Z}$ is symmetric, the matrices $\mathbf{P}$ and $\mathbf{Q}$ are also symmetric and commutative, that is,

$$\mathbf{PQ} = \mathbf{QP} \qquad (3\text{-}2\text{-}25)$$

Multiplying both the left- and right-hand sides of $\mathbf{PQ}$ and $\mathbf{QP}$ by $\mathbf{Q}^{-1}$ yields.

$$\mathbf{Q}^{-1}\mathbf{PQQ}^{-1} = \mathbf{Q}^{-1}\mathbf{QPQ}^{-1} \qquad (3\text{-}2\text{-}26)$$

Then

$$\mathbf{Q}^{-1}\mathbf{P} = \mathbf{PQ}^{-1} = \mathbf{S} \qquad (3\text{-}2\text{-}27)$$

The transpose $\tilde{S}$ of S is

$$\tilde{S} = \widetilde{Q^{-1}P} = \widetilde{PQ^{-1}} = Q^{-1}P = PQ^{-1} = S \qquad (3\text{-}2\text{-}28)$$

This means that the terms S_{ij} and S_{ji} of the S matrix are equal, and therefore the matrix S has a symmetry.

3-2-2 Unity Property

The unity property states that the sum of the products of each term of any one row or of any one column of the matrix S multiplied by its complex conjugate is unity; that is,

$$\sum_{i}^{n} S_{ij}S_{ij}^* = 1 \qquad \text{for } j = 1, 2, 3, \ldots, n \qquad (3\text{-}2\text{-}29)$$

From the principle of the conservation of energy, if the microwave devices are lossless and matched, the power input must be equal to the power output. The incident and reflected waves are related to the incident and reflected voltages by

$$a = \frac{V^+}{\sqrt{Z_0}} \qquad (3\text{-}2\text{-}30)$$

$$b = \frac{V^-}{\sqrt{Z_0}} \qquad (3\text{-}2\text{-}31)$$

It can be seen that

$$\text{incident power} = P_+ = \frac{1}{2} aa^* = \frac{1}{2}|a|^2 \qquad (3\text{-}2\text{-}32)$$

$$\text{reflected power} = P_- = \frac{1}{2} bb^* = \frac{1}{2}|b|^2 \qquad (3\text{-}2\text{-}33)$$

With no loss of generality, it is assumed that a wave of unit voltage is incident on port 1 of an n-port junction and that no voltage waves enter any of the other ports. Hence the power input is given by

$$P_{\text{in}} = a_1 a_1^* = |a_1|^2 \qquad (3\text{-}2\text{-}34)$$

which is equal to the power output leaving the ith port. That is,

$$P_{\text{in}} = a_1 a_1^* = P_{\text{out}} = \sum_{i}^{n} b_i b_i^* = b_1 b_1^* + b_2 b_2^* + \cdots + b_n b_n^* \qquad (3\text{-}2\text{-}35)$$

Since $b_i = S_{i1} a_1$, then

$$a_1 a_1^* = (S_{11}a_1)(S_{11}a_1)^* + (S_{21}a_1)(S_{21}a_1)^* + \cdots + (S_{n1}a_1)(S_{n1}a_1)^* \qquad (3\text{-}2\text{-}36)$$

Consequently

$$1 = S_{11}S_{11}^* + S_{21}S_{21}^* + \cdots + S_{n1}S_{n1}^* \qquad (3\text{-}2\text{-}37)$$

or

$$1 = \sum_i^n S_{ij}S_{ij}^* = \sum_i^n |S_{ij}|^2 \qquad \text{for } j = 1, 2, 3, \ldots \qquad (3\text{-}2\text{-}38)$$

Since S_{ij} is symmetric,

$$1 = \sum_j^n S_{ij}S_{ij}^* = \sum_j^n |S_{ij}|^2 \qquad \text{for } i = 1, 2, 3, \ldots \qquad (3\text{-}2\text{-}39)$$

For a lossy junction, the power dissipated at the junction is

$$P_{\text{diss}} = \frac{1}{2} \sum_{n=1}^n (a_n a_n^* - b_n b_n^*) \qquad (3\text{-}2\text{-}40)$$

It can be shown that

$$\sum_{n=1}^n a_n a_n^* = \tilde{\mathbf{a}}\mathbf{a}^* \qquad (3\text{-}2\text{-}41)$$

and

$$\sum_{n=1}^n b_n b_n^* = \tilde{\mathbf{a}}\mathbf{S}\mathbf{S}^*\mathbf{a}^* \qquad (3\text{-}2\text{-}42)$$

It should be noted that the right-hand terms of Eqs. (3-2-41) and (3-2-42) are 1×1 matrices, or just numbers. Hence, the power dissipated at the junction is given by

$$P_{\text{diss}} = \frac{1}{2} \tilde{\mathbf{a}}(1 - \mathbf{S}\overset{*}{\mathbf{S}})\mathbf{a}^* \qquad (3\text{-}2\text{-}43)$$

3-2-3 Zero Property

The zero property states that the sum of the products of each term of any row (or column) multiplied by the complex conjugate of the corresponding terms of any other row (or column) is zero.

$$\sum_i^n S_{ik}S_{ij}^* = 0 \qquad \text{for } k \approx j, \qquad \begin{array}{l} k = 1, 2, 3, \ldots, n \\ j = 1, 2, 3, \ldots, n \end{array} \qquad (3\text{-}2\text{-}44)$$

In general, the incident and reflected waves may exist at each of the n-ports. Then the incident power and reflected power for a lossless junction are

$$\sum_j^n a_j a_j^* = \sum_i^n b_i b_i^*$$

(3-2-45)

Substitution of Eq. (3-1-1) and its complex conjugate in Eq.(3-2-44) yields

$$\sum_j^n a_j a_j^* = \sum_j^n \left(\sum_i^n S_{ij} S_{ij}^* \right) a_j a_j^* + \sum_k' \sum_j \left(\sum_i S_{ik} S_{ij}^* \right) a_k a_j^*$$

$$+ \left[\sum_k' \sum_j \left(\sum_i S_{ik} S_{ij}^* \right) a_k a_j^* \right]^*$$

(3-2-46)

where the prime on $\sum'$ indicates that the terms of $k = j$ are not included in the sum. The first term on the right-hand side of Eq. (3-2-45) can be simplified by the use of Eq. (3-2-38). The last two terms on the right are of the form $(A + A^*)$, which is equal to twice the real part of A. Equation (3-2-45) can be simplified as

$$0 = 2 \, \text{Re} \sum_k' \sum_j \left(\sum_i S_{ik} S_{ij}^* \right) a_k a_j^*$$

(3-2-47)

Since the factor of $a_k a_j^*$ is nonzero,

$$\sum_i^n S_{ik} S_{ij}^* = 0$$

(3-2-48)

where $k \approx j$
$k = 1, 2, 3, \ldots, n$
$j = 1, 2, 3, \ldots, n$

For example, if only a_1 and a_2^* exist at port 1 and port 2, respectively, with all other ports terminated in their characteristic impedance, Eq. (3-2-48) becomes

$$S_{11} S_{12}^* + S_{21} S_{22}^* + S_{31} S_{32}^* + \cdots + S_{n1} S_{n2}^* = 0$$

(3-2-49)

Since S_{ij} is symmetric, it can be shown that

$$S_{11} S_{21}^* + S_{12} S_{22}^* + S_{13} S_{23}^* + \cdots + S_{1n} S_{2n}^* = 0$$

(3-2-50)

This has proved the zero property of the S parameters.

3-2-4 Phase-Shift Property

The phase-shift property states that if any of the terminal planes (or reference planes), say the kth port, is moved away from the junction by an electric distance $\beta_k \ell_k$, each of the coefficients S_{ij} involving k will be multiplied by the factor $e^{-j\beta_k \ell_k}$.

It is apparent that a change in the specified location of the terminal planes of an arbitrary junction will affect only the phase of the scattering coefficients of the junction. In matrix notation, the new S-parameter $\mathbf{S}'$ can be written as

$$\mathbf{S}' = \phi\mathbf{S}\phi \tag{3-2-51}$$

where $\mathbf{S}$ is the old scattering matrix, and

$$\phi = \begin{bmatrix} \phi_{11} & 0 & 0\cdot\cdot 0 \\ 0 & \phi_{22} & 0\cdot\cdot 0 \\ \cdots\cdots\cdots\cdots \\ 0 & 0\cdot\cdot\cdot 0 & \phi_{nn} \end{bmatrix} \tag{3-2-52}$$

where

$$\phi_{11} = \phi_{22} = \phi_{kk} = e^{-j\beta_k \ell_k} \qquad \text{for } k = 1, 2, 3, \cdots, n$$

It should be noted that while S_{ii} involves the product of two factors of $e^{-j\beta\ell}$, the new $\mathbf{S}'$ will be changed by a factor of $e^{-j2\beta\ell}$.

3-3 MASON'S SIGNAL-FLOW RULES

As described previously in this chapter, when frequencies are in the microwave region, it is very difficult to achieve short and open circuits for measuring the Z, Y, and H parameters. An alternative method is to use the S parameters in traveling waves for solving power gain at microwave frequencies. The two-port network of a microwave transistor amplifier is shown in Fig. 3-3-1 and the two S-parameter equations are expressed as

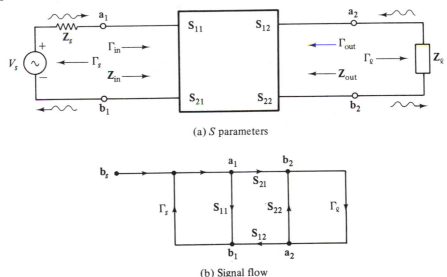

(a) S parameters

(b) Signal flow

Figure 3-3-1   Two-port network of a microwave amplifier.

$$b_1 = S_{11}a_1 + S_{12}a_2 \tag{3-3-1}$$

$$b_2 = S_{21}a_1 + S_{22}a_2 \tag{3-3-2}$$

The transfer function from b_s to b_2 can be derived by the nontouching loop rules of signal flow theory. The nontouching loop rules, which are often called *Mason's rules,* have the following terms:

1. *Path.* A path is a series of directed lines followed in sequence and in the same direction in such a way that no node is touched more than once. The value of the path is the product of all coefficients encountered enroute. In Fig. 3-3-1(b), there is only one path from b_s to b_2 and the value of the path is S_{21}. There are two paths from b_s to b_1, and the values are S_{11} and $S_{21}\Gamma_\ell S_{12}$.

2. *First-order loop.* A first-order loop is defined as the product of all coefficients along the paths starting from a node and moving in the direction of the arrows back to that original node without passing the same node twice. In Fig. 3-3-1(b), there are three first-order loops, and the values are $S_{11}\Gamma_s$, $S_{22}\Gamma_\ell$, and $S_{21}\Gamma_s S_{12}\Gamma_\ell$.

3. *Second-order loop.* A second-order loop is defined as the product of any two nontouching first-order loops. In Fig. 3-3-1(b), there is only one second-order loop and its value is $S_{11}\Gamma_s S_{22}\Gamma_\ell$.

4. *Third-order loop,* A third-order loop is the product of any three nontouching first-order loops. In Fig. 3-3-1(b), there is no third-order loop.

Then the transfer function for the ratio of a dependent variable in question to an independent variable of the source is expressed by

$$T = \frac{P_1[1 - \Sigma L(1)^1 + \Sigma L(2)^1 - \Sigma L(3)^1 + \cdots] + P_2[1 - \Sigma L(1)^2 + \Sigma L(2)^2 - \cdots] + P_3[1 \cdots]}{1 - \Sigma L(1) + \Sigma L(2) - \Sigma L(3) + \cdots} \tag{3-3-3}$$

where $P_1, P_2, P_3, \cdots$ are the various paths connecting these variables
  $\Sigma L(1), \Sigma L(2), \Sigma L(3), \cdots$ are the sums of all first-order, second-order, third-order, $\cdots$, loops, respectively,
  $\Sigma L(1)^1, \Sigma L(2)^1, \Sigma L(3)^1, \cdots$ are the sums of all first-order, second-order, third-order, $\cdots$, loops that do not touch the first path between the variables,
  $\Sigma L(1)^2, \Sigma L(2)^2, \Sigma L(3)^2, \cdots$ are the sum of all first-order, second-order, third-order, $\cdots$, loops that do not touch the second path
  From Fig. 3-3-1(b), the transfer function of b_2 over b_s is given by

$$\frac{b_2}{b_s} = \frac{S_{21}}{1 - S_{11}\Gamma_s - S_{22}\Gamma_\ell - S_{21}\Gamma_s S_{12}\Gamma_\ell + S_{11}\Gamma_s S_{22}\Gamma_\ell} \tag{3-3-4}$$

3-4 POWER-GAIN EQUATIONS

There are several power-gain equations which are derived from the two-port network of a microwave amplifier as shown in Fig. 3-3-1(b).

1. Transducer power gain
2. Available power gain
3. Power gain

3-4-1 Transducer Power Gain G_t

The transducer power gain of a microwave amplifier is defined as the ratio of the output power P_ℓ delivered to the Z_ℓ over the input power P_{avs} available from the source to the network. That is,

$$G_t = \frac{P_\ell}{P_{avs}} \tag{3-4-1}$$

where
$$P_\ell = P_{avn} \quad \text{when} \quad \Gamma_\ell = \Gamma_{out}^*$$
$$P_{avs} = P_{in} \quad \text{when} \quad \Gamma_{in} = \Gamma_s^*$$

P_{avn} = power available from the network

The power delivered to the load is the resultant of the power incident on the load minus the power reflected from the load.

$$P_\ell = \frac{1}{2}|\mathbf{b}_2|^2 - \frac{1}{2}|\mathbf{a}_2|^2 = \frac{1}{2}|\mathbf{b}_2|^2(1 - |\Gamma_\ell|^2) \tag{3-4-2}$$

where $\Gamma_\ell = (\mathbf{Z}_\ell - \mathbf{Z}_0)/(\mathbf{Z}_\ell + \mathbf{Z}_0)$ is the reflection coefficient of the load.

The power available from the source is given by

$$P_{avs} = \frac{\frac{1}{2}|\mathbf{b}_s|^2}{1 - |\Gamma_s|^2} \tag{3-4-3}$$

where $\Gamma_s = \dfrac{\mathbf{Z}_s - \mathbf{Z}_0}{\mathbf{Z}_s + \mathbf{Z}_0}$ is the reflection coefficient of the source

$\mathbf{b}_s$ = a function of $\mathbf{b}_2$ which is to be determined

Then the transducer power gain is expressed by

$$G_t = \frac{|\mathbf{b}_2|^2}{|\mathbf{b}_s|^2}(1 - |\Gamma_s|^2)(1 - |\Gamma_\ell|^2) \tag{3-4-4}$$

Substitution of Eq. (3-3-4) in Eq. (3-4-4) yields the transducer power gain as

$$G_t = \frac{(1 - |\Gamma_s|^2)|S_{21}|^2(1 - |\Gamma_\ell|^2)}{|(1 - S_{11}\Gamma_s)(1 - S_{22}\Gamma_\ell) - S_{12}S_{21}\Gamma_s\Gamma_\ell|^2} \tag{3-4-5}$$

Figure 3-4-1 shows a diagram for different reflection coefficients.

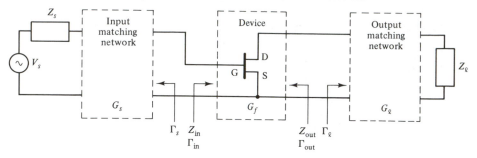

Figure 3-4-1   Reflection coefficients.

The transducer power gain Eq. (3-4-5) can also be expressed as

$$G_t = \frac{1 - |\Gamma_s|^2}{|1 - \Gamma_{in}\Gamma_s|^2} |S_{21}|^2 \frac{1 - |\Gamma_t|^2}{|1 - S_{22}\Gamma_t|^2} \tag{3-4-6}$$

$$G_t = \frac{1 - |\Gamma_s|^2}{|1 - S_{11}\Gamma_s|^2} |S_{21}|^2 \frac{1 - |\Gamma_t|^2}{|1 - \Gamma_{out}\Gamma_t|^2} \tag{3-4-7}$$

where

$$\Gamma_{in} = S_{11} + \frac{S_{12}S_{21}\Gamma_t}{1 - S_{22}\Gamma_t} = \frac{S_{11} - \Delta\Gamma_t}{1 - S_{22}\Gamma_t} \tag{3-4-8}$$

$$\Gamma_{out} = S_{22} + \frac{S_{12}S_{21}\Gamma_s}{1 - S_{11}\Gamma_s} = \frac{S_{22} - \Delta\Gamma_s}{1 - S_{11}\Gamma_s} \tag{3-4-9}$$

where $\Delta = S_{11}S_{22} - S_{12}S_{21}$.

There are three special cases of transducer power gain.

1. *Matched transducer power gain* ($\Gamma_s = \Gamma_t = 0$). When both the input and output networks are perfectly matched to the source impedance and the load impedance, respectively, the transducer power gain is

$$G_{tm} = |S_{21}|^2 \tag{3-4-10}$$

2. *Unilateral transducer power gain* ($|S_{12}|^2 = 0$). The unilateral transducer power gain G_{tu} is the forward power gain in a feedback amplifier having its reverse power gain set to zero ($|S_{12}|^2 = 0$) by adjusting a lossless reciprocal feedback network connected around the microwave amplifier. That is,

$$G_{tu} = \frac{1 - |\Gamma_s|^2}{|1 - S_{11}\Gamma_s|^2} |S_{21}|^2 \frac{1 - |\Gamma_t|^2}{|1 - S_{22}\Gamma_t|^2} \tag{3-4-11}$$

3. *Maximum unilateral transducer power gain.* The maximum unilateral transducer power gain is obtained when $\Gamma_s = S_{11}^*$ and $\Gamma_t = S_{22}^*$. Then

$$G_{tumax} = \frac{|S_{21}|^2}{(1 - |S_{11}|^2)(1 - |S_{22}|^2)} \qquad (3\text{-}4\text{-}12)$$

Example 3-4-1:   Transducer Power Gain of a GaAs MESFET Amplifier

A GaAs MESFET has the following S parameters measured with $V_{ds} = 4$ V and $I_{ds} = 30$ mA at 8 GHz with a 50-Ω reference:

$$S_{11} = 0.55 \;\underline{/158°} \qquad \Gamma_s = 0.20 \;\underline{/0°}$$

$$S_{12} = 0.01 \;\underline{/-5°} \qquad \Gamma_t = 0.33 \;\underline{/0°}$$

$$S_{21} = 1.95 \;\underline{/9°}$$

$$S_{22} = 0.46 \;\underline{/-148°}$$

Compute (a) the delta factor Δ, (b) the stability factor K, (c) the transducer power gain, (d) the matched transducer power gain, (e) the unilateral transducer power gain, and (f) the maximum unilateral transducer power gain.

Solution   (a) The delta factor is

$$\Delta = 0.55 \;\underline{/158°} \times 0.46 \;\underline{/-148°} - 0.01 \;\underline{/-5°} \times 1.95 \;\underline{/9°}$$

$$= 0.23 \;\underline{/9.87°}$$

(b) The stability factor K is

$$K = \frac{1 + |0.23|^2 - |0.55|^2 - |0.46|^2}{2|0.01 \times 1.95|}$$

$$= 13.5 > 1$$

(c) From Eq. (3-4-5) the transducer power gain is

$$G_t = \frac{(1 - |0.20|^2)|1.95|^2(1 - |0.33|^2)}{|(1 - 0.55 \;\underline{/158°} \times 0.20)(1 - 0.46 \;\underline{/-148°} \times 0.33) - 0.01 \;\underline{/-5°} \times 1.95 \;\underline{/9°} \times 0.20 \times 0.33|^2}$$

$$= 2.08 = 3.18 \text{ dB}$$

(d) Let $\Gamma_s = \Gamma_t = 0$; the matched transducer power gain is

$$G_{tm} = |S_{21}|^2 = |1.95|^2 = 3.80 = 5.8 \text{ dB}$$

(e) Let $|S_{12}|^2 = 0$; the unilateral transducer power gain is

$$G_{tu} = \frac{1 - |0.20|^2}{|1 - 0.55 \underline{/158°} \times 0.20|^2}|1.95|^2 \frac{1 - |0.33|^2}{|1 - 0.46 \underline{/-148°} \times 0.33|^2}$$

$$= 0.79 \times 3.80 \times 0.70$$

$$= 2.09 = 3.20 \text{ dB}$$

(f) The maximum unilateral transducer power gain is

$$G_{tu\max} = \frac{1}{1 - |0.55|^2}|1.95|^2 \frac{1}{1 - |0.45|^2}$$

$$= 1.43 \times 3.80 \times 1.27$$

$$= 6.88 = 8.37 \text{ dB}$$

3-4-2 Available Power Gain G_a

The available power gain is defined as the ratio of the power available from the network P_{avn} to the power available from the source P_{avs}. That is,

$$G_a = \frac{P_{avn}}{P_{avs}} = \frac{1 - |\Gamma_s|^2}{|1 - S_{11}\Gamma_s|^2} |S_{21}|^2 \frac{1}{1 - |\Gamma_{out}|^2} \tag{3-4-13}$$

where $P_{avn} = P_\ell$ when $\Gamma_\ell = \Gamma_{out}^*$. This can be done by substituting the following equation in Eq. (3-4-5).

$$\Gamma_\ell = \Gamma_{out}^* = \left(S_{22} + \frac{S_{12}S_{21}\Gamma_s}{1 - S_{11}\Gamma_s}\right)^*$$

Then the maximum available power gain for bilateral case is

$$G_{a\max} = \left|\frac{S_{21}}{S_{12}}\right| |K - (K^2 - 1)^{1/2}| \tag{3-4-14a}$$

where

$$K = \frac{1 + |\Delta|^2 - |S_{11}|^2 - |S_{22}|^2}{2|S_{12}S_{21}|} > 1 \tag{3-4-15}$$

with

$$\Delta = S_{11}S_{22} - S_{12}S_{21} \tag{3-4-16}$$

Hence the maximum available power gain is obtained only if the microwave amplifier is loaded with Γ_{sm} and $\Gamma_{\ell m}$ as reflection coefficients, where Γ_{sm} and $\Gamma_{\ell m}$ are the source and load reflection coefficients for maximum available power gain, respectively. The maximum frequency of oscillation is determined after the maximum available power gain is achieved.

When $K = 1$, the maximum stable gain is expressed as

$$G_{msg} = \frac{|S_{21}|}{|S_{12}|} \tag{3-4-14b}$$

Example 3-4-2:   Available Power Gain

A GaAs MESFET has the following S parameters measured at $V_{ds} = 5$ V and $I_{ds} = 40$ mA for 9 GHz with a 50-Ω reference:

$$S_{11} = 0.65 \; \underline{/-154°} \qquad \Gamma_s = 0.38\underline{/25°}$$

$$S_{12} = 0.02 \; \underline{/40°}$$

$$S_{21} = 2.04 \; \underline{/185°}$$

$$S_{22} = 0.55 \; \underline{/-30°}$$

Calculate   (a) the delta factor Δ, (b) the stability factor K, (c) the output reflection coefficient Γ_{out}, (d) the available power gain, (e) the maximum available power gain for bilateral case, and (f) the maximum stable power gain.

Solution   (a) The delta factor is

$$\Delta = 0.65 \; \underline{/-154°} \times 0.55 \; \underline{/-30°} - 0.02 \; \underline{/40°} \times 2.04 \; \underline{/185°}$$

$$= 0.32 \; \underline{/171°}$$

(b) The stability factor is

$$K = \frac{1 + |0.32|^2 - |0.65|^2 - |0.55|^2}{2|0.02 \times 2.04|} = 4.75 > 1$$

(c) The output reflection coefficient is

$$\Gamma_{out} = \frac{S_{22} - \Delta\Gamma_s}{1 - S_{11}\,\Gamma_s} = \frac{0.55 \; \underline{/-30°} - 0.32 \; \underline{/171°} \times 0.38 \; \underline{/25°}}{1 - 0.65 \; \underline{/-154°} \times 0.38 \; \underline{/25°}}$$

$$= 0.43 - j0.37 = 0.56 \; \underline{/-40.7°}$$

(d) The available power gain is

$$G_a = \frac{1 - |0.38|^2}{|1 - 0.65 \; \underline{/-154°} \times 0.38 \; \underline{/25°}\,|^2} \, |2.04|^2 \frac{1}{1 - |0.56|^2}$$

$$= 4.95 = 6.94 \text{ dB}$$

(e) The maximum available power gain in biateral case is

$$G_{amax} = \frac{|2.04|}{|0.02|} \, |4.75 - \sqrt{(4.75)^2 - 1}\,|$$

$$= 11.22 = 10.50 \text{ dB}$$

(f) The maximum stable power gain is

$$G_{msg} = \frac{|2.04|}{|0.02|} = 102 = 20 \text{ dB}$$

3-4-3 Operating Power Gain G_p

The power gain of a microwave amplifier is defined as the ratio of the power delivered to the load impedance P_ℓ over the power input to the network from the source P_{in}. That is,

$$G_p = \frac{P_\ell}{P_{in}} = \frac{1}{1 - |\Gamma_{in}|^2} |S_{21}|^2 \frac{1 - |\Gamma_\ell|^2}{|1 - S_{22}\Gamma_\ell|^2} \qquad (3\text{-}4\text{-}17)$$

where

$$\Gamma_{in} = S_{11} + \frac{S_{12}S_{21}\Gamma_\ell}{1 - S_{22}\Gamma_\ell} = \frac{S_{11} - S_{11}S_{22}\Gamma_\ell + S_{12}S_{21}\Gamma_\ell}{1 - S_{22}\Gamma_\ell}$$

$$= \frac{S_{11} - (S_{11}S_{22} - S_{12}S_{21})\Gamma_\ell}{1 - S_{22}\Gamma_\ell} \qquad (3\text{-}4\text{-}18)$$

$$= \frac{S_{11} - \Delta\Gamma_\ell}{1 - S_{22}\Gamma_\ell}$$

$$\Delta = S_{11}S_{22} - S_{12}S_{21} \qquad (3\text{-}4\text{-}19)$$

This equation can be expressed as

$$G_p = |S_{21}|^2 g_\ell \qquad (3\text{-}4\text{-}20)$$

where

$$g_\ell = \frac{1 - |\Gamma_\ell|^2}{\left(1 - \left|\dfrac{S_{11} - \Delta\Gamma_\ell}{1 - S_{22}\,\Gamma_\ell}\right|^2\right)|1 - S_{22}\Gamma_\ell|^2}$$

$$= \frac{1 - |\Gamma_\ell|^2}{|1 - S_{22}\Gamma_\ell|^2 - |S_{11} - \Delta\Gamma_\ell|^2} \qquad (3\text{-}4\text{-}21)$$

$$= \frac{1 - |\Gamma_\ell|^2}{1 - |S_{11}|^2 + |\Gamma_\ell|^2(|S_{22}|^2 - |\Delta|^2) - 2 \text{ Re } (\Gamma_\ell\, C_2)}$$

$$C_2 = S_{22} - \Delta S_{11}^* \qquad (3\text{-}4\text{-}22)$$

Example 3-4-3:   Operating Power Gain with Input Conjugate Matched

A GaAs MESFET has the following S parameters measured at $V_{ds} = 5$ V and $I_{ds} = 40$ mA for 8 GHz with a 50-Ω reference:

$$S_{11} = 0.45 \; \underline{/165°} \qquad \Gamma_{\ell} = 0.42 \; \underline{/140°}$$

$$S_{12} = 0.08 \; \underline{/10°}$$

$$S_{21} = 1.91 \; \underline{/10°}$$

$$S_{22} = 0.48 \; \underline{/15°}$$

Determine (a) the delta factor Δ, (b) the stability factor K, (c) the input reflection coefficient, (d) the operating power gain with the input conjugate matched, and (e) the power gain if the load is matched.

Solution   (a) The delta factor is

$$\Delta = 0.45 \; \underline{/165°} \times 0.48 \; \underline{/15°} - 0.08 \; \underline{/10°} \times 1.91 \; \underline{/10°}$$

$$= 0.094 \; \underline{/212°}$$

(b) The stability factor is

$$K = \frac{1 + |0.094|^2 - |0.45|^2 - |0.48|^2}{2|0.08 \times 1.91|} = 2.13 > 1$$

(c) The input reflection coefficient is

$$\Gamma_{in} = \frac{S_{11} - \Delta\Gamma_{\ell}}{1 - S_{22}\Gamma_{\ell}} = \frac{0.45 \; \underline{/165°} - 0.094 \; \underline{/212°} \times 0.42 \; \underline{/140°}}{1 - 0.48 \; \underline{/15°} \times 0.42 \; \underline{/140°}}$$

$$= 0.42 \; \underline{/169°}$$

(d) The operating power gain is

$$G_p = \frac{1}{1 - |0.42|^2}|1.91|^2 \frac{1 - |0.42|^2}{|1 - 0.48 \; \underline{/15°} \times 0.42 \; \underline{/140°}|}$$

$$= 3.07 = 4.88 \text{ dB}$$

(e) The power gain if $\Gamma_{\ell} = 0$ is

$$G_p = \frac{|1.91|^2}{1 - |0.45|^2} = 4.56 = 6.59 \text{ dB}$$

3-5 AMPLIFIER STABILITY

The stability of an amplifier is a very important consideration in a microwave circuit design. Stability or resistance to oscillation in a microwave circuit can be determined by the S parameters, the synthesized source, and the load impedances. Oscillations are possible in a two-port network if either the input or output port, or both, have negative resistance or when either $|\Gamma_{in}| > 1$ or $|\Gamma_{out}| > 1$. This occurs if $|S_{11}|$ or $|S_{22}|$ are greater than unity for a unilateral case ($|S_{12}| = 0$). However, even with negative resistance the amplifier might still be stable.

3-5-1 Types of Amplifier Stability

There are two types of stability:

1. *Conditional stability.* A network is conditionally stable if the real part of the input impedance Z_{in} and the output impedance Z_{out} is greater than zero for some positive real source and load impedances at a specific frequency.
2. *Unconditional stability.* A network is unconditionally stable if the real part of the input impedance Z_{in} and the output impedance Z_{out} is greater than zero for all positive real source and load impedances at a specific frequency.

It should be noted that positive real source and load impedance means that

$$|\Gamma_s| \leq 1 \quad \text{and} \quad |\Gamma_\ell| \leq 1$$

3-5-2 Stability Circles

The maximum power gain G_{max} that can be realized for a microwave amplifier without external feedback is defined as the forward power gain when the input and output are simultaneously and conjugately matched. Conjugately matched conditions mean that the reflection coefficient Γ_s of the source is equal to the conjugate of the input reflection coefficient Γ_{in}, and the reflection coefficient Γ_ℓ of the load is equal to the conjugate of the output reflection coefficient Γ_{out}. These are

$$\Gamma_s = \Gamma_{in}^* \quad \text{and} \quad \Gamma_\ell = \Gamma_{out}^*$$

For a microwave amplifier to be unconditionally stable, the magnitudes of S_{11}, S_{22}, Γ_{in}, and Γ_{out} must be smaller than unity and the transistor's inherent stability factor K must be greater than unity and positive. K is computed from

$$K = \frac{1 + |\Delta|^2 - |S_{11}|^2 - |S_{22}|^2}{2|S_{12}S_{21}|} > 1 \qquad (3\text{-}5\text{-}1)$$

where

$$|\Delta| = |S_{11}S_{22} - S_{12}S_{21}| < 1 \quad \text{or} \quad (1 - |S_{11}|^2) > |S_{12}S_{21}|$$
$$(1 - |S_{22}|^2) > |S_{12}S_{21}|$$

The input and output reflection coefficients are given by

$$\Gamma_{in} = S_{11} + \frac{S_{12}S_{21}\Gamma_\ell}{1 - S_{22}\Gamma_\ell} \tag{3-5-2}$$

$$\Gamma_{out} = S_{22} + \frac{S_{12}S_{21}\Gamma_s}{1 - S_{11}\Gamma_s} \tag{3-5-3}$$

The boundary conditions for stability are given by

$$|\Gamma_{in}| = 1 = \left| S_{11} + \frac{S_{12}S_{21}\Gamma_\ell}{1 - S_{22}\Gamma_\ell} \right| \tag{3-5-4}$$

$$|\Gamma_{out}| = 1 = \left| S_{22} + \frac{S_{12}S_{21}\Gamma_s}{1 - S_{11}\Gamma_s} \right| \tag{3-5-5}$$

Substitution of the real and imaginary values for the S parameters in Eqs. (3-5-4) and (3-5-5) yields the solutions of Γ_s and Γ_ℓ as

$$r_s \text{ (radius of } \Gamma_s \text{ circle)} = \frac{|S_{12}S_{21}|}{||S_{11}|^2 - |\Delta|^2|} \tag{3-5-6}$$

$$\mathbf{c}_s \text{ (center of } \Gamma_s \text{ circle)} = \frac{\mathbf{C}_s^*}{|S_{11}|^2 - |\Delta|^2} \tag{3-5-7}$$

$$r_\ell \text{ (radius of } \Gamma_\ell \text{ circle)} = \frac{|S_{12}S_{21}|}{||S_{22}|^2 - |\Delta|^2|} \tag{3-5-8}$$

$$\mathbf{c}_\ell \text{ (center of } \Gamma_\ell \text{ circle)} = \frac{\mathbf{C}_\ell^*}{|S_{22}|^2 - |\Delta|^2} \tag{3-5-9}$$

where

$$\Delta = S_{11}S_{22} - S_{12}S_{21} \tag{3-5-10}$$

$$\mathbf{C}_s = S_{11} - \Delta S_{22}^* \tag{3-5-11}$$

$$\mathbf{C}_\ell = S_{22} - \Delta S_{11}^* \tag{3-5-12}$$

The reflection coefficient of the source impedance required to conjugately match the input of the amplifier for maximum power gain is

$$\Gamma_{sm} = \mathbf{C}_s^* \left[\frac{B_s \pm \sqrt{B_s^2 - 4|\mathbf{C}_s|^2}}{2|\mathbf{C}_s|^2} \right] \tag{3-5-13}$$

where

$$B_s = 1 + |S_{11}|^2 - |S_{22}|^2 - |\Delta|^2 \tag{3-5-14}$$

The reflection coefficient of the load impedance required to conjugately match the output of a microwave device for maximum power gain is

$$\Gamma_{\ell m} = \mathbf{C}_\ell^* \left[\frac{B_\ell \pm \sqrt{B_\ell^2 - 4|\mathbf{C}_\ell|^2}}{2|\mathbf{C}_\ell|^2} \right] \tag{3-5-15}$$

where

$$B_\ell = 1 + |S_{22}|^2 - |S_{11}|^2 - |\Delta|^2 \tag{3-5-16}$$

If the computed values of B_s and B_ℓ are negative, the plus sign should be used in front of the radical in Eqs. (3-5-13) and (3-5-15). Conversely, if B_s and B_ℓ are positive, the negative sign should be used.

  Stability circules can be plotted directly on a Smith chart. These circles separate the output or input planes into stable and potentially unstable regions. A stability circle plotted on the output plane indicates the values of all loads that provide negative real input impedance, thereby causing the circuit to oscillate. A similar circle can be

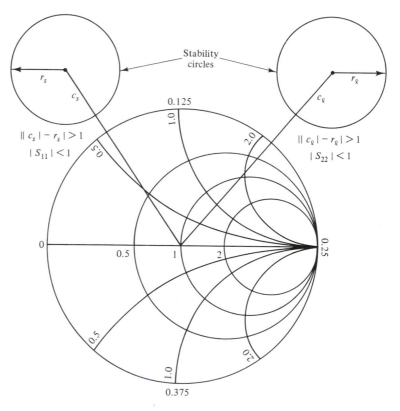

Figure 3-5-1 Unconditional stability circles.

plotted on the input plane, which indicates the values of all loads that provide negative real output impedance and again cause oscillation. A negative real impedance is defined as a reflection coefficient which has a magnitude that is greater than unity. The regions of instability occur within the circles whose centers and radii are expressed by Eqs. (3-5-6) through (3-5-9).

The stability criteria are:

1. *Stable:* $K > 1$ and $|\Delta| < 1$
   a. *Unconditionally stable:* $\|\mathbf{c}_s| - r_s| > 1$ for $|S_{22}| < 1$
                          $\|\mathbf{c}_\ell| - r_\ell| > 1$ for $|S_{11}| < 1$
   b. *Conditionally stable:* $\|\mathbf{c}_s| - r_s| < 1$ for $|S_{22}| < 1$
                          $\|\mathbf{c}_\ell| - r_\ell| < 1$ for $|S_{11}| < 1$
2. *Unstable (potentially):* $K > 1$ and $|\Delta| > 1$
                          $K < 1$ and $|\Delta| < 1$

In conclusion: for unconditional stability any passive source and load impedances in a two-port network must produce the stability circles completely outside the Smith chart, as shown in Fig. 3-5-1. If the stability circle is overlapped with the Smith chart, the stability is still conditional, as shown in Fig.3-5-2.

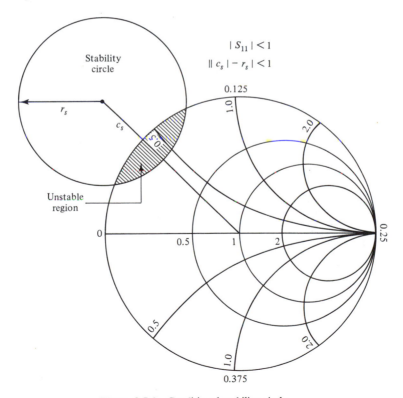

Figure 3-5-2 Conditional stability circle.

Example 3-5-1: Stability Circles of a Microwave Amplifier

A certain GaAs MESFET has the following S parameters measured at 9 GHz with a 50-Ω reference:

$$S_{11} = 0.64 \ \underline{/-170°}$$

$$S_{12} = 0.05 \ \underline{/15°}$$

$$S_{21} = 2.10 \ \underline{/30°}$$

$$S_{22} = 0.57 \ \underline{/-95°}$$

Compute (a) the delta factor Δ, and (b) the stability factor K. (c) Find the center and radius of the input stability circle and plot the circle. (d) Determine the center and radius of the output stability circle and plot the circle.

Solution   (a) The delta factor is

$$\Delta = S_{11}S_{22} - S_{12}S_{21}$$

$$= 0.64 \ \underline{/-170°} \times 0.57 \ \underline{/-95°} - 0.05 \ \underline{/15°} \times 2.10 \ \underline{/30°}$$

$$= 0.30 \ \underline{/111.45°}$$

$$|\Delta| = 0.30 < 1$$

$$|\Delta|^2 = 0.09$$

(b) The stability factor K is

$$K = \frac{1 + |\Delta|^2 - |S_{11}|^2 - |S_{22}|^2}{2|S_{12}S_{21}|}$$

$$= \frac{1 + |0.30|^2 - |0.64|^2 - |0.57|^2}{2 \times |0.05 \times 2.10|}$$

$$= 1.71 > 1$$

(c) The input stability circle is determined as follows:

$$C_s^* = S_{11}^* - \Delta^*S_{22} = 0.64 \ \underline{/170°} - 0.30 \ \underline{/-111.45°} \times 0.57 \ \underline{/-95°}$$

$$= 0.48 \ \underline{/176.52°}$$

The center of the input stability circle is

$$\mathbf{c}_s = \frac{C_s^*}{|S_{11}|^2 - |\Delta|^2} = \frac{0.48 \ \underline{/176.42°}}{|0.64|^2 - |0.30|^2} = 1.50 \ \underline{/176.42°}$$

The radius is

$$r_s = \frac{|S_{12}S_{21}|}{||S_{11}|^2 - |\Delta|^2|} = \frac{|0.05 \times 2.10|}{|0.64|^2 - |0.30|^2} = 0.33$$

(d) The output stability circle is found as follows:

$$C_\ell^* = S_{22}^* - \Delta^* S_{11} = 0.57 \underline{/95°} - 0.30 \underline{/-111.45°} \times 0.64 \underline{/-170°}$$
$$= 0.39 \underline{/103.32°}$$

The center of the output stability circle is

$$c_\ell = \frac{C_\ell^*}{|S_{22}|^2 - |\Delta|^2} = \frac{0.39 \underline{/103.32°}}{|0.57|^2 - |0.30|^2} = 1.70 \underline{/103.32°}$$

The radius is

$$r_\ell = \frac{|S_{12}S_{21}|}{||S_{22}|^2 - |\Delta|^2|} = \frac{0.05 \times 2.10}{|0.57|^2 - |0.30|^2} = 0.48$$

Both the input and output stability circles are completely outside the Smith chart, as shown in Fig. 3-5-3, so the amplifier is unconditionally stable.

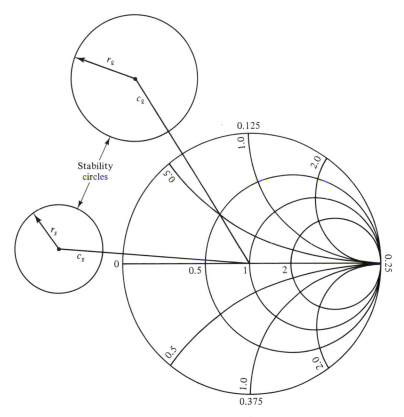

Figure 3-5-3 Stability circles for Example 3-5-1.

3-6 CONSTANT-GAIN CIRCLES

3-6-1 Unilateral Case ($|S_{12}| = 0$)

It is obvious that for $\Gamma_s = S_{11}^*$ or $\Gamma_\ell = S_{22}^*$, the power gain G_s or G_ℓ is equal to maximum, respectively. It is also clear that for $|\Gamma_s| = 1$ or $|\Gamma_\ell| = 1$, the power gain G_s or G_ℓ has a value zero. For any arbitrary value of G_s or G_ℓ between these extremes of zero and $G_{s\max}$ or $G_{\ell\max}$, solutions for Γ_s or Γ_ℓ lie on a circle.

For $0 < G_s < G_{s\max}$ (dB)

$$g_s = \frac{1 - |\Gamma_s|^2}{|1 - S_{11}\Gamma_s|^2} \qquad \text{(numerical value)} \qquad (3\text{-}6\text{-}1)$$

It is convenient to plot these circles on a Smith chart. The circles have their centers located on the vector drawn from the center of the Smith chart to the point S_{11}^* or S_{22}^*. The distance from the center of the Smith chart to the center of the constant-gain circle along the vector $\mathbf{S}_{11}^*$ or $\mathbf{S}_{22}^*$ is given by

$$d_s = \frac{g_{ns}|S_{11}|}{1 - |S_{11}|^2(1 - g_{ns})} \qquad (3\text{-}6\text{-}2)$$

The radius of the constant-gain circle is expressed by

$$r_s = \frac{\sqrt{1 - g_{ns}}\,(1 - |S_{11}|^2)}{1 - |S_{11}|^2(1 - g_{ns})} \qquad (3\text{-}6\text{-}3)$$

where g_n is the normalized gain value for the gain circle g_s or g_ℓ, respectively. That is,

$$g_{ns} = \frac{g_s}{g_{s\max}} = g_s(1 - |S_{11}|^2) \qquad (3\text{-}6\text{-}4)$$

where $0 \le g_{ns} \le 1$.

The S parameters of the signal side should be replaced by the S parameters of the load side in Eqs. (3-6-1) through (3-6-4), respectively, for the constant-gain circles of the load side. Any value of Γ_s and Γ_ℓ along a specific gain circle would result in a power gain of that amount of specific decibels.

Example 3-6-1:   Constant-Gain Circles for Unilateral Case

A certain microwave transistor has the following S parameters measured at 4 GHz with a 50-Ω reference:

$$S_{11} = 0.707 \,\underline{/-155°}$$

$$S_{12} = 0$$

$$S_{21} = 5.00 \,\underline{/180°}$$

$$S_{22} = 0.510 \,\underline{/-20°}$$

Plot (a) the input constant-gain circles for 2, 1, 0, −1 dB, and (b) the output constant-gain circles for the same decibel levels.

Solution   (a) Compute the normalized gain values for different power gain levels and then calculate the centers and radii of the gain circles. The values are tabulated in Table 3-6-1. The maximum input power gain is

$$G_{smax} = \frac{1}{1 - |S_{11}|^2} = \frac{1}{1 - |0.707|^2}$$

$$= 2 = 3 \text{ dB}$$

TABLE 3-6-1 COMPUTED VALUES

G_s (dB)	2	1	0	−1
g_s	1.59	1.26	1.00	0.79
g_{ns}	0.80	0.63	0.50	0.40
d_s	0.63	0.55	0.47	0.40
r_s	0.25	0.38	0.47	0.55
G_ℓ (dB)		1	0	−1
g_ℓ		1.26	1.00	0.79
$g_{n\ell}$		0.93	0.74	0.58
d_ℓ		0.48	0.41	0.33
r_ℓ		0.20	0.41	0.54

The maximum output network gain is

$$G_{\ell max} = \frac{1}{1 - |S_{22}|^2} = \frac{1}{1 - |0.51|^2}$$

$$= 1.35 = 1.3 \text{ dB}$$

There is no 2-dB constant-gain circle for the output network.
For 2-dB gain:

$$g_{ns} = g_s(1 - |S_{11}|^2) = 1.59 \, (1 - |0.707|^2)$$

$$= 0.80$$

$$d_s = \frac{g_{ns}|S_{11}|}{1 - |S_{11}|^2(1 - g_{ns})} = \frac{0.80 \times |0.707|}{1 - |0.707|^2(1 - 0.80)}$$

$$= 0.63$$

$$r_s = \frac{\sqrt{1 - g_{ns}} \, (1 - |S_{11}|^2)}{1 - |S_{11}|^2(1 - g_{ns})}$$

$$= \frac{\sqrt{1 - 0.80}(1 - |0.707|^2)}{1 - |0.707|^2(1 - 0.80)}$$

$$= 0.25$$

(b) The input and output constant-gain circles are shown in Fig. 3-6-1.

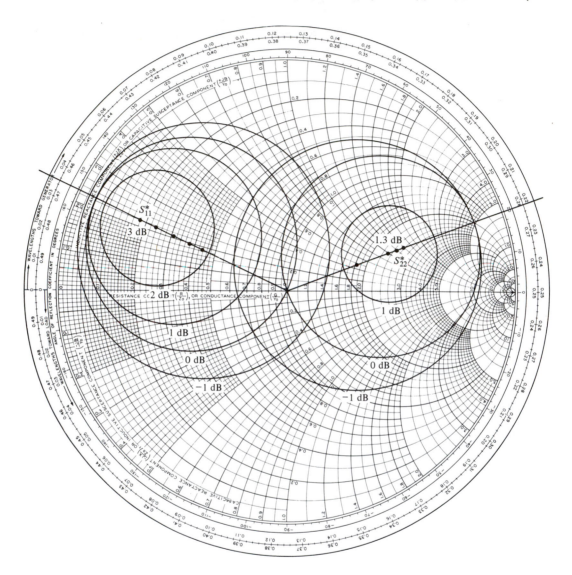

Figure 3-6-1 Constant-gain circles for the input network.

3-6-2 Unilateral Figure of Merit

When $|S_{12}|$ is not zero but is very small and assumed to be zero, the error that occurred may be determined by the unilateral figure of merit. From Eqs. (3-4-5) and (3-4-11), the magnitude ratio of the transducer power gain to the unilateral transducer power gain is

$$\frac{G_t}{G_{tu}} = \frac{1}{|1 - X|^2} \tag{3-6-5}$$

where

$$X = \frac{S_{12}S_{21}\Gamma_s\Gamma_t}{(1 - S_{11}\Gamma_s)(1 - S_{22}\Gamma_t)} \qquad (3\text{-}6\text{-}6)$$

The boundary condition for the ratio is

$$\frac{1}{(1 + |X|)^2} < \frac{G_t}{G_{tu}} < \frac{1}{(1 - |X|)^2} \qquad (3\text{-}6\text{-}7)$$

When $\Gamma_s = S_{11}^*$ and $\Gamma_t = S_{22}^*$ the unilateral transducer power gain reaches a maximum value and, in this case, the maximum error introduced when using $G_{tu\,max}$ is bounded by

$$\frac{1}{(1 + M)^2} < \frac{G_t}{G_{tu\,max}} < \frac{1}{(1 - M)^2} \qquad (3\text{-}6\text{-}8)$$

where

$$M = \frac{|S_{12}||S_{21}||S_{11}||S_{22}|}{(1 - |S_{11}|^2)(1 - |S_{22}|^2)} \qquad (3\text{-}6\text{-}9)$$

M is called the *unilateral figure of merit*. Normally, the value of the unilateral figure of merit varies with frequency because of its dependence on the S parameters. Its value is very small, around 0.03 or the -15-dB range. The maximum error is equal to

$$\text{maximum error} = \pm\frac{1}{(1 + M)^2} \quad \text{dB} \qquad (3\text{-}6\text{-}10)$$

Example 3-6-2: Unilateral Figure of Merit

A GaAs MESFET has the following S parameters measured at $V_{ds} = 3$ V and $I_{ds} = 30$ mA for 8 GHz with a 50-Ω reference:

$$S_{11} = 0.52 \,/\!-145° \qquad \Gamma_s = 0.43 \,/70°$$

$$S_{12} = 0.03 \,/20° \qquad \Gamma_t = 0.45 \,/80°$$

$$S_{21} = 2.56 \,/170°$$

$$S_{22} = 0.48 \,/\!-20°$$

Calculate (a) the error factor X, (b) the gain error ratio G_t/G_{tu}, (c) the maximum error factor M, and (d) the maximum gain error ratio $G_t/G_{tu\,max}$.

Solution Check the delta factor and stability factor:

$$\Delta = 0.52 \,/\!-145° \times 0.48 \,/\!-20° - 0.03 \,/20° \times 2.56 \,/170°$$

$$= 0.168 \,/197°$$

$$K = \frac{1 + |0.168|^2 - |0.52|^2 - |0.48|^2}{2|0.03 \times 2.56|} = 3.53 > 1$$

(a) From Eq. (3-6-6) the error factor X is

$$X = \frac{0.03 \underline{/20°} \times 2.56 \underline{/170°} \times 0.43 \underline{/70°} \times 0.45 \underline{/80°}}{(1 - 0.52\underline{/-145°} \times 0.43 \underline{/70°})(1 - 0.48 \underline{/-20°} \times 0.45 \underline{/80°})}$$

$$= 0.017 \underline{/340°}$$

(b) From Eq. (3-6-7) the gain error ratio is

$$\frac{1}{(1 + |0.017|)^2} < \frac{G_t}{G_{tu}} < \frac{1}{(1 - |0.017|)^2}$$

$$0.97 < \frac{G_t}{G_{tu}} < 1.03$$

$$-0.13 \text{ dB} < \frac{G_t}{G_{tu}} < +0.13 \text{ dB}$$

(c) From Eq. (3-6-9) the error factor M is

$$M = \frac{|0.03||2.56||0.52||0.48|}{(1 - |0.52|^2)(1 - |0.48|^2)} = 0.04$$

(d) From Eq. (3-6-8) the maximum gain error is

$$\frac{1}{(1 + 0.04)^2} < \frac{G_t}{G_{tumax}} < \frac{1}{(1 - 0.04)^2}$$

$$0.92 < \frac{G_t}{G_{tumax}} < 1.09$$

$$-0.36 \text{ dB} < \frac{G_t}{G_{tumax}} < +0.37 \text{ dB}$$

3-6-3 Bilateral Case ($|S_{12}| \neq 0$)

The bilateral case occurs when S_{12} cannot be neglected. The transducer power-gain equation, as shown in Eq. (3-4-6), is

$$G_t = \frac{1 - |\Gamma_s|^2}{|1 - \Gamma_{in}\Gamma_s|^2} |S_{21}|^2 \frac{1 - |\Gamma_t|^2}{|1 - S_{22}\Gamma_t|^2} \tag{3-6-11}$$

where

$$\Gamma_{in} = S_{11} + \frac{S_{12}S_{21}\Gamma_t}{1 - S_{22}\Gamma_t} \tag{3-6-12}$$

It can be seen that Γ_{in} is a function of Γ_t and it is a tedious work to plot a constant-gain circle for this case. In practice, the constant operating power-gain circles are often plotted for the bilateral case of the stability considertion of the microwave amplifier design.

3-7 CONSTANT OPERATING POWER-GAIN CIRCLES (BILATERAL CASE)

Because the operating power gain is independent of the source impedance in bilateral case, its gain circles for both unconditionally stable and potentially unstable microwave devices can easily be plotted. There are two cases to be studied: unconditionally stable and potentially unstable.

3-7-1 Unconditionally Stable

From the operating power-gain equation (3-4-20) we can write

$$G_p = |S_{21}|^2 g_p \tag{3-7-1}$$

where

$$g_p = \frac{1 - |\Gamma_\ell|^2}{\left(1 - \left|\dfrac{S_{11} - \Delta\Gamma_\ell}{1 - S_{22}\Gamma_\ell}\right|^2\right)\left|1 - S_{22}\Gamma_\ell\right|^2}$$

$$= \frac{1 - |\Gamma_\ell|^2}{|1 - S_{22}\Gamma_\ell|^2 - |S_{11} - \Delta\Gamma_\ell|^2}$$

$$= \frac{1 - |\Gamma_\ell|^2}{1 - |S_{11}|^2 + |\Gamma_\ell|^2(|S_{22}|^2 - |\Delta|^2) - \mathrm{Re}\,(\Gamma_\ell C_\ell)} \tag{3-7-2}$$

$$\Delta = S_{11}S_{22} - S_{12}S_{21} \tag{3-7-3}$$

$$C_\ell = S_{22} - \Delta S_{11}^* \tag{3-7-4}$$

In order to find the centers of the constant operating power-gain circles from the center of the Smith chart and the radii of the circles, substituting $\Gamma_\ell = U_\ell + jV_\ell$ in Eq. (3-7-2) and letting $\mathrm{Re}\,(\Gamma_\ell C_\ell) = U_\ell\,\mathrm{Re}\,C_\ell - V_\ell\,\mathrm{Im}\,C_\ell$, we obtain

$$U_p = \frac{g_p\,\mathrm{Re}\,(C_\ell^*)}{1 + g_p(|S_{21}|^2 - |\Delta|^2)} \tag{3-7-5}$$

and

$$V_p = \frac{g_p\,\mathrm{Im}\,(C_\ell^*)}{1 + g_p(|S_{22}|^2 - |\Delta|^2)} \tag{3-7-6}$$

The radii of the circles are given by

$$r_{p\ell} = \frac{(1 - 2K|S_{12}S_{21}|g_p + |S_{12}S_{21}|^2 g_p^2)^{1/2}}{|1 + g_p(|S_{22}|^2 - |\Delta|^2)|} \tag{3-7-7}$$

where

$$K = \frac{1 + |\Delta|^2 - |S_{11}|^2 - |S_{22}|^2}{2|S_{12}S_{21}|} \qquad (3\text{-}7\text{-}8)$$

The distance from the center of the Smith chart to the center of the circle is found to be

$$d_{p\ell} = (U_p + V_p)^{1/2} = \frac{g_p C_\ell^*}{|1 + g_p(|S_{22}|^2 - |\Delta|^2)|} \qquad (3\text{-}7\text{-}9)$$

where

$$\mathbf{C}_\ell^* = |C_\ell^*| \; \underline{/\theta_\ell}.$$

The centers of the power-gain circles are located along a line with an angle

$$\theta_\ell = \tan^{-1}\!\left(\frac{\operatorname{Im} \mathbf{C}_\ell^*}{\operatorname{Re} \mathbf{C}_\ell^*}\right) \qquad (3\text{-}7\text{-}10)$$

The maximum operating power gain occurs when $r_p = 0$. This means that from Eq. (3-7-7)

$$g_{p,\max}^2 |S_{12}S_{21}|^2 - 2K|S_{12}S_{21}|g_{p,\max} + 1 = 0$$

and the maximum value of g_p is

$$g_{p,\max} = \frac{1}{|S_{12}S_{21}|}(K - \sqrt{K^2 - 1}) = \frac{G_{p,\max}}{|S_{21}|^2} \qquad (3\text{-}7\text{-}11)$$

where the maximum operating power gain is given by

$$G_{p,\max} = \frac{|S_{21}|}{|S_{12}|}(K - \sqrt{K^2 - 1}) \qquad (3\text{-}7\text{-}12)$$

where the minus sign is used with inherent stability.

For a given power gain G_p, the load reflection coefficient Γ_ℓ is chosen from the constant operating power-gain circle. $G_{p,\max}$ approaches when Γ_ℓ is chosen at the distance where $g_{p,\max} = G_{p,\max}/|S_{21}|^2$. The maximum output power gain is obtained when a conjugate match is realized at the input side ($\Gamma_s = \Gamma_{in}^*$). When $\Gamma_s = \Gamma_{in}^*$ the input power is equal to the maximum available input power. Therefore, the operating power gain and the maximum transducer power gain are equal. The values of Γ_s and Γ_ℓ that result in $G_{p,\max}$ are identical to Γ_{sm} and $\Gamma_{\ell m}$, respectively.

Similarly, the distance from the center of the Smith chart to the center of the input constant operating power-gain circle is given by

$$d_{ps} = \frac{g_p C_s^*}{|1 + g_p(|S_{11}|^2 - |\Delta|^2)|} \qquad (3\text{-}7\text{-}13)$$

where

$$C_s^* = S_{11}^* - \Delta^* S_{22} = |C_s^*| \; \underline{/\theta_s} \tag{3-7-14}$$

$$\theta_s = \tan^{-1}\left(\frac{\text{Im } C_s^*}{\text{Re } C_s^*}\right) \tag{3-7-15}$$

$$\Delta = S_{11}S_{22} - S_{12}S_{21} \tag{3-7-16}$$

The radius of the circle is expressed by

$$r_{ps} = \frac{(1 - 2K|S_{12}S_{21}|g_p + |S_{12}S_{21}|^2 g_p^2)^{1/2}}{|1 + g_p(|S_{11}|^2 - |\Delta|^2)|} \tag{3-7-17}$$

where

$$K = \frac{1 + |\Delta|^2 - |S_{11}|^2 - |S_{22}|^2}{2|S_{12}S_{21}|} \tag{3-7-18}$$

Example 3-7-1:   Constant Operating Power-Gain Circles

A certain GaAs MESFET has the following S parameters measured at 8 GHz with a 50-Ω resistance:

$$S_{11} = 0.26 \; \underline{/-55°}$$

$$S_{12} = 0.08 \; \underline{/80°}$$

$$S_{21} = 2.14 \; \underline{/65°}$$

$$S_{22} = 0.82 \; \underline{/-30°}$$

Plot the power-gain circles.

Solution

1. Compute the delta factor Δ and the stability factor K:

$$\Delta = 0.26 \; \underline{/-55°} \times 0.82 \; \underline{/-30°} - 0.08 \; \underline{/80°} \times 2.14 \; \underline{/65°}$$

$$= 0.36 \; \underline{/-62.7°}$$

$$|\Delta| = 0.36 < 1$$

$$|\Delta|^2 = 0.13$$

$$K = \frac{1 + |0.36|^2 - |0.26|^2 - |0.82|^2}{2 \times |0.08 \times 2.14|} = 1.15 > 1$$

2. The maximum operating power gain is

$$G_{p.\max} = \frac{2.14}{0.08} [1.15 - \sqrt{(1.15)^2 - 1}]$$

$$= 15.52 = 11.91 \text{ dB}$$

3. Compute $g_{p,max}$.

$$g_{p,max} = \frac{15.52}{|2.14|^2} = 3.39$$

4. The distance of the circle center is computed as

$$C_{\ell}^* = S_{22}^* - \Delta^* S_{11} = 0.82 \; \underline{/30°} - 0.36 \; \underline{/62.7°} \times 0.26 \; \underline{/-55°}$$

$$= 0.73 \; \underline{/32.83°}$$

$$d_{p\ell} = \frac{3.39 \times 0.73 \; \underline{/32.83°}}{|1 + 3.39(|0.82|^2 - 0.13)|} = 0.87 \; \underline{/32.83°}$$

5. The radius of the circle is computed as

$$r_{p\ell} = \frac{\sqrt{1 - 2 \times 1.15|0.08 \times 2.14| \times 3.39 + |0.08 \times 2.14|^2 (3.39)^2}}{|1 + 3.39 \; [|0.82|^2 - |0.36|^2]|}$$

$$= 0.00$$

6. Compute the distances and the radii of the circles for the power gains of 10, 8, and 6 dB.

7. The computed values are tabulated in Table 3-7-1.

TABLE 3-7-1　COMPUTED VALUES

(dB)	$G_{p,max}$ (Numerical)	g_p	d_p	r_p
11.91	15.52	3.39	0.87	0
10	10	2.18	0.73	0.25
8	6.31	1.38	0.58	0.41
6	3.98	0.87	0.43	0.56

8. Plot the power-gain circles on the Smith chart as shown in Fig. 3-7-1.

9. The normalized load impedance for 11.91-dB power gain is read from the plot at Γ_ℓ as

$$z_\ell = 0.80 + j3.20$$

The load impedance is $Z_\ell = 40 + j160 \; \Omega$.

10. For each value of the distance d_p used for the power gain desired, maximum output power occurs with a conjugate match at the input for which $\Gamma_{in}^* = \Gamma_s$. Then from Eq. (3-4-8), we have

$$\Gamma_s = \left(0.26 \; \underline{/-55°} + \frac{0.08 \; \underline{/80°} \times 2.14 \; \underline{/65°} \times 0.87 \; \underline{/32.83°}}{1 - 0.82 \; \underline{/-30°} \times 0.87 \; \underline{/32.83°}} \right)^*$$

$$= 0.31 \; \underline{/-76.87°}$$

Then read z_s at Γ_s as

$$z_s = 0.95 - j0.65$$

and the input impedance is

$$Z_s = 47.5 - j32.5 \ \Omega$$

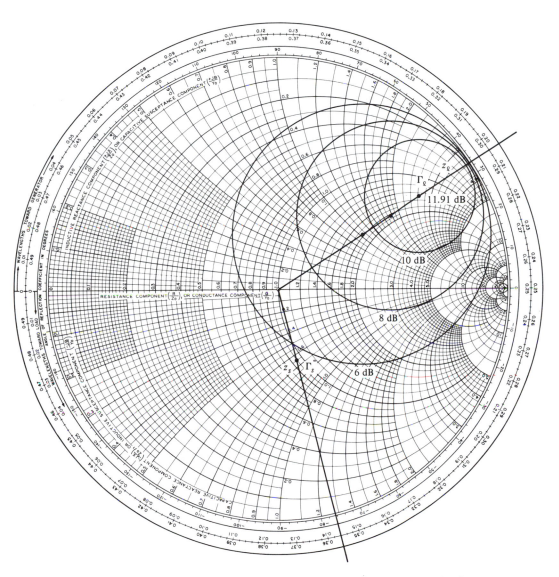

Figure 3-7-1 Constant operating power-gain circles for Example 3-7-1.

3-7-2 Potentially Unstable

When the microwave device is potentially unstable, the output stability circle is first drawn on the Smith chart. After the power gain is chosen, a power operating power-gain circle is plotted. A value of d_p at a point on the power-gain circle within the stable region is selected. The normalized load impedance can be read from the Smith chart.

Design Procedures

1. Draw the constant operating power-gain circle for a given power gain G_p in decibels.
2. Draw the output stability circles.
3. Choose Γ_ℓ in the stable region.
4. Compute Γ_{in} and determine if a conjugate match at the input is possible.
5. Draw the input stability circle and determine if $\Gamma_s = \Gamma_{in}^*$ is within the input stable region.
6. If $\Gamma_s = \Gamma_{in}^*$ is not in the stable region (or in the stable region but very close to the input stability circle), a new value of Γ_s must be selected arbitrarily or a value of G_p is chosen again. Be careful to ensure that the values of Γ_ℓ and Γ_s should not be too close to their respective stability circles, because oscillation may occur when the input and output circles are not matched.

Design Example 3-7-2:   Operating Power-Gain Circles for Potential Stability (Bilateral Case)

A certain GaAs MESFET has the following S parameters measured at 9 GHz with a 50-Ω resistance reference:

$$S_{11} = 0.45 \; \underline{/-60°}$$

$$S_{12} = 0.09 \; \underline{/70°}$$

$$S_{21} = 2.50 \; \underline{/74°}$$

$$S_{22} = 0.80 \; \underline{/-50°}$$

Design an amplifier for an operating power gain of 10 dB.

Solution

1. Compute the delta factor Δ and the stability factor K.

$$\Delta = 0.45 \; \underline{/-60°} \times 0.80 \; \underline{/-50°} - 0.09 \; \underline{/70°} \times 2.50 \; \underline{/74°}$$

$$= 0.20 \; \underline{/-70.7°}$$

$$|\Delta| = 0.20 < 1$$

$$|\Delta|^2 = 0.04$$

$$K = \frac{1 + 0.04 - |0.45|^2 - |0.80|^2}{2 \times |0.09 \times 2.50|}$$

$$= 0.44 < 1$$

The device is potentially unstable because $K < 1$.

2. Compute the maximum stable power gain at $K = 1$.

$$G_{msp} = \frac{|S_{21}|}{|S_{12}|} = \frac{2.5}{0.09} = 27.78 = 14.44 \text{ dB}$$

A 10-dB power gain is possible.

3. Compute d_p and r_p for the 10-dB operating power-gain circle.

$$g_p = \frac{G_p}{|S_{21}|^2} = \frac{10}{|2.5|^2} = 1.60$$

$$C_{\ell}^* = S_{22}^* - \Delta^* S_{11} = 0.80 \ \underline{/50°} - 0.20 \ \underline{/70.7°} \times 0.45 \ \underline{/-60°}$$

$$= 0.73 \ \underline{/54.55°}$$

$$d_{p\ell} = \frac{1.60 \times 0.73 \ \underline{/54.55°}}{|1 + 1.6 \times (|0.8|^2 - |0.20|^2)|} = 0.60 \ \underline{/54.55°}$$

$$r_{p\ell} = \frac{\sqrt{1 - 2 \times 0.44 \times |0.09 \times 2.5| \times 1.6 + |0.09 \times 2.5|^2 \times (1.6)^2}}{|1 + 1.6(|0.80|^2 - |0.20|^2)|}$$

$$= 0.46$$

The 10-dB power gain circle is plotted on the Smith chart shown in Fig. 3-7-2.

4. Compute d_ℓ and r_ℓ for the output stability circle.

$$C_\ell^* = 0.73 \ \underline{/54.55°}$$

$$C_\ell = \frac{0.73 \ \underline{/54.55°}}{|0.80|^2 - |0.20|^2} = 1.22 \ \underline{/54.55°}$$

$$r_\ell = \frac{|0.09 \times 2.50|}{||0.80|^2 - |0.20|^2|} = 0.38$$

The output stability circle is plotted on the same Smith chart.

5. Since $|S_{11}| < 1$, the stable region is the region outside the output stability circle. The load reflection coefficient Γ_ℓ is chosen on the 10-dB power-gain circle at the location A. Then

$$\Gamma_\ell = 0.38 \ \underline{/104°}$$

$$z_\ell = 0.70 + j0.50$$

$$Z_\ell = 35 + j25 \ \Omega$$

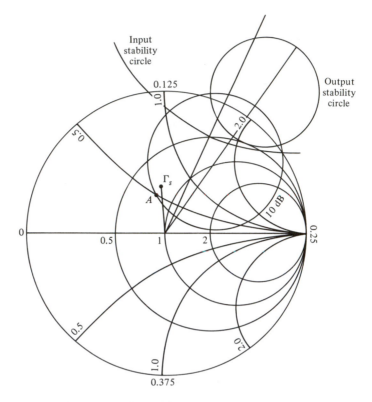

Figure 3-7-2. Potential stability for Example 3-7-2.

6. Compute Γ_s for a possible conjugate match.

$$\Gamma_s = \Gamma_{in}^* = \left(0.45 \underline{/-60^\circ} + \frac{0.09 \underline{/70^\circ} \times 2.50 \underline{/74^\circ}}{1 - 0.80 \underline{/-50^\circ} \times 0.38 \underline{/104^\circ}}\right)^*$$

$$= 0.30 \underline{/-266.2^\circ}$$

7. Compute d_s and r_s for the input stability circle.

$$\mathbf{C}_s^* = 0.45 \underline{/60^\circ} - 0.20 \underline{/70.7^\circ} \times 0.80 \underline{/-50^\circ}$$

$$= 0.35 \underline{/76.37^\circ}$$

$$\mathbf{c}_s = \frac{0.35 \underline{/76.37^\circ}}{|0.45|^2 - |0.20|^2} = 2.19 \underline{/76.37^\circ}$$

$$r_s = \frac{0.09 \times 2.50}{||0.45|^2 - |0.20|^2|} = 1.41$$

The input stability circle is ploted on the same Smith chart.

8. From Fig. 3-7-2, Γ_s is a stable source reflection coefficient because it is located in the stable region outside the input stability circle.

3-8 CONSTANT-NOISE-FIGURE CIRCLES

Noise is a problem in a two-port amplifier. Even there is no input signal, a small output noise voltage is present. In an amplifier, the output noise power is the sum of the amplified noise power from the input signal plus the noise power generated by the amplifier itself. The sensitivity of a detection system is determined by its signal-to-noise ratio. There are two main noise sources:

1. *Thermal noise.* Thermal noise, also called *Johnson noise,* is the random fluctuation of the electron motion by thermal agitation in a conductor. When a noise resistor is equal to the load resistor, the mean-square thermal voltage at a temperature T with a bandwidth B is given by

$$\overline{V_n^2} = 4KTBR_n \qquad (3\text{-}8\text{-}1)$$

where $K = 1.38 \times 10^{-23}$ J/°K is the Boltzmann constant
   T = absolute temperature, degrees Kelvin
   B = noise bandwidth, hertz
   R_n = noise resistance, ohms
Similarly, the mean-square thermal current is

$$\overline{I_n^2} = 4KTBG \qquad (3\text{-}8\text{-}2)$$

where G is the noise conductance. The maximum available noise power from a resistor R_n can be written as

$$P_{n\max} = \frac{V_n^2}{4R_n} = KTB \qquad (3\text{-}8\text{-}3)$$

This type of noise distribution is usually called *white noise.*

2. *Shot noise.* Shot noise, also called *Schottky noise,* is the fluctuation in the number of carriers in a current and it is present in all active devices because of the discrete nature of the electron flow. The mean-square shot noise current is expressed by

$$\overline{I_{sn}^2} = 2gI_{dc}B \qquad (3\text{-}8\text{-}4)$$

where $g = 1.60 \times 10^{-19}$ C is the electron charge
   I_{dc} = direct current flowing through the device
   B = bandwidth, hertz
It should be observed that the same noise power exists in a given bandwidth regardless of the center frequency. Such a distribution, which gives the same noise per unit bandwidth anywhere in the spectrum, is called *white noise.*

One important consideration of microwave amplifier design is the noise figure. The *noise figure* is defined as the signal-to-noise ratio at the input to the signal-to-noise ratio at the output. That is,

$$F = \frac{S_{in}/N_{in}}{S_o/N_o} = \frac{S_{in}N_o}{S_oN_{in}} \tag{3-8-5}$$

The noise figure of a two-port microwave amplifier is given by

$$F = F_{min} + \frac{r_n}{g_s}|y_s - y_o|^2$$

$$= F_{min} + \frac{r_n}{g_s}[(g_s - g_o)^2 + (b_s - b_o)^2] \tag{3-8-6}$$

where F_{min} = minimum noise figure, which is a function of the device operating
frequency and current

$r_n = R_n/Z_o$ is the normalized noise resistance of the two ports

$y_s = g_s + jb_s$ is the normalized source admittance

$y_o = g_o + jb_o$ is the normalized optimum source admittance which results
in the minimum noise figure

The normalized source admittance can be written in terms of the source reflection
coefficient Γ_s as

$$y_s = \frac{1 - \Gamma_s}{1 + \Gamma_s} \tag{3-8-7}$$

Similarly, the normalized optimum source admittance can be expressed as

$$y_o = \frac{1 - \Gamma_o}{1 + \Gamma_o} \tag{3-8-8}$$

where Γ_o is the optimum source reflection coefficient which results in the minimum
noise figure.

Substitution of Eqs. (3-8-7) and (3-8-8) in Eq. (3-8-6) yields the noise-figure
equation as

$$F = F_{min} + \frac{4r_n|\Gamma_s - \Gamma_o|^2}{(1 - |\Gamma_s|^2)|1 + \Gamma_o|^2} \tag{3-8-9}$$

The resistance r_n can be found by measuring F for $\Gamma_s = 0$ when a 50-Ω source
resistance is used. Then

$$r_n = (F_{\Gamma_s=0} - F_{min})\frac{|1 + \Gamma_o|^2}{4|\Gamma_o|^2} \tag{3-8-10}$$

To determine the noise-figure circle for a given noise figure F_i, we define a noise
figure parameter N_i first as

$$N_i = \frac{|\Gamma_s - \Gamma_o|^2}{1 - |\Gamma_s|^2} = \frac{F_i - F_{min}}{4r_n}|1 + \Gamma_o|^2 \tag{3-8-11}$$

Equation (3-8-11) can be rewritten as

$$(\Gamma_s - \Gamma_o)(\Gamma_s^* - \Gamma_o^*) = N_i(1 - |\Gamma_s|^2) = |\Gamma_s - \Gamma_o|^2 \qquad (3\text{-}8\text{-}12)$$

or

$$|\Gamma_s|^2(1 + N_i) + |\Gamma_o|^2 - 2 \operatorname{Re}(\Gamma_s \Gamma_o^*) = N_i \qquad (3\text{-}8\text{-}13)$$

Let's multiply both sides of Eq. (3-8-13) by a factor $(1 + N_i)$; we have

$$|\Gamma_s|^2(1 + N_i)^2 + |\Gamma_o|^2 - 2(1 + N_i)\operatorname{Re}(\Gamma_s \Gamma_o^*) = N_i^2 + N_i(1 - |\Gamma_o|^2)$$

and

$$|\Gamma_s(1 + N_i) - \Gamma_o|^2 = N_i^2 + N_i(1 - |\Gamma_o|^2)$$

Then

$$\left| \Gamma_s - \frac{\Gamma_o}{1 + N_i} \right|^2 = \frac{N_i^2 + N_i(1 - |\Gamma_o|^2)}{(1 + N_i)^2} \qquad (3\text{-}8\text{-}14)$$

Equation (3-8-14) represents a family of circles in terms of N_i. The circle centers and radii are given, respectively, by

$$c_{Fi} = \frac{\Gamma_o}{1 + N_i} \qquad (3\text{-}8\text{-}15)$$

$$r_{Fi} = \frac{1}{1 + N_i}[N_i^2 + N_i(1 - |\Gamma_o|^2)]^{1/2} \qquad (3\text{-}8\text{-}16)$$

From Eqs. (3-8-11), (3-8-15), and (3-8-16), $N_i = 0$ when $F_i = F_{min}$, and the center of the F_{min} circle with zero radius is located at Γ_o on the Smith chart. The centers of the other noise-figure circles lie along the optimum source reflection coefficient Γ_o vector. If a given source impedance is located along a specific noise circle, that impedance would result in specific noise figure in decibels at that point.

Example 3-8-1:   Noise-Figure Circles

A certain GaAs MESFET has the following noise-figure parameters measured at $V_{ds} = 5$ V, $I_{ds} = 20$ mA, with a 50-Ω resistance for a frequency of 9 GHz.

$$F_{min} = 2 \text{ dB}$$

$$\Gamma_o = 0.485 \underline{/155°}$$

$$R_n = 4 \ \Omega$$

Plot the noise-figure circles for given values of F_i at 2.5, 3.0, 3.5, 4.0, and 5.0 dB.

Solution

1. From Eqs. (3-8-11), (3-8-15), and (3-8-16), the values of N_i, $\mathbf{c}_{Fi}$, and r_{Fi} for F_i at 2.5 dB are computed as follows:

$$N_i = \frac{1.78 - 1.59}{4(4/50)}\,|1 + 0.485\,\underline{/155°}\,|^2 = 0.21$$

$$\mathbf{c}_{Fi} = \frac{0.485\,\underline{/155°}}{1 + 0.21} = 0.40\,\underline{/155°}$$

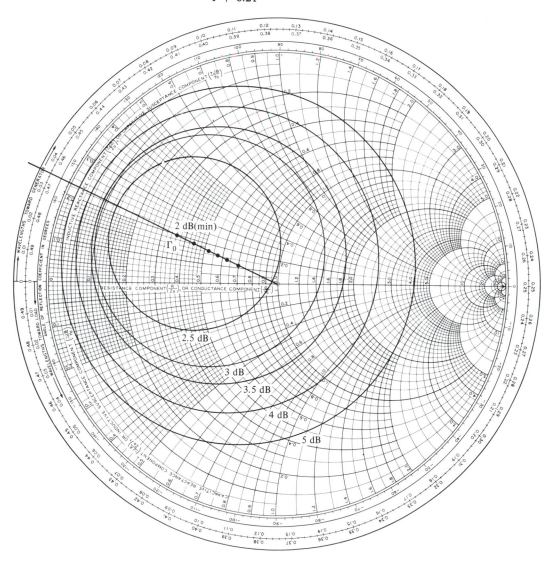

Figure 3-8-1 Noise-figure circles for Example 3-8-1.

$$r_{Fi} = \frac{1}{1 + 0.21} [(0.21)^2 + 0.21(1 - |0.485|^2)]^{1/2}$$

$$= 0.37$$

2. Similarly, the values of N_i, c_{Fi}, and r_{Fi} for F_i at 3, 3.5, 4, and 5 dB are also computed.
3. All values are tabulated in Table 3-8-1.

Table 3-8-1   VALUES OF NOISE-FIGURE CIRCLES

F_i(dB)	2.5	3	3.5	4	5
f_i	1.78	2	2.24	2.5	3.16
N_i	0.21	0.45	0.71	1	1.72
c_{Fi}	0.40 /155°	0.33 /155°	0.28 /155°	0.24 /155°	0.18 /155°
r_{Fi}	0.37	0.51	0.55	0.66	0.76

4. The noise-figure circles are plotted in Fig. 3-8-1.

PROBLEMS

3-4 Power Gain

3-4-1. Derive Eqs. (3-4-6) and (3-4-7) from Eq. (3-4-5).

3-4-2. The NE694 GaAs MESFET has the following S parameters measured at $V_{ds} = 7$ V and $I_{ds} = 30$ mA at 8 GHz:

$$S_{11} = 0.68 \,/{-167°} \qquad \Gamma_s = 0.15 \,/0°$$
$$S_{12} = 0.04 \,/45° \qquad \Gamma_t = 0.20 \,/0°$$
$$S_{21} = 1.55 \,/8°$$
$$S_{22} = 0.73 \,/{-113°}$$

Compute:
a. The delta factor Δ.
b. The stability factor K.
c. The maximum unilateral transducer power gain.
d. The maximum available power gain.

3-4-3. The DXL 3501A-P100F GaAs FET has the following S parameters measured at $V_{ds} = 8$ V and $I_{ds} = 0.5 I_{dss}$ at 10 GHz:

$$S_{11} = 0.67 \,/{-176°} \qquad \Gamma_s = 0.10 \,/0°$$
$$S_{12} = 0.06 \,/{-15°} \qquad \Gamma_t = 0.18 \,/0°$$
$$S_{21} = 1.02 \,/{-3°}$$
$$S_{22} = 0.81 \,/{-115°}$$

Calculate:
(a) The delta factor Δ.
(b) The stability factor K.

(c) The maximum unilateral transducer power gain.
(d) The maximum available power gain.

3-4-4. The S parameters of a certain GaAs MESFET measured at 5 GHz with a 50-Ω resistance matching the input and output are:

$$S_{11} = 0.55 \; \underline{/25°}$$
$$S_{12} = 0.6 \; \underline{/-180°}$$
$$S_{21} = 5.0 \; \underline{/180°}$$
$$S_{22} = 0.35 \; \underline{/-20°}$$

(a) Calculate the maximum available power gain G_{max}.
(b) Compute the unilateral power gain G_u for $\Gamma_s = 0.1 \; \underline{/0°}$ and $\Gamma_\ell = 0.1 \; \underline{/0°}$.
(c) Determine the maximum unilateral power gain $G_{u max}$.

3-5 Amplifier Stability

3-5-1. A certain microwave GaAs FET has the following S parameters measured at $V_{ds} = 3$ V and $I_{ds} = 10$ mA at 4 GHz:

$$S_{11} = 0.89 \; \underline{/-50°}$$
$$S_{12} = 0.06 \; \underline{/66°}$$
$$S_{21} = 3.26 \; \underline{/141°}$$
$$S_{22} = 0.58 \; \underline{/-24°}$$

(a) Calculate the delta factor Δ.
(b) Find the stability factor K.
(c) Compute the center and radius of the input stability circle and plot the circle.
(d) Determine the center and radius of the output stability circle and plot the circle.

3-5-2. A GaAs MESFET has the following S parameters measured at $V_{ds} = 6$ V and $I_{ds} = 0.50 I_{dss}$ at 4 GHz:

$$S_{11} = 0.42 \; \underline{/-54°}$$
$$S_{12} = 0.90 \; \underline{/80°}$$
$$S_{21} = 2.40 \; \underline{/63°}$$
$$S_{22} = 0.92 \; \underline{/-41°}$$

(a) Compute the delta factor Δ.
(b) Calculate the stability factor K.
(c) Find the center and radius of the input stability circle and plot the circle.
(d) Determine the center and radius of the output stability circle and plot the circle.

3-6 Constant-Gain Circles

3-6-1. A certain GaAs MESFET has the following S parameters measured at $V_{ds} = 8$ V and $I_{ds} = 0.50 I_{dss}$ at 12 GHz:

$$S_{11} = 0.65 \; \underline{/-157°}$$
$$S_{12} = 0$$

$$S_{21} = 2.03 \ \underline{/25°}$$
$$S_{22} = 0.76 \ \underline{/-75°}$$

Plot:

(a) The input constant-gain circles for 3, 2, 1, 0, −1, and −2 dB.

(b) The output constant-gain circles for the same decibel levels.

3-6-2. The NE720 GaAs MESFET has the following S parameters measured at $V_{ds} = 4$ V and $I_{ds} = 30$ mA at 4 GHz:

$$S_{11} = 0.70 \ \underline{/-127°}$$
$$S_{12} = 0$$
$$S_{21} = 3.13 \ \underline{/76°}$$
$$S_{22} = 0.47 \ \underline{/-30°}$$

Plot:

(a) The input constant-gain circles for 3, 2, 1, 0, −1, and −2 dB.

(b) The output constant-gain circles for the same decibel levels.

3-6-3. Derive Eqs. (3-6-2) and (3-6-3) from Eqs. (3-6-1) and (3-6-4).

3-7 Constant-Power-Gain Circles

3-7-1. The NE868898-7 GaAs FET has the following S parameters measured at $V_{ds} = 9$ V and $I_{ds} = 1.2$ A at 7 GHz:

$$S_{11} = 0.42 \ \underline{/155°}$$
$$S_{12} = 0.13 \ \underline{/-10°}$$
$$S_{21} = 2.16 \ \underline{/25°}$$
$$S_{22} = 0.51 \ \underline{/-70°}$$

(a) Compute the delta factor Δ.

(b) Calculate the stability factor K.

(c) Find the maximum operating power gain.

(d) Plot the power-gain circles for 8 and 6 dB.

3-7-2. A certain GaAs MESFET has the following parameters measured at $V_{ds} = 7$ V and $I_{ds} = 50$ mA at 4 GHz:

$$S_{11} = 0.40 \ \underline{/-135°}$$
$$S_{12} = 0.09 \ \underline{/21°}$$
$$S_{21} = 2.30 \ \underline{/63°}$$
$$S_{22} = 0.45 \ \underline{/-68°}$$

(a) Calculate the delta factor Δ.

(b) Find the stability factor K.

(c) Plot the power-gain circles for 7 and 5 dB.

3-7-3. A microwave transistor has the following S parameters with a 50-Ω resistance at 3 GHz:

$$S_{11} = 0.40 \; \underline{/-60°}$$
$$S_{12} = 0.10 \; \underline{/55°}$$
$$S_{21} = 2.00 \; \underline{/40°}$$
$$S_{22} = 0.91 \; \underline{/-43°}$$

(a) Compute the delta factor Δ.
(b) Calculate the stability factor K.
(c) Find the maximum stable power gain at $K = 1$.
(d) Plot the power-gain circle for 10-dB gain.
(e) Plot the input and output stability circles.
(f) Design an amplifier for an operating power gain of 10 dB by choosing a proper stable source reflection coefficient from the plot.

3-8- Constant-Noise-Figure Circles

3-8-1. The HP HXTR-6101 microwave transistor has the following parameters at 4 GHz with a reference resistance of 50 ohms.

Minimum-noise figure	$F_{min} = 2.5$ dB
Optimum source reflection coefficient for F_{min}	$\Gamma_o = 0.475 \; \underline{/155°}$
Equivalent noise resistance	$R_n = 3.5 \; \Omega$

Plot the noise-figure circles for given values of F_i at 2.5, 3.0, 3.5, 4.0, and 5.0 dB.

3-8-2. A certain GaAs MESFET has the following parameters at 8 GHz with a reference resistance of 50 ohms.

Minimum-noise figure	$F_{min} = 2.5$ dB
Optimum source reflection coefficient for F_{min}	$\Gamma_o = 0.52 \; \underline{/160°}$
Equivalent noise resistance	$R_n = 4.2 \; \Omega$

Plot the noise-figure circles for given values of F_i at 2.5, 3.0, 3.5, 4.0, and 4.5 dB.

Chapter 4

Small-Signal and Narrowband Amplifier Design

4-0 INTRODUCTION

In general, the microwave amplifiers can be operated in Class A, B, and C modes. In Class A, the collector or drain current is never cut off; in Class B, the device is biased approximately at the cutoff point; and in Class C, the device is biased beyond cutoff. However, the microwave amplifiers are usually classified into small-signal mode and large-signal mode. In small-signal mode, the optimum load admittance of a GaAs MESFET amplifier is equal to the drain conductance ($\partial I_{ds} / \partial V_{ds}$). In large-signal operation, the input signal level will increase and the gate voltage V_{gs} will enter the nonlinear region. The output current, which is the drain current I_{ds}, will then reach its upper limit and the load-line gradient must increase to increase the output power. Figures 4-0-1 and 4-0-2 show the two types of operations. I_{dss} refers to the saturation drain at $V_{gs} = 0$.

The dc-biasing operating point of a microwave amplifier depends on its specific applications. The safe operating region for a GaAs MESFET amplifier and a microwave silicon transistor can be determined, respectively, by their operating voltage and current as tabulated in Table 4-0-1.

Specifically, the safe operating point for a GaAs MESFET can be determined by:

1. Maximum drain–source voltage V_{ds}
2. Maximum drain current I_{ds}
3. Maximum input signal power to the gate
4. Maximum power dissipation at maximum junction temperature, 175°C

123

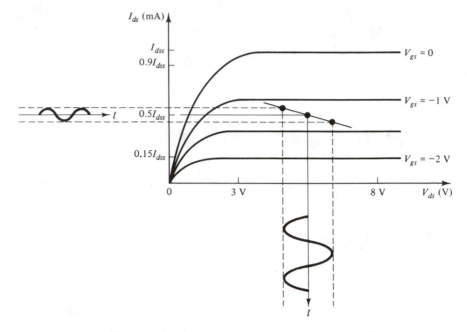

Figure 4-0-1. Small-signal operation of microwave amplifier.

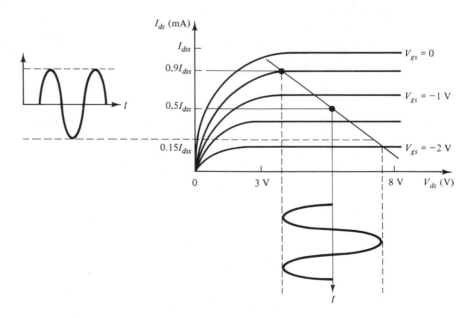

Figure 4-0-2. Large-signal operation of microwave amplifier.

TABLE 4-0-1 SAFE OPERATING POINT FOR GaAs MESFET AND SILICON MICROWAVE TRANSISTOR

Device	Voltage	Current	Applications
GaAs MESFET	V_{ds} (V)	I_{ds} (mA)	
(I_{dss} = 80 mA)	$\geq$ 10	40	High power
	5	80	High gain
	3.5	10	Low noise
	$\geq$ 10	40	Low distortion
	$\geq$ 8	0	Class B
Silicon microwave transistor	V_{CE} (V)	I_{CE} (mA)	
	20	25	High power
	10	10	High gain
	10	3	Low noise
	$\geq$ 20	25	Low distortion
	$\geq$ 20	0	Class B
	$\geq$ 28	0	Class C

Specifically, the safe operating point for a silicon microwave transistor can be determined by:

1. Maximum collector–emitter voltage V_{CE}
2. Maximum collector current I_{CE}
3. Secondary breakdown
4. Maximum power dissipation at maximum junction temperature, 200°C

Microwave amplifiers are commonly used in radar, communications, navigational, and industrial electronic systems. Their characteristics are listed in Table 4-0-2.

TABLE 4-0-2 CHARACTERISTICS OF COMMONLY USED MICROWAVE AMPLIFIERS

Device	Frequency range (GHz)	Characteristics
BARITT	Above 8	Low noise, low power
GaAs MESFET	2–20	Low noise, high gain, medium power
Gunn diode	Above 10	Low noise, medium power
IMPATT diode	Up to 100	CW and pulse modes
Silicon transistor	Up to 3	Low noise, high gain, high power
Tunnel diode	Above 20	Low noise, low power
TRAPATT diode	2–4	High noise, high power

4-1 dc-BIASING CIRCUITS

The design of dc-biasing circuits for microwave amplifiers is as important as the design of the matching networks for the amplifiers because the amplifiers' high gain, high power, high efficiency, and low noise also depend on the dc-biasing circuits. The purpose of dc-biasing circuit design is to provide a proper quiescent point and hold that point constant over temperature and device-parameter variations.

4-1-1 dc-Biasing Circuits for Microwave GaAs MESFETs

There are two types of dc-biasing circuits for microwave GaAs MESFET amplifiers: passive (or self) and active dc-biasing circuits.

Passive dc-biasing circuits.   There are three types of passive dc power supply circuits commonly used for microwave FET amplifiers.

Type 1:   Bipolar Power Supply. Figure 4-1-1 shows a dc-biasing circuit that requires two power supplies. For GaAs MESFET, the range of the gate voltage requires that

$$V_p < V_{gs} < 0 \qquad (4\text{-}1\text{-}1)$$

where V_p is the pinch-off voltage. The gate voltage can be expressed as

$$V_{gs} = V_p\left(1 - \sqrt{\frac{I_{ds}}{I_{dss}}}\right) \qquad (4\text{-}1\text{-}2)$$

where I_{ds} = drain current
      I_{dss} = saturation drain current at $V_{gs} = 0$

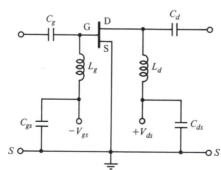

Figure 4-1-1. Bipolar power supply circuit.

A caution must be taken to prevent the burnout of a GaAs MESFET if the drain is biased positive before the gate. The proper turn-on method is first to apply a negative bias to the gate and then apply the positive drain voltage. Alternatively, in order to turn on both power supplies simultaneously, a long RC time constant circuit and short RC time constant circuit must be added to the V_{ds} and V_{gs} supply, respectively. When the source is connected directly to the ground terminal, the source inductance can be made relatively small. By doing so, low noise, high gain, high power, and high efficiency can be achieved at higher frequencies.

Type 2:   One Power Supply. Figure 4-1-2 shows two dc-biasing circuits that require only one power supply for each circuit.

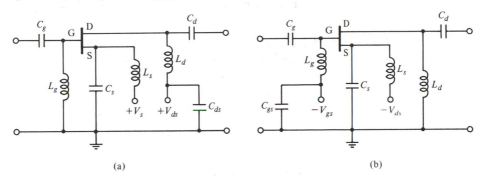

(a)   (b)

Figure 4-1-2   One-power supply circuits.

One positive power supply. The circuit as shown in Fig. 4-1-2(a) employs only one positive power supply. The gate voltage V_{gs} must be negatively biased before any drain voltage V_{ds} is applied and the source voltage V_s must be applied first before V_{ds} is applied.

One negative power supply. The circuit as shown in Fig. 4-1-2(b) uses only one negative power supply. The gate voltage V_{gs} must be turned on before the source voltage V_s is on.

These types of biasing circuits need very good microwave bypass capacitors at the source; this may cause a problem at higher frequencies because any small series source impedance could cause a high noise and possible oscillations.

Type 3:   Unipolar Power Supply. Figure 4-1-3 shows two dc-biasing circuits that require only one power source. Part (a) shows that $V_s = I_{ds}R_s <$t V_{ds} and part

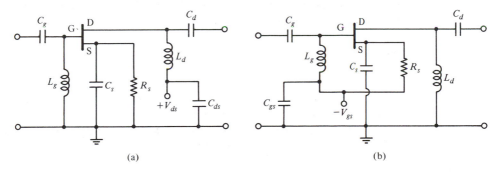

(a)   (b)

Figure 4-1-3   Unipolar power supply circuits.

(b) implies that $V_s = -I_{ds}R_s > -V_{gs}$. These types of biasing circuits use a source resistor to provide the turn-on and turn-off transient protection. However, the amplifier efficiency and noise figure may be degraded because the source resistor will dissipate some power and generate noise. Also, the source bypass capacitor may cause low-frequency oscillations.

In all dc-biasing circuits, the element values can be specified as follows:

1. All inductors, L_g, L_d, and L_s are RF chokes (RFCs) which can be made of two or three turns of No. 36 enameled wire on a 0.1-in. air core.
2. The value of R_s is adjusted to provide the right V_s for a proper quiescent point and transient protection.
3. All source bypass capacitors, C_s, are 0.01 μF at 100-V disk.
4. All other capacitors could have the values of 0.01 μF at 100-V disk.

Also in all dc-biasing circuits, it is sometimes necessary to couple the capacitors with shunt zener diodes. The zener diodes can provide additional protection against transients, reversing biasing, and overvoltage. Figure 4-1-4 shows a dc-biasing circuit with shunt zener diodes for transient protection.

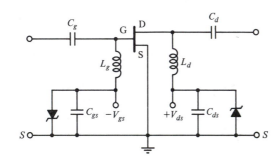

Figure 4-1-4. dc-biasing circuit with zener diodes.

Example 4-1-1:   Passive dc-Biasing Circuit

A passive dc-biasing circuit is to be designed for the HFET-1101 GaAs MESFET. The circuit requires a very good source bypassing circuit.

Solution

1. The passive dc-biasing circuit to be designed is shown in Fig. 4-1-5.
2. The RF chokes (RFCs) for L_g and L_d can be made of two or three turns of No. 36 enameled wire on a 0.1-in. air core.
3. R_s and R_d are 5-kΩ 1-W potentiometers. R_s is adjusted to supply the correct drain current I_{ds} and R_d is adjusted to provide the proper drain–source voltage V_{ds}.
4. All capacitors are 1-nF high-Q Johanson 50S41Q102MB capacitors.

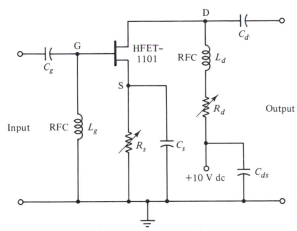

Figure 4-1-5. Passive dc-biasing circuit to be designed.

Active dc-biasing circuits.   Besides the aforementioned passive dc-biasing circuits, several active dc-biasing circuits can also be used for the GaAs-MESFET amplifier. Figure 4-1-6 shows an active dc-biasing circuit for a common-source GaAs MESFET. The quiescent point is controlled by R_{B2} and R_E. R_{B2} is adjusted to provide for proper V_{ds} and R_E is adjusted for proper I_{ds}. The best bias circuit for GaAs-MESFET amplifiers must therefore satisfy the following conditions. The MESFET source must have a good ground connection and transient protection. When one MESFET in a chain fails, the other MESFETs must not be affected.

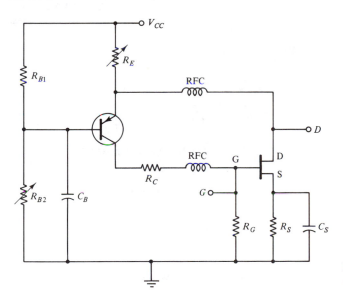

Figure 4-1-6. Active dc-biasing circuit for GaAs MESFET.

Example 4-1-2:   Active dc-Biasing Circuit

An active dc-biasing circuit is to be designed for the HFET-1101 GaAs MESFET. The biasing circuit requires a dual power supply.

Solution

1. The active dc-biasing circuit to be designed is shown in Fig. 4-1-7.
2. Capacitors C_g and C_d are 1-nF high-Q Johanson 50S41Q102MB.
3. Capacitor C_B is 5-nF at 250-V disk.

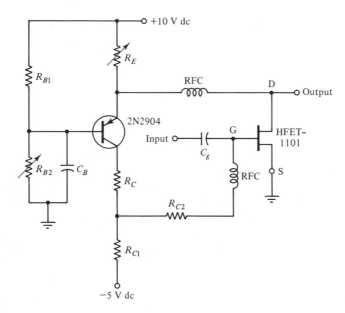

Figure 4-1-7   Active dc-biasing circuit to be designed.

4. The values of the resistors are:

$$R_{B1} = 38\text{-k}\Omega \ \frac{1}{4}\text{-W resistor}$$

$$R_{B2} = 100\text{-k}\Omega \ 1\text{-W potentiometer}$$

$$R_E = 5\text{-k}\Omega \ 1\text{-W potentiometer}$$

$$R_C = 1\text{-k}\Omega \ \frac{1}{4}\text{-W resistor}$$

$$R_{C1} = 10\text{-k}\Omega \ \frac{1}{4}\text{-W resistor}$$

$$R_{C2} = 1\text{-M}\Omega \ \frac{1}{4}\text{-W resistor}$$

5. RFCs: The RFC chokes can be made of two or three turns of No. 36 enameled wire on a 0.1-in. air core.

4-1-2 dc-Biasing Circuits for Microwave Silicon Transistors

The collector current I_C of a microwave silicon transistor is a temperature-dependent result of its reverse collector current I_{CBO}, dc-current gain h_{FE}, and base–emitter voltage V_{BE}. Most microwave transistor circuits for a high power gain or low noise figure often require that the emitter lead be dc grounded as close to the package as possible so that the emitter series feedback is kept at a minimum. At microwave frequencies, the emitter bypass capacitor R_E at the design frequency often introduces low-frequency instability and gives rise to bias oscillations. At low frequencies, a reverse collector current I_{CBO} flows through the reverse-biased p-n junction of a transistor. Classically, this leakage current is doubled for every 10°C temperature rise in a silicon semiconductor junction. However, microwave silicon transistors have a more complicated reverse current flow at microwave frequencies. A small component of this current flow is the conventional I_{CBO} term, but the major contributor is a surface current that flows across the top of the silicon crystal lattice and is linearly increased with temperature rise. Therefore, the total reverse current is made up of I_{CBO} and surface components, and it increases with temperature rise at a rate of 0.5%/°C much less than expected.

Each type of microwave transistor amplifier may require a different dc-biasing circuit. For example, a microwave transistor biased for high linear output power must hold its quiescent point such that the 1-dB compression point is not degraded with temperature rise and so that the maximum power dissipation in device is not exceeded with increasing temperature.

Mathematically, the collector current I_C of a microwave silicon transistor can be expressed as

$$I_C = f(I_{CBO}, h_{FE}, V_{BE}) \qquad (4\text{-}1\text{-}3)$$

Because the three parameters I_{CBO}, h_{FE}, and V_{BE} are temperature dependent, the changing collector current can be written as

$$\Delta I_C = S_{ICBO}\,\Delta I_{CBO} + S_{hFE}\,\Delta h_{FE} + S_{VBE}\,\Delta V_{BE} \qquad (4\text{-}1\text{-}4)$$

where $S_{ICBO} = \left.\dfrac{\Delta I_C}{\Delta I_{CBO}}\right|_{\substack{\Delta h_{FE}=0 \\ \Delta V_{BE}=0}}$ is the stability factor of the reverse collector current I_{CBO}

$S_{hFE} = \left.\dfrac{\Delta I_C}{\Delta h_{FE}}\right|_{\substack{\Delta I_{CBO}=0 \\ \Delta V_{BE}=0}}$ is the stability factor of the dc common-emitter current gain h_{FE}

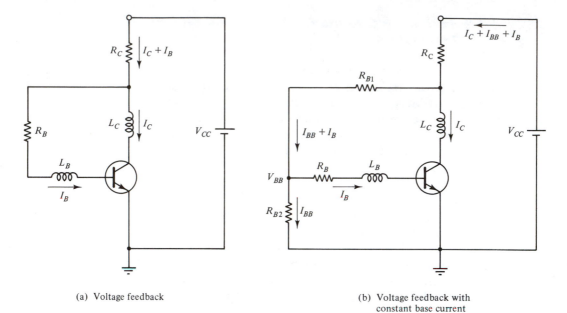

(a) Voltage feedback

(b) Voltage feedback with
constant base current

(c) Dc-biasing circuit with
bypass-emitter resistor

Figure 4-1-8 Passive dc-biasing circuits for microwave silicon transistor.

$$S_{VBE} = \frac{\Delta I_C}{\Delta V_{BE}}\Bigg|_{\substack{\Delta I_{CBO}=0 \\ \Delta h_{FE}=0}} \quad \text{is the stability factor of the base–emitter voltage } V_{BE}$$

The stability factors can be calculated from Eq. (4-1-4) for a specific dc-biasing circuit. Then the variations of the collector current I_C with temperature can be predicted. In practical circuit design, the values of the resistors in the dc-biasing circuit can be determined from the stability factor and the quiescent point.

There are two types of dc-biasing circuits for microwave silicon transistors: passive (or self) and active dc-biasing circuits.

Passive dc-biasing circuits.   Figure 4-1-8 shows three configurations of passive dc-biasing circuits for microwave silicon transistors. The two grounded-emitter dc-biasing circuits as shown in Fig. 4-1-8(a) and (b) are often used at microwave frequencies. The circuit in Fig. 4-1-8(b) produces lower values of resistance and is more compatible with thin- or thick-film resistor values. The circuit of Fig. 4-1-8(c) has a bypass-emitter resistor and can be used at the lower microwave frequencies. The bypass-emitter resistor provides excellent stability.

Design Example 4-1-3:   Passive dc-Biasing Circuits Design

A certain microwave silicon transistor has the following parameters:

$$V_{CC} = 30 \text{ V} \qquad I_{BB} = 1.5 \text{ mA} \qquad \text{Quiescent point:}$$

$$V_{BB} = 3 \text{ V} \qquad I_{CBO} = 0 \qquad V_{CE} = 10 \text{ V}$$

$$V_{BE} = 1 \text{ V} \qquad h_{FE} = 50 \qquad I_C = 15 \text{ mA}$$

To hold the quiescent point constant with temperature changing, it is necessary to design passive dc-biasing circuits as shown in Fig. 4-1-8.

Solution

1. Design a passive dc-biasing circuit with voltage feedback.

$$I_B = \frac{I_C}{h_{FE}} = \frac{15 \times 10^{-3}}{50} = 0.3 \text{ mA}$$

$$R_B = \frac{V_{CE} - V_{BE}}{I_B} = \frac{10 - 1}{0.3 \times 10^{-3}} = 30 \text{ k}\Omega$$

$$R_C = \frac{V_{CC} - V_{CE}}{I_C + I_B} = \frac{30 - 10}{15 \times 10^{-3} + 0.3 \times 10^{-3}} = 1.3 \text{ k}\Omega$$

2. Design a passive dc-biasing circuit with voltage feedback and constant base current.

$$R_B = \frac{V_{BB} - V_{BE}}{I_B} = \frac{3 - 1}{0.3 \times 10^{-3}} = 6.67 \text{ k}\Omega$$

$$R_{B1} = \frac{V_{CE} - V_{BB}}{I_{BB} + I_B} = \frac{10 - 3}{1.5 \times 10^{-3} + 0.3 \times 10^{-3}} = 3.89 \text{ k}\Omega$$

$$R_{B2} = \frac{V_{BB}}{I_{BB}} = \frac{3}{1.5 \times 10^{-3}} = 2 \text{ k}\Omega$$

$$R_C = \frac{V_{CC} - V_{CE}}{I_C + I_{BB} + I_B} = \frac{30 - 10}{15 \times 10^{-3} + 1.5 \times 10^{-3} + 0.3 \times 10^{-3}}$$

$$= 1.19 \text{ k}\Omega$$

3. Design a passive dc-biasing circuit with a bypass emitter resistor.

$$R_E = \frac{V_{BB} - V_{BE}}{I_B} = \frac{3 - 1}{0.3 \times 10^{-3}} = 6.67 \text{ k}\Omega$$

Active dc-biasing circuits. The purpose of a biasing circuit is to hold the quiescent point constant. A commonly used active dc-biasing circuit is shown in Fig. 4-1-9, where Q_1 is a p-n-p transistor and Q_2 is a microwave silicon transistor. The active dc-biasing circuit is to maintain a relatively constant quiescent point over a wide temperature range. The quiescent point can be adjusted by R_{B2} and R_E. R_{B2} is adjusted to provide the proper voltage V_{CE} and R_E is adjusted to supply the collector current I_C. The two RF chokes (RFCs), L_E and L_C, can be made of two or three turns of No. 36 enameled wire on a 0.1-in. air core. R_{B1} may be a 100-$\Omega\frac{1}{4}$-W resistor. The bypass capacitors may be the type of 0.01-μF at 100-V disk.

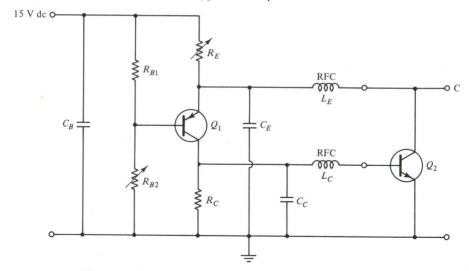

Figure 4-1-9 Active dc-biasing circuit for microwave silicon transistor.

From the standpoint of temperature compensation, passive biasing circuits are inexpensive to build and can provide satisfactory results. However, when the dc current gain variation is large, automatic compensation can be achieved only an active biasing circuit. Furthermore, the active biasing circuit can provide better operating-point stability, especially for low-noise or high-power amplifiers. The active biasing circuit is actually a feedback loop that senses the collector current of the microwave transistor and adjusts the base current to hold the collector current constant.

4-1-3 Biasing-Circuit Design

The selection of the dc quiescent point Q in a microwave amplifier circuit depends on particular applications. The dc-biasing circuit of a GaAs MESFET must provide a stable quiescent point. The source resistor R_s is expressed as

$$R_s = -\frac{V_{gs}}{I_{ds}} = -\frac{V_p}{I_{ds}}\left(1 - \sqrt{\frac{I_{ds}}{I_{dss}}}\right) \tag{4-1-5}$$

Equation (4-1-5) indicates that the negative feedback resistor R_s decreases the effect of variations of I_{ds} with respect to temperature and I_{dss}.

In general, the Class A linear amplifier can provide low noise, low power, and high power gain, but the Class B or Class AB amplifier can generate higher power and high efficiency. The selection of the dc quiescent point for a GaAs MESFET amplifier can be classified into three operations:

1. *Low-power and low-noise operation.* Low-power and low-noise amplifier operates a relatively low drain–source voltage V_{ds} and current I_{ds}, where I_{ds} is usually equal to $0.15I_{dss}$. Under this condition, the quiescent point Q is selected at the point 1, as shown in Fig. 4-1-10. This type of operation is a typical Class A mode.

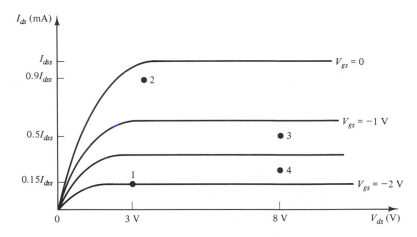

Figure 4-1-10   dc quiescent points for GaAs-MESFET amplifier.

2. *High-gain and low-noise operation.* For high-gain and low-noise operation, the drain voltage remains the same level as Class A and the drain current I_{ds} is increased to $0.90I_{ds}$ for high-power gain. The dc quiescent point Q may be selected at point 2, as shown in Fig. 4-1-10.

3. *High power and high efficiency.* As the power output is increased, the drain voltage V_{ds} must be also increased. On the other hand, the drain current I_{ds} must be decreased to maintain linear operation in Class A. The dc quiescent

point Q is recommended at point 3 for $V_{ds} \simeq 8 - 10$ V and $I_{ds} = 0.51 I_{dss}$, as shown in Fig. 4-1-10. For high-efficiency operation, the drain current I_{ds} must be decreased and the bias voltage V_{ds} should be shifted to point 4, as shown in Fig. 4-1-10. This type of operation is a typical Class AB or B.

4-2 SMALL-SIGNAL AMPLIFIER DESIGN

As described in Section 4-0, the small-signal amplifier operates with a small drain current, say $I_{ds} = 0.50 I_{dss}$, at the midrange of the drain voltage V_{ds}, say 5 V. It is a typical Class A linear operation.

Design Procedure.   The design procedures for a small-signal microwave amplifier are stated as follows:

1. List the specifications of the microwave amplifier to be designed, such as frequency, power output, and power gain.
2. Select a device that will meet the specifications.
3. Measure the S parameters of the selected device for the required specifications.
4. Design the input and output networks for the amplifier.

Figure 4-2-1 shows a microwave amplifier circuit to be designed.

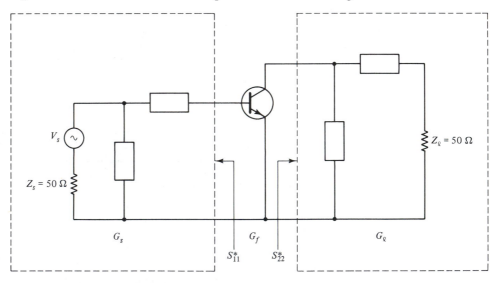

Figure 4-2-1   Microwave amplifier circuit to be designed.

Design Example 4-2-1:   Small-Signal Microwave Amplifier Design

A microwave transistor has the following S parameters measured with a 50-Ω resistance at $V_{ce} = 4$ V and $I_c = 0.50 I_{cmax}$ for a frequency of 3 GHz.

$$S_{11} = 0.707 \; \underline{/-155°}$$

$$S_{12} = 0$$

$$S_{21} = 4 \; \underline{/180°}$$

$$S_{22} = 0.51 \; \underline{/-20°}$$

Design the input and output matching networks of the amplifier for a power gain of 15 dB at the frequency of 3 GHz.

Design Steps

Step 1. The forward transducer power gain is $10 \, \ell og \, |4|^2 = 12$ dB. For a total of 15 dB gain, 2 dB may come from the input network and 1 dB from the output network.

Step 2. Plot the input and output constant-gain circles. The values of the constant-gain circles are calculated from Eqs. (3-6-1) through (3-6-4) and they are tabulated in Tables 4-2-1 and 4-2-2. The constant-gain circles are plotted in Fig. 4-2-2.

TABLE 4-2-1 VALUES FOR THE IN-PUT CONSTANT-GAIN CIRCLES

G_s(dB)	2	1	0
g_s	1.59	1.26	1.00
g_{ns}	0.80	0.63	0.50
d_s	0.63	0.55	0.47
r_s	0.25	0.38	0.47

TABLE 4-2-2 VALUES FOR THE OUTPUT CONSTANT-GAIN CIRCLES

G_ℓ(dB)	1	0
g_ℓ	1.26	1.00
$g_{n\ell}$	0.93	0.74
d_ℓ	0.48	0.41
r_ℓ	0.20	0.41

Step 3. The output matching network. We start the load impedance of 50 Ω at the center of the Smith chart and proceed along the constant unit resistance circle in the counterclockwise direction until we arrive at the constant-conductance circle toward S_{22}^*, but stop at the 1-dB gain circle. The first element from the load is a series capacitor.

$$jX = -j1.45 \times 50 = -j72.50 \; \Omega$$

$$C_\ell = \frac{1}{\omega X} = \frac{1}{2\pi \times 3 \times 10^9 \times 72.5} = 0.73 \text{ pF}$$

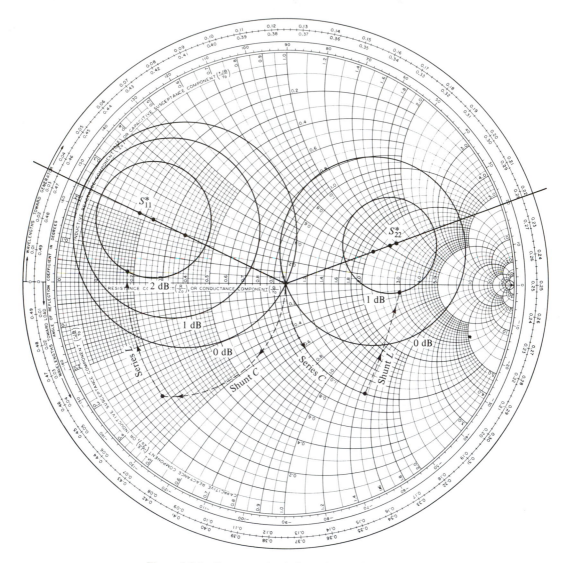

Figure 4-2-2 Constant-gain circles and network elements.

The second element from the load is a shunt inductor

$$jB = -\frac{j0.43}{50} = -j8.60 \text{ mmho}$$

$$L_\ell = \frac{1}{\omega B} = \frac{1}{2\pi \times 3 \times 10^9 \times 8.6 \times 10^{-3}}$$

$$= 6.17 \text{ nH}$$

Step 4. Determine the input matching network. Similarly, the matched source terminal is located at the center of the Smith chart. We follow the constant unity conductance circle in the clockwise direction and then proceed along the constant-resistance circle toward the point at S_{11}^* but stop at the 2-dB gain circle as shown on the Smith chart. The first element from the source is a shunt capacitor.

$$jB = \frac{j2.10}{50} = j42.00 \text{ mmhos}$$

$$C_s = \frac{B}{\omega} = \frac{42.0 \times 10^{-3}}{2\pi \times 3 \times 10^9} = 2.23 \text{ pF}$$

The second element from the source is a series inductor:

$$jX = j0.41 \times 50 = j20.5 \ \Omega$$

$$L_s = \frac{X}{\omega} = \frac{21.50}{2\pi \times 3 \times 10^9} = 1.09 \text{ nH}$$

Step 5. The circuit designed for the amplifier is shown in Fig. 4-2-3.

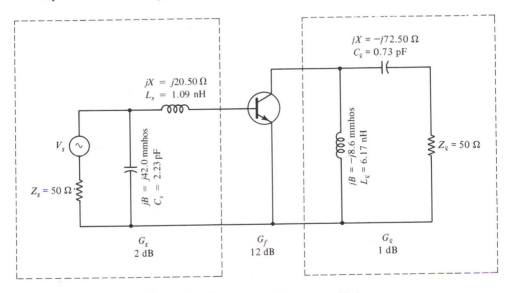

Figure 4-2-3   Designed circuit for the amplifier.

4-3 HIGH-GAIN AMPLIFIER DESIGN

For high-gain operation, the drain current I_{ds} must be high, say $I_{ds} = 0.90 I_{dss}$, and the drain voltage may be set at 4 or 6 V for a GaAs-MESFET amplifier. The input and the output matching circuits must be conjugately matched to the device with $\Gamma_s = S_{11}^*$ and $\Gamma_\ell = S_{22}^*$ for maximum transducer power gain. Figure 4-3-1 shows a GaAs-MESFET amplifier circuit and its matching networks to be designed for maximum power gain.

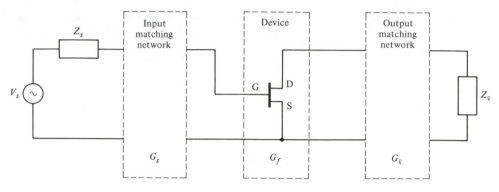

(a) Microwave amplifier matching circuit

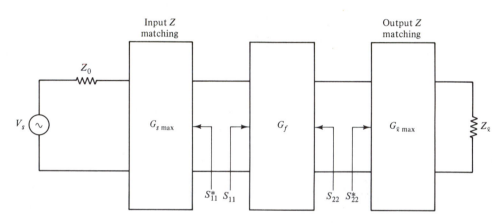

(b) Matching circuits for maximum power gain

Figure 4-3-1   Microwave amplifier circuits to be designed.

The unilateral transducer power gain G_{tu} as indicated by Eq. (3-4-11) is

$$G_{tu} = \frac{1 - |\Gamma_s|^2}{|1 - S_{11}\Gamma_s|^2} |S_{21}|^2 \frac{1 - |\Gamma_\ell|^2}{|1 - S_{22}\Gamma_\ell|^2}$$

$$= g_s\, g_f\, g_\ell \quad \text{(numerical values)} \tag{4-3-1}$$

$$= G_s + G_f + G_\ell \quad \text{dB}$$

where $g_s = \dfrac{1 - |\Gamma_s|^2}{|1 - S_{11}\Gamma_s|^2}$ is the additional power gain or loss resulting from the input impedance-matching network between the device and source

$g_f = |S_{21}|^2$ is the forward power gain of the device with the input and output terminated in matching loads

$g_\ell = \dfrac{1 - |\Gamma_\ell|^2}{|1 - S_{22}\Gamma_\ell|^2}$ is the additional power gain or loss due to the impendance-matching network between the output of the device and the load

Maximum unilateral transducer gain can be accomplished by choosing impedance-matching networks such that $\Gamma_s = S_{11}^*$, $Z_s = Z_0$, $\Gamma_\ell = S_{22}^*$, and $Z_\ell = Z_0$, as shown in Fig. 4-3-1(b). Then Eq. (4-3-1) can be written

$$G_{tumax} = \frac{1}{1 - |S_{11}|^2} |S_{21}|^2 \frac{1}{1 - |S_{22}|^2}$$

$$= g_{smax}\, g_f\, g_{\ell max} \qquad \text{(numerical values)} \tag{4-3-2}$$

$$= G_{smax} + G_f + G_{\ell max} \qquad \text{dB}$$

where
$$g_{smax} = \frac{1}{1 - |S_{11}|^2}$$
$$g_f = |S_{21}|^2$$
$$g_{\ell max} = \frac{1}{1 - |S_{22}|^2}$$

Design Procedure

1. List the specifications of the microwave amplifier to be designed, such as frequency, power gain, and power output.
2. Choose a device that will meet the specifications.
3. Measure the S parameters of the device chosen against the specifications.
4. Check the stability conditions.
5. Plot the constant-gain circles.
6. Compute the unilateral figure of merit.
7. Check the error range for unilateral assumption.
8. Design the input and the output matching networks for maximum power gain.

Design Example 4-3-1:   High-Gain Microwave Amplifier Design

A GaAs MESFET has the following S parameters measured with a 50-Ω resistance at $V_{ds} = 4$ V, $I_{ds} = 0.90I_{dss}$ for 9 GHz.

$$S_{11} = 0.55 \;\underline{/-150°}$$

$$S_{12} = 0.04 \;\underline{/20°}$$

$$S_{21} = 2.82 \;\underline{/180°}$$

$$S_{22} = 0.45 \;\underline{/-30°}$$

Design the input and the output matching networks for maximum power gain at 9 GHz as shown in Fig. 4-3-2.

Design Steps

Step 1. Calculate the maximum transducer power gain with the assumption of $|S_{12}| = 0$.

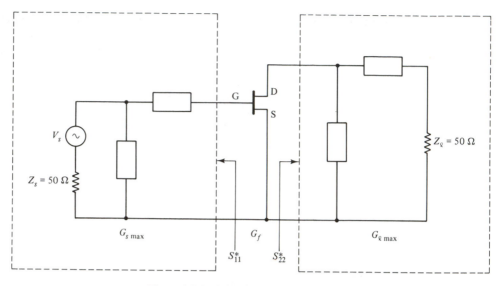

Figure 4-3-2 Desired amplifier circuits.

$$G_{tu\,max} = \frac{1}{1 - |S_{11}|^2} |S_{21}|^2 \frac{1}{1 - |s_{22}|^2}$$

$$= \frac{1}{1 - |0.55|^2} |2.82|^2 \frac{1}{1 - |0.45|^2}$$

$$= 1.43 \times 7.95 \times 1.25 = 14.21 = 11.52 \text{ dB}$$

$$= 1.55 \text{ dB} + 9 \text{ dB} + 0.97 \text{ db} = 11.52 \text{ dB}$$

The maximum transducer power gain is 11.52 dB.

Step 2. Check the stability conditions. The delta factor is

$$\Delta = S_{11}S_{22} - S_{12}S_{21}$$

$$= 0.55 \, \underline{/-150°} \times 0.45 \, \underline{/-30°} - 0.04 \, \underline{/20°} \times 2.82 \, \underline{/180°}$$

$$= 0.25 \, \underline{/-180°} - 0.11 \, \underline{/200°}$$

$$= 0.16 \, \underline{/165°}$$

$$\Delta = 0.16 < 1$$

$$|\Delta|^2 = 0.025$$

The stability factor is

$$K = \frac{1 + |\Delta|^2 - |S_{11}|^2 - |S_{22}|^2}{2|S_{12}S_{21}|}$$

$$= \frac{1 + |0.16|^2 - |0.55|^2 - |0.45|^2}{2|0.04 \times 2.82|}$$

$$= 2.28 > 1$$

The device is unconditionally stable.

Step 3. Plot the constant-gain circles. The constant-gain circles in this example are located at S_{11}^* and S_{22}^* and they are 1.55 dB and 0.97 dB, respectively.

Step 4. Compute the unilateral figure of merit from Eq. (3-6-9).

$$M = \frac{|S_{12}| |S_{21}| |S_{11}| |S_{22}|}{(1 - |S_{11}|^2)(1 - |S_{22}|^2)}$$

$$= \frac{|0.04| |2.82| |0.55| |0.45|}{(1 - |0.55|^2)(1 - |0.45|^2)}$$

$$= 0.05$$

From Eq. (3-6-8), the error at the frequency of 9 GHz is

$$\frac{1}{(1 + 0.05)^2} < \frac{G_t}{G_{tu\text{max}}} < \frac{1}{(1 - 0.05)^2}$$

$$0.91 < \frac{G_t}{G_{tu\text{max}}} < 1.11$$

$$-0.41 \text{ dB} < \frac{G_t}{G_{tu\text{max}}} < +0.45 \text{ dB}$$

The maximum error at 9 GHz is from $+0.45$ to -0.41 dB. In this case, the error is small enough to justify the unilateral assumption.

Step 5. Determine the output matching network. We start from the center of the Smith chart and proceed along the constant unit resistance circle in the counterclockwise direction until we arrive at the constant-conductance circle that intersects the point representing S_{22}^*, as shown in Fig. 4-3-3.

The first element from the load is a series capacitor.

$$jX = -j1.2 \times 50 = -j60.0 \ \Omega$$

$$C_t = \frac{1}{\omega X} = \frac{1}{2\pi \times 9 \times 10^9 \times 60} = 0.29 \text{ pF}$$

Then we add an inductive susceptance along the constant-conductance circle so that the impedance looking into the matching network will be equal to S_{22}^*.

The second element from the load is a shunt inductor:

$$jB = \frac{-j0.73}{50} = -j14.60 \times 10^{-3} \text{ mho}$$

$$L_t = \frac{1}{\omega B} = \frac{1}{2\pi \times 9 \times 10^9 \times 14.60 \times 10^{-3}}$$

$$= 1.21 \text{ nH}$$

Step 6. Find the input matching circuit. Similarly, the procedures can be applied at the input side in the clockwise direction and the results are a shunt capacitor and a series inductor.

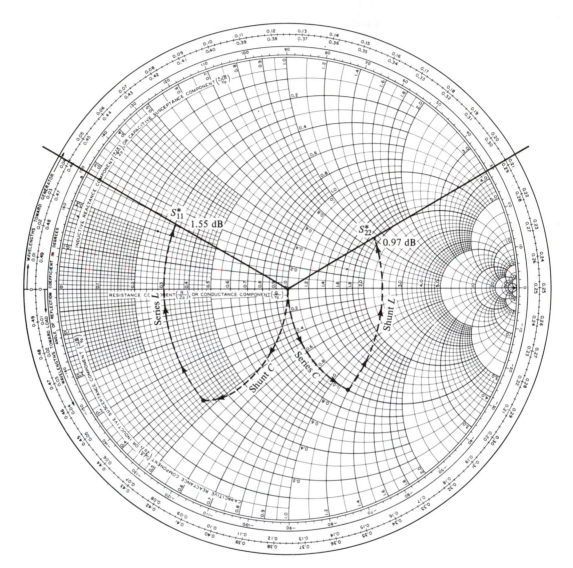

Figure 4-3-3 Graphical solutions on Smith chart.

The first element from the source is a shunt capacitor.

$$jB = \frac{j1.66}{50} = j33.32 \times 10^{-3}$$

$$C_s = \frac{B}{\omega} = \frac{33.32 \times 10^{-3}}{2\pi \times 9 \times 10^9} = 0.59 \text{ pF}$$

The second element from the source is a series inductor.

$$jX = j0.71 \times 50 = j35.50 \ \Omega$$

$$L_s = \frac{X}{\omega} = \frac{35.50}{2\pi \times 9 \times 10^9} = 0.63 \ \text{nH}$$

Step 7. Sketch the designed matching networks (Fig. 4-3-4).

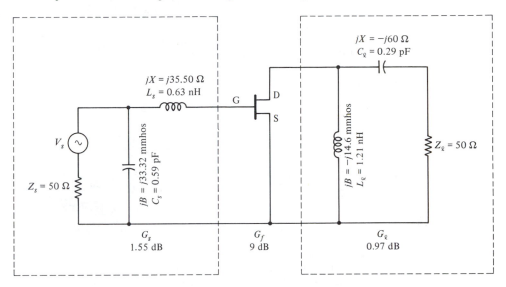

Figure 4-3-4   Designed matching networks.

4-4 LOW-NOISE AMPLIFIER DESIGN

In electronic receiver systems, the maximum power gain and minimum-noise figure of an amplifier are equally important. However, from design practices, maximum power gain and minimum-noise figure cannot be achieved simultaneously in most cases. Therefore, high gain and low noise are the trade-off choice.

Design Procedure

1. Plot the constant-gain and constant-noise circles on the same Smith chart.
2. Compute the maximum power gain for the amplifier to be designed.
3. Choose a source power-gain circle to intercept with a low-noise circle for compensation.
4. Determine the input and the output matching networks.

Design Example 4-4-1:   Design a Low-Noise Amplifier

A certain microwave transistor has the following scattering and noise parameters measured with a 50-Ω resistance at $V_{ce} = 4$ V and $I_{ce} = 30$ mA for 1 GHz.

$$S_{11} = 0.707 \ \underline{/-155°} \qquad F_{\min} = 3 \ \text{dB}$$

$$S_{12} = 0.0 \qquad\qquad \Gamma_o = 0.45 \;\underline{/180°}$$

$$S_{21} = 5.00 \;\underline{/180°} \qquad R_n = 4\Omega$$

$$S_{22} = 0.51 \;\underline{/-20°}$$

where Γ = optimum source reflection coefficient for F_{min}
　　　R_n = equivalent noise resistance
Design the input and output matching networks of the microwave amplifier for a power gain of 16 dB and low-noise figure of 3.5 dB.

Design Steps

Step 1. Calculate the maximum transducer power gain.

$$G_{tumax} = \frac{1}{1 - |S_{11}|^2} \, |S_{21}|^2 \, \frac{1}{1 - |S_{22}|^2}$$

$$= \frac{1}{1 - |0.707|^2} \, |5|^2 \frac{1}{1 - |0.51|^2}$$

$$= 2 \times 25 \times 1.35 = 67.50 = 18.30 \text{ dB}$$

$$= 3 \text{ dB} + 14 \text{ dB} + 1.30 \text{ dB} = 18.30 \text{ dB}$$

It is projected that in order to have a total power gain of 16 dB, 1.22 dB must come from the input matching network and 0.78 from the output matching network in addition to the 14-dB transducer power gain.

Step 2. Plot the 3.5-dB constant-noise-figure circle.

$$F_i = 3.5 \text{ dB} = 2.24$$

$$N_i = \frac{F_i - F_{min}}{4r_n} \, |1 + \Gamma_o|^2 = \frac{2.24 - 2}{4(4/50)} |1 + 0.45 \;\underline{/180°}\,|^2$$

$$= 0.23$$

$$\mathbf{c}_{Fi} = \frac{\Gamma_o}{1 + N_i} = \frac{0.45 \underline{/180°}}{1 + 0.23} = 0.37 \underline{/180°}$$

$$r_{Fi} = \frac{1}{1 + N_i}[N_i^2 + N_i(1 - |\Gamma_o|^2)]^{1/2}$$

$$= \frac{1}{1 + 0.23}[(0.23)^2 + 0.23\,(1 - |0.45|^2)]^{1/2}$$

$$= 0.39$$

The 3.5-dB noise circle is plotted on the Smith chart in Fig. 4-4-1.

Step 3. Plot the input 1.22-dB power-gain circle.

$$G_s = 1.22 \text{ dB} = 1.32$$

$$g_{ns} = g_s(1 - |S_{11}|^2) = 1.32(1 - |0.707|^2) = 0.66$$

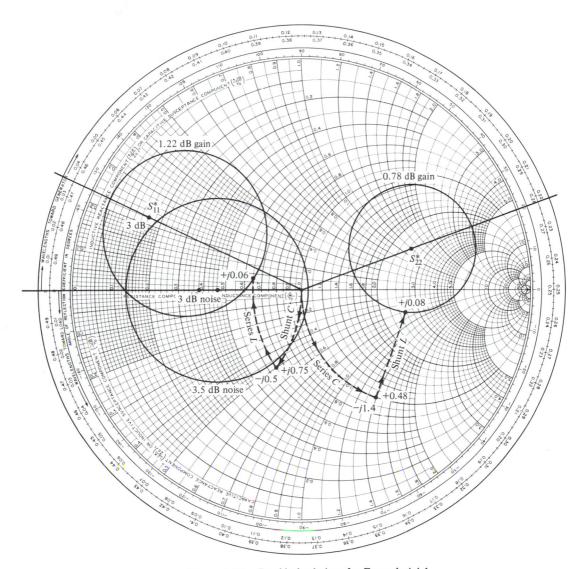

Figure 4-4-1 Graphical solutions for Example 4-4-1.

$$d_s = \frac{g_{ns}|S_{11}|}{1 - |S_{11}|^2(1 - g_{ns})} = \frac{0.66 \times |0.707|}{1 - |0.707|^2(1 - 0.66)}$$

$$= 0.56$$

$$r_s = \frac{\sqrt{1 - g_{ns}}(1 - |s_{11}|^2)}{1 - |S_{11}|^2(1 - g_{ns})} = \frac{\sqrt{1 - 0.66}\,(1 - |0.707|^2)}{1 - |0.707|^2(1 - 0.66)}$$

$$= 0.35$$

The 1.22-dB power-gain circle is plotted on the same Smith chart.

Step 4. Plot the output 0.78-dB power-gain circle.

$$G_\ell = 0.78 \text{ dB} = 1.20$$

$$g_{n\ell} = g_\ell(1 - |S_{22}|^2) = 1.20(1 - |0.51|^2) = 0.89$$

$$d_\ell = \frac{g_{n\ell}|S_{22}|}{1 - |S_{22}|^2(1 - g_{n\ell})} = \frac{0.89|0.51|}{1 - |0.51|^2(1 - 0.89)}$$

$$= 0.47$$

$$r_\ell = \frac{\sqrt{1 - g_{n\ell}}(1 - |S_{22}|^2)}{1 - |S_{22}|^2(1 - g_{n\ell})} = \frac{\sqrt{1 - 0.89}\,(1 - |0.51|^2)}{1 - |0.51|^2(1 - 0.89)}$$

$$= 0.25$$

The 0.78-dB power-gain circle is plotted on the same Smith chart.

Step 5. Determine the output matching network. We follow the constant unit resistance circle from the center of the Smith chart in the counterclockwise direction and then proceed along the constant-conductance circle toward the point S_{22}^*, but stop at the 0.78-dB power-gain circle. The first element from the load is a series capacitor.

$$jX = -j1.40 \times 50 = -j70.0 \ \Omega$$

$$C_\ell = \frac{1}{\omega X} = \frac{1}{2\pi \times 10^9 \times 70} = 2.27 \text{ pF}$$

The second element from the load is a shunt inductor.

$$jB = -\frac{j0.40}{50} = -j8.00 \times 10^{-3} \text{ mho}$$

$$L_\ell = \frac{1}{\omega B} = \frac{1}{2\pi \times 10^9 \times 80.0 \times 10^{-3}} = 19.89 \text{ nH}$$

Step 6. Find the input matching network. Similarly, we start from the center of the Smith chart and proceed along the constant unit conductance circle in the clockwise direction and arrive at the constant resistance circle that passes the 3.5-dB noise circle and ends at the 1.22-dB power-gain circle. The first element from the source is a shunt capacitor.

$$jB = \frac{j0.75}{50} = j15 \times 10^{-3} \text{ mho}$$

$$C_s = \frac{B}{\omega} = \frac{15 \times 10^{-3}}{2\pi \times 10^9} = 2.39 \text{ pF}$$

The second element from the source is a series inductor.

$$jX = j0.56 \times 50 = j28.0 \ \Omega$$

$$L_s = \frac{X}{\omega} = \frac{28}{2\pi \times 10^9} = 4.40 \text{ nH}$$

(*Note:* The input matching network is not unique.)

Step 7. Draw the designed matching networks for the amplifier (Fig. 4-4-2).

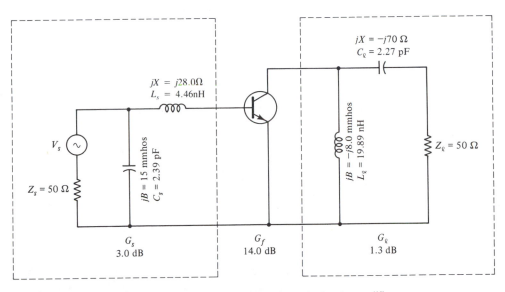

Figure 4-4-2 Designed matching networks for the amplifier.

4-5 NARROWBAND AMPLIFIER DESIGN

A narrowband amplifier traditionally has a bandwidth of less than 10%. The design procedures for this type of microwave amplifier are similar to the small-signal amplifier design except that the scattering, noise-figure, and power-gain parameters must be measured at the center frequency of the narrow bandwidth. There are two design methods available to deal with either maximum power gain or minimum-noise figure at a time.

4-5-1 Narrowband Amplifier Design for Maximum Power Gain

From Section 3-5-2, the reflection coefficient of the source impedance required to conjugately match the input of the amplifier for maximum power gain is Γ_{sm}, as shown in Eq. (3-5-13). Similarly, the reflection coefficient of the load impedance required to conjugately match the output of the amplifier is $\Gamma_{\ell m}$, as shown in Eq. (3-5-15). Actually, these two parameters are measured and supplied by the device manufacturers. The designer can use these data Γ_{sm} and $\Gamma_{\ell m}$ to design the input and output matching networks as shown in Fig. 4-5-1.

Input matching network.   From transmission-line theory, the source reflection coefficient for maximum power gain is given by

$$\Gamma_{sm} = \frac{Z_{sin} - Z_o}{Z_{sin} + Z_o} \qquad (4\text{-}5\text{-}1)$$

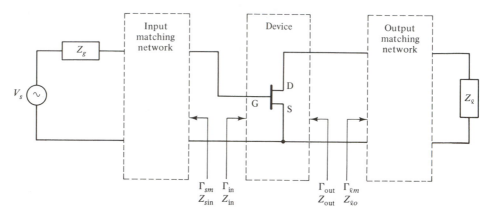

Figure 4-5-1. Matching networks of narrowband amplifier to be designed.

where Z_{sin} = source input impedance

Z_o = characteristic impedance

The source equivalent impedance for maximum power gain is expressed as

$$Z_{sin} = Z_o \frac{1 + \Gamma_{sm}}{1 - \Gamma_{sm}} \tag{4-5-2}$$

where

$$\Gamma_{sm} = |\Gamma_{sm}| \cos \underline{/\Gamma_{sm}} + j|\Gamma_{sm}| \sin \underline{/\Gamma_{sm}} \tag{4-5-3}$$

is the source reflection coefficient for maximum available power gain at minimum-noise figure F_{min}.

Substitution of Eq. (4-5-3) in Eq. (4-5-2) and then multiplying the numerator and denominator of the resultant by the denominator's conjugate results in the source equivalent impedance for maximum power gain at F_{min} as

$$Z_{sin} = Z_o \frac{1 - |\Gamma_{sm}|^2 + j2|\Gamma_{sm}| \sin \underline{/\Gamma_{sm}}}{1 + |\Gamma_{sm}|^2 - 2|\Gamma_{sm}| \cos \underline{/\Gamma_{sm}}} \tag{4-5-4}$$

Output matching network. Similarly, the load equivalent impedance for maximum power gain is

$$Z_{to} = Z_o \frac{1 - |\Gamma_{\ell m}|^2 + j2|\Gamma_{\ell m}| \sin \underline{/\Gamma_{\ell m}}}{1 + |\Gamma_{\ell m}|^2 - 2|\Gamma_{\ell m}| \cos \underline{/\Gamma_{\ell m}}} \tag{4-5-5}$$

where $\Gamma_{\ell m}$ is the load reflection coefficient for maximum power gain. The main purposes of the input and output matching networks are to provide adequate impedances for maximum power gain.

Design Procedure

1. Convert the source and load reflection coefficients Γ_{sm} and $\Gamma_{\ell m}$ to the source and load equivalent impedances.

2. Determine the source and load equivalent admittances from the impedances computed.
3. Realize the susceptance components with either a short or an open circuit of a three-eighths-wavelength stub.
4. Realize the conductance components with quarter-wavelength transformers.

Design Example 4-5-1:   Narrowband Amplifier Design for Maximum Power Gain

A certain GaAs-MESFET amplifier is to be designed for use in the frequency range 5.8 to 6.2 GHz. Its center frequency is 6 GHz. The various measured parameters at 6 GHz with a 50-Ω resistance are

$$\Gamma_{sm} = 0.761 \ \underline{/177°}$$

$$\Gamma_{\ell m} = 0.719 \ \underline{/104°}$$

$$G_{a\max} = 12 \text{ dB}$$

Design (a) the input and (b) the output matching networks with 50-Ω reference for maximum power gain.

Design Steps: (a) Input matching network

Step 1. Determine the source equivalent impedance from Eq. (4-5-4).

$$Z_{sin} = \frac{50(1 - |0.761|^2) + j2 \times 50 \times 0.761 \sin (177°)}{1 + |0.761|^2 - 2|0.761| \cos (177°)}$$

$$= 6.77 + j1.28$$

Step 2. Convert Z_{sin} to Y_{sin}.

$$Y_{sin} = \frac{1}{Z_{sin}} = \frac{1}{6.77 + j1.28} = 0.15 - j0.02$$

Step 3. Use a three-eighths-wavelength stub. An open-circuited stub looks like a shunt admittance $Y = jY_o \tan (\beta\ell)$. Therefore, an open-circuited stub that is three-eighths-wavelength long looks like a shunt inductor of susceptance $-jY_o$. Thus

$$Z_{o1} = \frac{1}{Y_o} = \frac{1}{\text{Im} [Y_{sin}]} = \frac{1}{0.02} = 50.0 \ \Omega$$

Step 4. Use a quarter-wave transformer. In order to transform the source resistance of 50 Ω to the source equivalent conductance of 0.15 mho, a quarter-wave transformer needs the follwing characteristic impedance:

$$Z_{o2} = \sqrt{\frac{50}{0.15}} = 18.26 \ \Omega$$

(b) Output matching network

Step 1. Determine the load equivalent impedance from Eq. (4-5-5).

$$Z_{to} = \frac{50(1 - |0.719|^2) + j2 \times 50 \times 0.719 \sin (104°)}{1 + |0.719|^2 - 2 \times 0.719 \cos (104°)}$$

$$= 12.83 + j37.30$$

Step 2. Convert Z_{to} to Y_{to}:

$$Y_{to} = \frac{1}{Z_{to}} = \frac{1}{12.83 + j37.30} = 0.008 - j0.024$$

Step 3. Use a three-eighths-wavelength stub. An open-circuited stub that is three-eighths-wavelength long looks like a shunt inductor of susceptance $-jY_o$. Thus

$$Z_{o1} = \frac{1}{Y_o} = \frac{1}{\mathrm{Im}\,[Y_{to}]} = \frac{1}{0.024} = 41.67 \ \Omega$$

Step 4. Use a quarter-wave transformer. A quarter-wave transformer is neeed to transform the load equivalent impedance to the load resistance 50 Ω.

$$Z_{o2} = \sqrt{\frac{50}{0.008}} = 79 \ \Omega$$

Step 5. Draw the designed matching networks for the amplifier (Fig. 4-5-2).

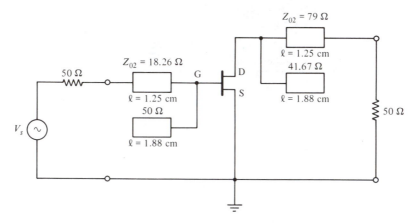

Figure 4-5-2   Designed matching networks for Example 4-5-1.

4-5-2 Narrowband Amplifier Design for Minimum-Noise Figure

For some particular applications, the minimum-noise figure of a microwave amplifier is more important than the power gain, such as the preamplifier. The basic equations of the source and load equivalent impedances for the minimum-noise figure can be written as

$$Z_{sn} = \frac{Z_o(1 - |\Gamma_o|^2) + j2Z_o\,|\Gamma_o|\sin\,\underline{/\Gamma_o}}{1 + |\Gamma_o|^2 - 2\,|\Gamma_o|\cos\,\underline{/\Gamma_o}} \tag{4-5-6}$$

where

Γ_o = optimum source reflection coefficient for minimum noise $F_{\min}$

and

$$Z_{\ell n} = \frac{Z_o(1 - |\Gamma_\ell|^2) + j2Z_o|\Gamma_\ell| \sin \underline{/\Gamma_\ell}}{1 + |\Gamma_\ell|^2 - 2|\Gamma_\ell| \cos \underline{/\Gamma_\ell}} \qquad (4\text{-}5\text{-}7)$$

where Γ_ℓ = load reflection coefficient for minimum noise F_{min}.

Design Procedure.   The design procedures for a minimun-noise narrowband amplifier are similar to those for a maximum-gain narrowband amplifier.

Design Example 4-5-2:   Narrowband Amplifier Design for Minimum-Noise Figure

The same GaAs MESFET used in Example 4-5-1 has the noise parameters measured at the center frequency of 6 GHz as

$$F_{min} = 3 \text{ dB}$$

$$R_n = 10 \ \Omega$$

$$\Gamma_o = 0.541 \ \underline{/140°}$$

$$\Gamma_\ell = 0.543 \ \underline{/105°}$$

Design (a) the input and (b) the output matching networks with a 50-Ω reference for a minimum-noise figure, as shown in Fig. 4-5-3.

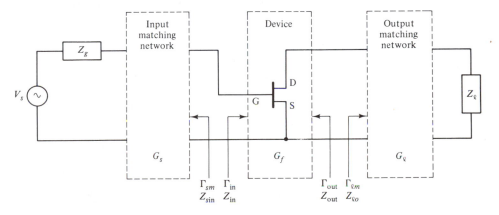

Figure 4-5-3.   Matching networks of narrowband amplifier to be designed.

Design Steps: (a) Input matching network

Step 1. Determine the source equivalent impedance for minimum-noise figure from Eq. (4-5-6):

$$Z_{sn} = \frac{50(1 - |0.541|^2) + j2 \times 50 \times 0.541 \sin (140°)}{1 + |0.541|^2 - 2|0.541| \cos (140°)}$$

$$= 16.75 + j16.40 \ \Omega$$

Step 2. Convert Z_{sn} to Y_{sn}.

$$Y_{sn} = \frac{1}{Z_{sn}} = \frac{1}{16.75 + j16.40} = 0.031 - j0.030$$

Step 3. Use a three-eighths-wavelength stub. An open-circuited stub looks like a shunt admittance $Y = jY_o \tan(\beta\ell)$. Therefore, an open-circuited stub that is three-eighths-wavelength long looks like a shunt inductor of susceptance $-jY_o$. Thus

$$Z_{o1} = \frac{1}{Y_o} = \frac{1}{\text{Im}[Y_{sn}]} = \frac{1}{0.030} = 33.33 \ \Omega$$

Step 4. Use a quarter-wave transformer. To transform the load resistance of 50 Ω to the load equivalent conductance 0.031 mho, a quarter-wave transformer needs the following characteristic impedance:

$$Z_{o2} = \sqrt{\frac{50}{0.031}} = 40.16 \ \Omega$$

(b) Output matching network

Step 1. Determine the load equivalent impedance for the minimum-noise figure from Eq. (4-5-7).

$$Z_{\ell n} = \frac{50(1 - |0.543|^2) + j2 \times 50 \times 0.543 \sin(105°)}{1 + |0.543|^2 - 2|0.543| \cos(105°)}$$
$$= 22.61 + j33.41 \ \Omega$$

Step 2. Convert $Z_{\ell n}$ to $Y_{\ell n}$.

$$Y_{\ell n} = \frac{1}{Z_{\ell n}} = \frac{1}{22.61 + j33.41} = 0.014 - j0.021 \ \Omega$$

Step 3. Use a three-eighths-wavelength stub. The output matching network is similar to the input matching circuit. An open-circuited stub that is three-eighths-wavelength

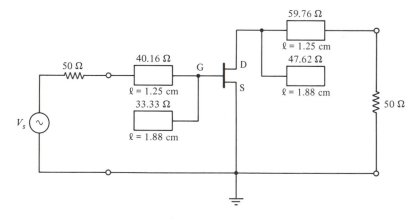

Figure 4-5-4　Designed matching networks for Example 4-5-2.

long looks like a shunt inductor of a susceptance $-jY_o$. Then

$$Z_{o1} = \frac{1}{Y_o} = \frac{1}{\text{Im}\,[Y_{tn}]} = \frac{1}{0.021} = 47.62\ \Omega$$

Step 4. Use a quarterwave transformer. Since the load resistance is 50 Ω, a quarter-wave transformer needs the following characteristic impedance to match the load equivalent impedance to the load resistance:

$$Z_{o2} = \sqrt{\frac{50}{0.014}} = 59.76\ \Omega$$

Step 5. Draw the designed matching networks for the amplifier (Fig. 4-5-4).

PROBLEMS

4-1 dc-Biasing Circuits

4-1-1. The quiescent point of a GaAs MESFET is at

$$V_{ds} = 5\ \text{V} \qquad V_{dd} = 30\ \text{V}$$
$$I_{ds} = 14\ \text{mA} \qquad V_{gs} = -1\ \text{V}$$

A passive dc-biasing circuit is to be designed for the MESFET. The circuit requires a unipolar power supply and should have a source bypass resistor. Determine:
(a) The drain Resistance R_d.
(b) The source bypass resistance R_s.

4-1-2. A microwave silicon transistor has the following parameters:

$V_{CC} = 25\ \text{V}$	$I_{BB} = 1\ \text{mA}$	Quiescent point:
$V_{BB} = 2\ \text{V}$	$I_{CBO} = 0$	$V_{CE} = 8\ \text{V}$
$V_{BE} = 1\ \text{V}$	$h_{FE} = 50$	$I_C = 10\ \text{mA}$

Design:
(a) A passive dc-biasing circuit with voltage feedback.
(b) A passive dc-biasing circuit with voltage feedback and constant base current.
(c) A passive dc-biasing circuit with a bypass-emitter resistor.

4-1-3. A GaAs MESFET has the following parameters:

$$I_{dss} = 15\ \text{mA}$$
$$I_{ds} = 2.25\ \text{mA}$$
$$V_p = -7\ \text{V}$$

Compute:
(a) The gate–source voltage V_{gs}.
(b) The dc-biasing source resistor R_s.

4-2 Small-Signal Amplifier Design

4-2-1. The DXL 2501A-P70 GaAs FET has the following S parameters measured at $V_{ds} = 6$ V and $I_{ds} = 40$ mA at 6 GHz with a 50-Ω resistance:

$$S_{11} = 0.69 \;/-159°$$
$$S_{12} = 0.0$$
$$S_{21} = 2.20 \;/46°$$
$$S_{22} = 0.74 \;/-67°$$

The GaAs FET is to be used for an amplifier. Design the input and output matching networks of the amplifier for a power gain of 10 dB at 6 GHz.

4-2-2. The DXL 3501 A (chip) GaAs FET has the following S parameters measured at V_{ds} = 8 V and I_{ds} = $0.5I_{dss}$ at 10 GHz with a 50-Ω resistance:

$$S_{11} = 0.66 \;/-143°$$
$$S_{12} = 0.0$$
$$S_{21} = 1.26 \;/46°$$
$$S_{22} = 0.74 \;/-59°$$

The GaAs FET is to be used for an amplifier. Deisgn the input and output matching networks of the amplifier for a power gain of 5 dB at 10 GHz.

4-2-3. The NE41137 GaAs FET has the following S parameters measured at V_{ds} = 5 V and I_{ds} = 10 mA at 3 GHz with a 50-Ω resistance:

$$S_{11} = 0.38 \;/-169°$$
$$S_{12} = 0.0$$
$$S_{21} = 1.33 \;/-39°$$
$$S_{22} = 0.95 \;/-66°$$

The GaAs FET is to be used for an amplifier. Design the input and output matching networks of the amplifier for a power gain of 6 dB at 3 GHz.

4-2-4. The S parameters of a GaAs MESFET are measured with a 50-Ω resistance at 3 GHz as

$$S_{11} = 0.60 \;/-150°$$
$$S_{12} = 0.07 \;/-180°$$
$$S_{21} = 2.50 \;/-40°$$
$$S_{22} = 0.40 \;/-15°$$

(a) Calculate and plot the input and output constant-power-gain circles on a Smith chart.

(b) Overlap two Smith charts to facilitate the design procedures. Use the original chart to read the impedance and the overlaid chart to read the admittance.

(c) Determine the capacitance and inductance of the output matching network.

(d) Find the capacitance and inductance of the input matching network.

4-2-5. The S parameters of a certain MESFET are measured with a 50-Ω line at 5 GHz as

$$S_{11} = 0.45 \;/-160°,$$
$$S_{21} = 7.00 \;/180°$$
$$S_{22} = 0.35 \;/-25°$$

The MESFET is to be used as an amplifier in an amplifier circuit. The problem is to design two networks to match the input and output impedances of the MESFET. The load resistance and the output impedance of the signal source are considered to be 50 Ω.

(a) Determine the capacitance in picofarads and the inductance in millihenrys of the output matching network.

(b) Find the capacitance in picofarads and the inductance in millihenrys of the input matching network.

(c) Draw the complete circuit diagram for the amplifier designed.

4-3 High-Gain Amplifier Design

4-3-1. The NE868495-4 GaAs power FET has the following S parameters measured at $V_{ds} =$ 9 V and $I_{ds} = 600$ mA at 5 GHz with a 50-Ω line:

$$S_{11} = 0.45 \ \underline{/-163°}$$
$$S_{12} = 0.04 \ \underline{/40°}$$
$$S_{21} = 2.55 \ \underline{/-106°}$$
$$S_{22} = 0.46 \ \underline{/-65°}$$

The GaAs FET is to be used for a high-gain amplifier. Design the input and output matching networks of the amplifier for maximum power gain at 5 GHz.

4-3-2. The NE868898-7 GaAs power FET has the following S parameters measured at $V_{ds} =$ 9 V and $I_{ds} = 1.2$ A at 6.8 GHz with a 50-Ω line:

$$S_{11} = 0.49 \ \underline{/-175°}$$
$$S_{12} = 0.02 \ \underline{/5°}$$
$$S_{21} = 2.65 \ \underline{/50°}$$
$$S_{22} = 0.56 \ \underline{/-45°}$$

The GaAs FET is to be used for a high-gain amplifier. Design the input and output matching networks of the amplifier for maximum power gain at 6.8 GHz.

4-3-3. The DXL 3501A-P100F GaAs medium-power FET has the following S parameters measured at $V_{ds} = 8$ V and $I_{ds} = 0.5I_{dss}$ at 5 GHz with a 50-Ω line:

$$S_{11} = 0.60 \ \underline{/-135°}$$
$$S_{12} = 0.09 \ \underline{/21°}$$
$$S_{21} = 2.30 \ \underline{/63°}$$
$$S_{22} = 0.54 \ \underline{/-68°}$$

The GaAs FET is to be used for a power amplifier. Design the input and output matching networks of the amplifier for maximum power gain at 5 GHz.

4-3-4. The DXL 3503A (chip) Ku-band medium-power GaAs FET has the following S parameters measured at $V_{ds} = 6$ V and $I_{ds} = 0.5I_{dss}$ at 18 GHz with a 50-Ω line:

$$S_{11} = 0.64 \ \underline{/-160°}$$
$$S_{12} = 0.08 \ \underline{/127°}$$

$$S_{21} = 0.81 \ \underline{/23°}$$
$$S_{22} = 0.77 \ \underline{/-78°}$$

The GaAs FET is to be used for a high-gain amplifier. Design the input and output matching networks of the amplifier for maximum power gain at 18 GHz.

4-3-5. The DXL 3608A (chip) X-band GaAs power FET has the following S parameters measured at $V_{ds} = 5$ V and $I_{ds} = 115$ mA at 8 GHz with a 50-Ω line:

$$S_{11} = 0.82 \ \underline{/-142°}$$
$$S_{12} = 0.04 \ \underline{/77°}$$
$$S_{21} = 1.75 \ \underline{/68°}$$
$$S_{22} = 0.56 \ \underline{/-60°}$$

The GaAs FET is to be used for high-gain amplifier. Design the input and output matching networks for the power amplifier for maximum power gain at 8 GHz measured with a 50-Ω line.

4-3-6. A certain GaAs MESFET has the following parameters measured with a 50-Ω line:

$$S_{11} = 0.66 \ \underline{/-160°} \qquad \Gamma_{in} = 0.65 \ \underline{/160°}$$
$$S_{12} = 0.05 \ \underline{/150°} \qquad \Gamma_{out} = 0.60 \ \underline{/30°}$$
$$S_{21} = 2.20 \ \underline{/25°}$$
$$S_{22} = 0.55 \ \underline{/-30°}$$

(a) Calculate the stability factor K.
(b) Compute the maximum unilateral transducer power gain in decibels.
(c) Design an input matching network N_{in}.
(d) Design an output matching network N_{out}.
(e) Sketch the complete matching networks.

4-4 Low-Noise Amplifier Design

4-4-1. The NE71083 low-noise GaAs FET has the following S parameters measured at $V_{ds} = 3$ V and $I_{ds} = 10$ mA at 6 GHz with a 50-Ω line:

$$S_{11} = 0.44 \ \underline{/-103°} \qquad F_{min} = 3 \text{ dB}$$
$$S_{12} = 0.07 \ \underline{/20°} \qquad \Gamma_o = 0.51 \ \underline{/180°}$$
$$S_{21} = 2.28 \ \underline{/81°} \qquad R_n = 5 \ \Omega$$
$$S_{22} = 0.62 \ \underline{/-77°}$$

Design the input and output matching networks of the amplifier for power gain of 10 dB and noise figure of 4.5 dB.

4-4-2. The NE67383 Ku-band low-noise GaAs FET has the following S parameters measured at $V_{ds} = 3$ V and $I_{ds} = 10$ mA at 16 GHz with a 50-Ω line:

$$S_{11} = 0.57 \ \underline{/-111°} \qquad F_{min} = 2.5 \text{ dB}$$
$$S_{12} = 0.01 \ \underline{/-31°} \qquad \Gamma_o = 0.48 \ \underline{/175°}$$
$$S_{21} = 2.09 \ \underline{/-42°} \qquad R_n = 4 \ \Omega$$
$$S_{22} = 0.47 \ \underline{/-69°}$$

Design the input and output matching networks of the amplifier for power gain of 9 dB and a low-noise figure of 4.5 dB.

4-4-3. The NE21889 X-band low-noise GaAs FET has the following S parameters measured at $V_{ds} = 4$ V and $I_{ds} = 30$ mA at 10 GHz with a 50-Ω line:

$$S_{11} = 0.40 \ \underline{/-125°} \qquad F_{min} = 3 \text{ dB}$$
$$S_{12} = 0.01 \ \underline{/20°} \qquad \Gamma_o = 0.43 \ \underline{/170°}$$
$$S_{21} = 1.58 \ \underline{/-5°} \qquad R_n = 4.5 \ \Omega$$
$$S_{22} = 0.64 \ \underline{/-80°}$$

Design the input and output matching networks of the amplifier for power gain of 6 dB and a low-noise figure of 3.5 dB.

4-4-4. The DXL 2503 Ku-band low-noise GaAs FET has the follwing S parameters measured at $V_{ds} = 3.5$ V and $I_{ds} = 30$ mA at 14 GHz with a 50-Ω line:

$$S_{11} = 0.46 \ \underline{/-143°} \qquad F_{min} = 2.5 \text{ dB}$$
$$S_{12} = 0.01 \underline{/98°} \qquad \Gamma_o = 0.55\underline{/160°}$$
$$S_{21} = 1.70\underline{/59°} \qquad R_n = 5 \ \Omega$$
$$S_{22} = 0.40 \ \underline{/-84°}$$

Design the input and output matching networks of the amplifier for a maximum power gain of 6.4 dB and a low-noise figure of 3 dB.

4-4-5. The DXL 1503A-P70 X-band low-noise GaAs FET has the following S parameters measured at $V_{ds} = 3.5$ V and $I_{ds} = 12$ mA at 12 GHz with a 50-Ω line:

$$S_{11} = 0.48 \ \underline{/-130°} \qquad F_{min} = 2.5 \text{ dB}$$
$$S_{12} = 0.01 \ \underline{/92°} \qquad \Gamma_o = 0.45 \ \underline{/175°}$$
$$S_{21} = 2.22 \ \underline{/75°} \qquad R_n = 3.5 \ \Omega$$
$$S_{22} = 0.52 \ \underline{/-65°}$$

Design the input and output matching networks of the amplifier for a maximum power gain of 9 dB and a low-noise figure of 3 dB.

4-5 Narrowband Amplifier Design

4-5-1. The HP HXTR-6101 microwave transistor has the following parameters at 4 GHz:

S parameters	Gain parameters	Noise parameters
$S_{11} = 0.552 \ \underline{/169°}$	$K = 1.012$	$F_{min} = 2.5$ dB
$S_{12} = 0.049 \ \underline{/23°}$	$G_{amax} = 14.7$ dB	$\Gamma_o = 0.475 \ \underline{/166°}$
$S_{21} = 1.68 \ \underline{/26°}$	$\Gamma_{sm} = 0.941 \ \underline{/-154°}$	$R_n = 3.5 \ \Omega$
$S_{22} = 0.839 \ \underline{/-67°}$	$\Gamma_{\ell m} = 0.979\underline{/70°}$	

The transistor is to be used for a low-noise narrowband amplifier. Design the input and output matching networks with a 50-Ω reference for a minimum-noise amplifier.

4-5-2. A certain GaAs power MESFET has the following parameters measured at 8 GHz with a 50-Ω reference:

$$\Gamma_{sm} = 0.721 \ \underline{/160°}$$

$$\Gamma_{\ell m} = 0.703 \underline{/115°}$$
$$G_{a\max} = 12 \text{ dB}$$
$$F_{\min} = 3 \text{ dB}$$
$$R_n = 10 \text{ }\Omega$$

The MESFET is to be used for a high power-gain narrowband amplifier. Design the input and output matching networks with the 50-Ω reference for a maximum power-gain narrowband amplifier.

Chapter 5

Balanced Amplifier Design and Power-Combining Techniques

5-0 INTRODUCTION

In Chapter 4, small-signal and narrowband amplifier designs were discussed. To obtain high power and broadband, the simplest and easiest method is to use power-combining techniques in amplifier design. For example, the balanced amplifier incorporated with two hybrid couplers is commonly used in microwave integrated circuits for high power and broadband components. For future airborne phased-array radar and spaceborne communications systems applications, a large number of identical components are required and the balanced amplifier is the most desirable candidate for achieving the goals projected. In this chapter several commonly used microwave power combiners/dividers, such as Lange coupler and Wilkinson power combiner, and the balanced amplifier design are discussed.

5-1 LANGE COUPLERS

Lange couplers are interdigitated microstrip couplers consisting of four parallel strip-lines with alternate lines tied together. A single ground plane, a single dielectric, and a single layer of metallization are used. This type of couplers has four ports and is used for a power combiner [1]. Figure 5-1-1 shows the schematic diagram of a Lange coupler. A signal wave incident in port 1 couples equal power into ports 2 and 4, but none into port 3. There are two basic types: 180° hybrid and 90° (quadrature) hybrid. The latter is also called a 3-dB directional coupler.

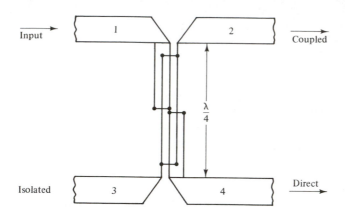

Figure 5-1-1 Lange coupler.

5-1-1 Basic Equations of Lange Couplers

The structure of a Lange coupler described in Fig. 5-1-1 can be rearranged to a four-port network as shown in Fig. 5-1-2.

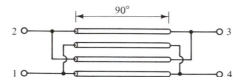

Figure 5-1-2 Interdigitated Lange coupler.

When the four-port network is used as a Lange coupler, the phase angle for one-quarter wavelength at the center frequency of the operating band is 90°, so this type of coupler is also called a *quadrature coupler*. The current–voltage relation at the center frequency is given by the following matrix [2]:

$$
\begin{bmatrix} I_1 \\ I_2 \\ I_3 \\ I_4 \end{bmatrix} = \begin{bmatrix} 0 & 0 & jN & jM \\ 0 & 0 & jM & jN \\ jN & jM & 0 & 0 \\ jM & jN & 0 & 0 \end{bmatrix} = \begin{bmatrix} V_1 \\ V_2 \\ V_3 \\ V_4 \end{bmatrix} \tag{5-1-1}
$$

where

$$
M = \frac{k}{2} Y_{11} + \left(\frac{k}{2} - 1 \right) \frac{Y_{12}^2}{Y_{11}}
$$

$$
N = (k - 1) Y_{12} \tag{5-1-2}
$$

$$
k = \text{number of lines}
$$

The admittance $Y_{m(m+1)}$ can be expressed in terms of self and mutual capacitances and phase velocity as

$$Y_{m(m+1)} = -C_{m(m+1)}v_p \qquad (5\text{-}1\text{-}3)$$

and

$$Y_{mm} = (C_{m0} + C_{(m-1)m} + C_{m(m+1)})\, v_p \qquad (5\text{-}1\text{-}4)$$

where $C_{m(m+1)}$ = mutual capacitance per unit length between lines m and $m+1$
C_{m0} = capacitance per unit length between line m and ground
v_p = phase velocity
If the spacings between lines are identical, the capacitances are reduced to

$$C_{m(m+1)} = C_{12} \qquad (5\text{-}1\text{-}5)$$

$$C_{m0} = \begin{cases} C_{10} & \text{if } m = 1, k \\ C_{20} & \text{if } m \neq 1, k \end{cases} \qquad (5\text{-}1\text{-}6)$$

$$C_{20} \simeq C_{10} - \frac{C_{10}C_{12}}{C_{10} + C_{12}} \qquad (5\text{-}1\text{-}7)$$

Substitution of Eqs. (5-1-5) through (5-1-7) into Eqs. (5-1-3) and (5-1-4) yields the following relations:

$$Y_{mm} = Y_{11} \qquad \text{if } m = 1, k \qquad (5\text{-}1\text{-}8)$$

$$= Y_{11} + \frac{Y_{12}^2}{Y_{11}} \qquad \text{if } m \neq 1, k \qquad (5\text{-}1\text{-}9)$$

For perfect match and perfect isolation, the terminated admittance at the four ports should follow the conditions

$$Y_0^2 = M^2 - N^2 \qquad (5\text{-}1\text{-}10)$$

$$= Y_{0o}Y_{0e} \qquad (5\text{-}1\text{-}11)$$

$$(M^2 - N^2) > 0$$

where Y_{0o} = odd-mode admittance
Y_{0e} = even-mode admittance
If port 1 is fed by an input power, the power-coupling ratios are determined by the following equations:

$$\frac{P_2}{P_{1in}} = \frac{N^2}{M^2} \qquad (5\text{-}1\text{-}12)$$

$$\frac{P_3}{P_{1in}} = 0 \qquad (5\text{-}1\text{-}13)$$

$$\frac{P_4}{P_{\text{lin}}} = \frac{M^2 - N^2}{M^2} \qquad (5\text{-}1\text{-}14)$$

Equation (5-1-13) indicates that port 3 is perfectly isolated from port 1, so its power is zero, as expected.

In practice, the coupling coefficient K and the normalized odd-mode impedance z are given by [3]

$$K = \frac{(k - 1)(1 - R^2)}{(k - 1)(1 + R^2) + 2R} \qquad (5\text{-}1\text{-}15)$$

and

$$z = \frac{Z_{0o}}{Z_0} \frac{\sqrt{R[(k - 1) + R][(k - 1)R + 1]}}{1 + R} \qquad (5\text{-}1\text{-}16)$$

where

$$R = \frac{Z_{0o}}{Z_{0e}} \qquad (5\text{-}1\text{-}17)$$

The relationships among K, z, and R are plotted in Figures 5-1-3, 5-1-4, and 5-1-5.

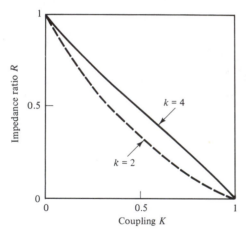

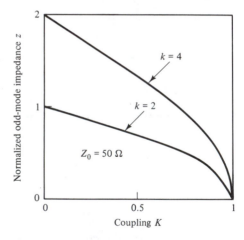

Figure 5-1-3 Impedance ratio versus coupling coefficient. (After Adolph Presser [3]. Copyright © 1978 IEEE. Reprinted by permission of the IEEE, Inc.)

Figure 5-1-4 Normalized odd-mode impedance versus coupling coefficient. (After Adolph Presser [3]. Copyright © 1978 IEEE. Reprinted by permission of the IEEE, Inc.)

5-1-2 Design Example

The procedures for designing a Lange coupler are given as follows:

1. Determine the desired coupling coefficient K.
2. Choose the characteristic impedance Z_0.

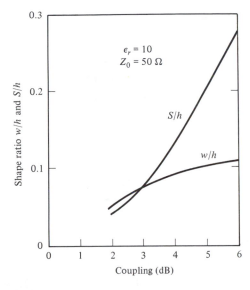

Figure 5-1-5 Nominal shape ratios versus coupling coefficient. (After Adolph Presser [3]. Copyright © 1978 IEEE. Reprinted by permission of the IEEE, Inc.)

3. Calculate the strip length ℓ for a quarter wavelength at the center frequency of the operating band.

4. Find the gap spacing S and the strip width w for a given substrate height h and with a relative dielectric constant ϵ_{r1}.

Design Example 5-1-1:   Lange Coupler

A four-strip Lange coupler is to be used in a balanced amplifier on an alumina substrate for power division at the center frequency of 10 GHz. The terminal characteristic impedance is 50 Ω, the relative dielectric constant of alumina is 10, and the coupling coefficient K is 3 dB, as shown in Fig. 5-1-6. Determine (a) the strip length ℓ; (b) the impedance ratio R; (c) the normalized odd-mode impedance z; (d) the strip width w, spacing between strips S, and the substrate height h; and (e) the impedances Z_{0o} and Z_{0e}.

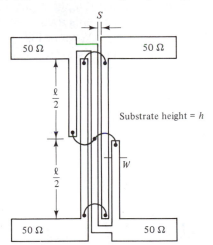

Figure 5-1-6 Lange coupler to be designed. (After Adolph Presser [3]. Copyright © 1978 IEEE. Reprinted by permission of the IEEE, Inc.)

Solution   (a) The strip length is

$$\ell = \frac{\lambda}{4} = \frac{300 \times 10^8}{4 \times 10^{10}} = 0.75 \text{ cm} = 7.5 \text{ mm}$$

(b) From Eq. (5-1-15) the impedance ratio R is found as

$$0.5 = \frac{(4-1)(1-R^2)}{(4-1)(1+R^2)+2R}$$

$$R = 0.48$$

(c) From Eq. (5-1-16) the normalized impedance is

$$z = \frac{50}{50} \frac{\sqrt{0.48(3+0.48)(3 \times 0.48 + 1)}}{1 + 0.48}$$

$$= 1.36$$

(d) From the plots as shown in Figs. 5-1-3 and 5-1-5, we find that

$$\frac{S}{h} = 0.075 \qquad \epsilon_r = 10$$

$$\frac{w}{h} = 0.08 \qquad Z_0 = 50 \ \Omega$$

Let $h = 25$ mils; then $w = 2$ mils and $S = 1.9$ mils.

(e) From Appendix B we obtain

$$Z_{0o} = 50 \ \Omega$$

$$Z_{0e} = 150 \ \Omega$$

5-2 BALANCED AMPLIFIER DESIGN

One of the commonly used microwave solid-state amplifier integrated circuits for high power and broadband is the balanced amplifier configuration [4]. A balanced amplifier consists of two 3-dB and 90° Lange couplers as shown in Fig. 5-2-1. The input power is split into two parts, so this type of amplifier is also called a two-way combined amplifier.

As shown in Fig. 5-2-1, two GaAs MESFETs are connected at the ports of two 3-dB quadrature Lange couplers. This design allows the input and output amplifier ports to be mismatched, but the combined amplifier will appear matched if the parallel stages are balanced. The total reflection coefficients, power gain, and the voltage-standing-wave ratio (VSWR) can be expressed as

$$S_{11} = \frac{1}{2}(S_{11a} - S_{11b}) \qquad\qquad (5\text{-}2\text{-}1)$$

$$S_{22} = \frac{1}{2}(S_{22a} - S_{22b}) \qquad\qquad (5\text{-}2\text{-}2)$$

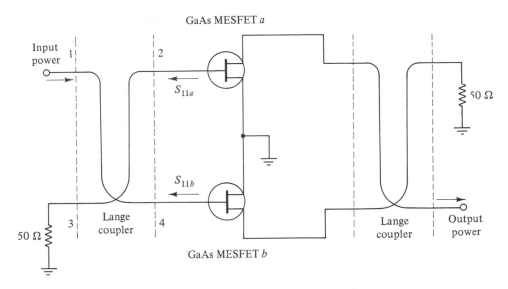

Figure 5-2-1   Balanced amplifier with Lange couplers.

$$\text{reverse power loss} = |S_{12}|^2 = \frac{1}{4}|S_{12a} + S_{12b}|^2 \qquad (5\text{-}2\text{-}3)$$

$$\text{forward power gain} = |S_{21}|^2 = \frac{1}{4}|S_{21a} + S_{21b}|^2 \qquad (5\text{-}2\text{-}4)$$

$$\text{VSWR(input)} = \frac{1 + |S_{11}|}{1 - |S_{11}|} \qquad (5\text{-}2\text{-}5a)$$

and

$$\text{VSWR(output)} = \frac{1 + |S_{22}|}{1 - |S_{22}|} \qquad (5\text{-}2\text{-}5b)$$

where a and b indicate the two GaAs MESFETs and 1 and 2 refer to the input and output ports of the devices.

If the two GaAs MESFETs are identical, the input and output reflection coefficients are reduced to half of the corresponding reflections of the two GaAs MESFETs, the total power gain is equal to the individual power gain, and the VSWRs at the input and output are unity. If one of the two GaAs MESFETs fails, the power loss of the balanced amplifier would be only 6 dB. This type of power loss, called *graceful degradation*, means that the output power is reduced but not completely lost. The balanced amplifier configuration is by far the most common design in a modern microwave integrated circuit. Its advantages are:

1. *Good input and output VSWRs.* If the reflection coefficients of the two active devices are very close and their VSWRs could be much improved.

2. *Good stability*. The reflections from the two active devices each terminate in a 50-Ω load, which usually guarantees stability.

3. *High reliability*. If one active device fails, the amplifier power is reduced but not completely lost. Consequently, the amplifier reliability is enhanced.

4. *Low tuning work*. The tuning work for a balanced amplifier is much less than in designing and tuning an unbalanced amplifier. The tuning labor is even less for a multistage balanced amplifier than for a conventional amplifier because each stage is isolated by a Lange coupler.

5. *High output power*. The linear output power is increased by 3 dB in comparison with a one-way amplifier.

In practical design, it is often necessary to insert a matching network in front of and after the GaAs MESFET to ensure a minimum reflection from the mismatched load or source, as shown in Fig. 5-2-2. This type of design practice has the advantage over manual tuning work when nonidentical GaAs MESFETs are used in the circuit.

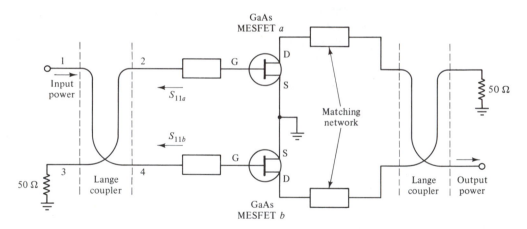

Figure 5-2-2   Balanced amplifier circuit with matching networks.

Example 5-2-1:   GaAs-MESFET Balanced Amplifier

A GaAs MESFET has the following parameters:
    MESFET *a*:

Reflection coefficients	$S_{11a} = 0.81 \ \underline{/53°}$
	$S_{22a} = 0.55 \ \underline{/20°}$
Forward transmission coefficient	$S_{21a} = 6.22 \ \underline{/180°}$

MESFET b:

Reflection coefficients $S_{11b} = 0.79 \; \underline{/55°}$

$S_{22b} = 0.50 \; \underline{/22°}$

Forward transmission $S_{21b} = 6.10 \; \underline{/178°}$
coefficient

Calculate (a) the input and output reflection coefficients of the balanced amplifier, and (b) the input and output VSWRs. Determine (c) the power gain in decibels for the balanced amplifier, and (d) the power loss in decibels if one MESFET fails. (e) Calculate the linear output of power capability in comparison with two MESFETs in series.

Solution (a) Using Eqs. (5-2-1) and (5-2-2), the input and output reflection coefficients are

$$|S_{11}| = \frac{1}{2}|(0.81 \; \underline{/53°} - 0.79 \; \underline{/55°})| = 0.02$$

$$|S_{22}| = \frac{1}{2}|(0.55 \; \underline{/20°} - 0.50 \; \underline{/22°})| = 0.03$$

(b) The input and output VSWRs are

$$\text{VSWR(input)} = \frac{1 + |S_{11}|}{1 - |S_{11}|} = \frac{1 + 0.02}{1 - 0.02} = 1.04$$

$$\text{VSWR(output)} = \frac{1 + |S_{22}|}{1 - |S_{22}|} = \frac{1 + 0.03}{1 - 0.03} = 1.06$$

(c) Using Eq. (5-2-4), the power gain is

$$\text{gain} = \frac{1}{4}|6.22 \; \underline{/180°} + 6.10 \; \underline{/178°}|^2 = \frac{1}{4} \times 151.82$$

$$= 37.96 = 15.8 \text{ dB}$$

(d) If one MESFET fails, the power loss is

$$\text{loss} = 10 \; \ell og\left(\frac{1}{4}\right) = -6 \text{ dB}$$

(e) The linear output power is increased by

$$P_{out} = 10 \; \ell og \, (2) = 3 \text{ dB}$$

5-3 CHIP CHARACTERIZATION

As indicated by Eqs. (5-2-1) and (5-2-2), the two-way amplifier will be balanced and its VSWRs will be unity if and only if the two GaAs MESFETs are identical. In practical design, however, the characteristic S parameters of the two GaAs MESFETs are not actually measured and they are not the same. When their characteristics are different, the amplifier will not be balanced and manual tuning work is needed to bring the amplifier on balance. Therefore, for mass production, it is required to characterize the GaAs-MESFET chips before placing them in the circuits in order to increase the hybrid reproducibility, minimize the tuning work, reduce production cost, and to improve the module performance.

There are three types of characterization for GaAs-MESFET chips:

1. *Dc and low-frequency tests*. The dc and low-frequency tests can be carried out individually for each chip on a wafer by a standard test fixture such as the automatic wafer prober model 1032 made by the Rucker-Kolls Company, shown in Fig. 5-3-1.

2. *RF packaged-chip tests*. Packaged chips are those which are ribbon bonded to bring the gate, source, and drain contacts out to bonding pads over the semi-insulating substrate after the chips are removed from the wafer. In practice, the S parameters at microwave frequencies are measured on a sample packaged chip; it is then assumed that the 100 chips around the sample chip would have the same S parameters. This assumption is not true because the wafer yield is very low, about 6 to 10%, due to lattice imperfections and fabrication defects. The packaged chip test can be made using the Maury Microwave MT950 test fixture shown in Fig. 5-3-2.

Figure 5-3-1 dc test fixture. (Reprinted by permission of Rucker-Kolls Co.)

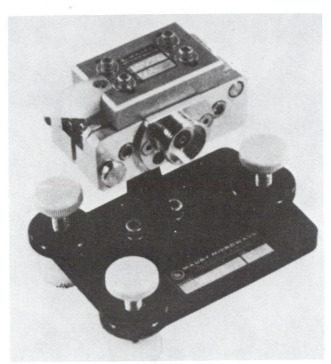

Figure 5-3-2 RF packaged-chip test fixture. (Reprinted by permission of Maury Microwave Corporation.)

3. *RF chip tests*. There are two types of RF chip characterization: destructive and nondestructive tests. After the chip is removed from a wafer and wire bonded, an RD chip test is made on the chip. This type of test is called a *destructive test* because the tested chip cannot be used again. The test fixture for destructive testing is the same as for the packaged-chip test. The nondestructive chip test is made to each chip either on the wafer or after the chip is removed from the wafer without wire bonding. So far the nondestructive test is not successful because the test fixture for nondestructive testing is not yet widely available. However, Cascade Microwave has introduced a microwave wafer probing test fixture, shown in Fig. 5-3-3.

5-4 POWER-COMBINING TECHNIQUES

Microwave power combining and dividing are sometimes very desirable in radar and space communications systems when a large amount of power cannot be supplied by a single power source. In higher microwave frequencies the power output of a microwave power source decreases rapidly according to the inverse frequency squared. To achieve a high power output, power combiners are needed to combine several single small-power sources. Power combining with a circulator in an IMPATT amplifier is an example. On the other hand, when the input power is too high beyond the power

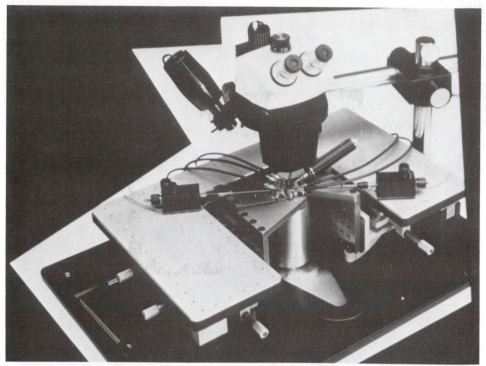

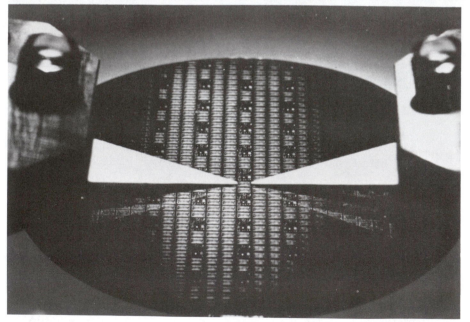

Figure 5-3-3   RF chip test fixtures. (Reprinted by permission of Cascade Microtech.)

limit of a single device capability, a power divider is necessary. The balanced amplifier incorporated with two Lange couplers is an example of the use of power-dividing and power-combining techniques.

In general, there are two categories of microwave power combiners and dividers: those that combine the output of N devices with multiple stages and those that combine the output of N devices in a single stage. The former category is simpler and more widely used, such as the binary combining structures (tree or corporate type) and nonbinary combining structures (chain or serial type). The latter category, also called an *N-way combiner*, includes resonant-cavity combining structures and nonresonant-cavity combining structures. All these power combiners and dividers are tabulated in Table 5-4-1, where it is seen that all microwave power combiners and dividers can be subdivided into four types:

1. Binary combiner/divider structures
2. Nonbinary combiner/divider structures
3. Resonant-cavity structures
4. Nonresonant-cavity structures

TABLE 5-4-1   MICROWAVE POWER COMBINERS/DIVIDERS

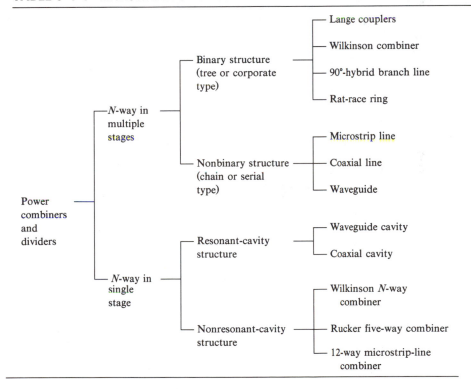

5-4-1 Binary Combiner/Divider Structures

The number of devices combined in this type of power combining structure is binary. This means that the device number must be equal to $2N$, where N is a positive integer. This type, also called *tree* or *corporate structure*, is shown in Fig. 5-4-1. The commonly used two-way combiners or dividers in the tree or corporate structure are shown in Fig. 5-4-2.

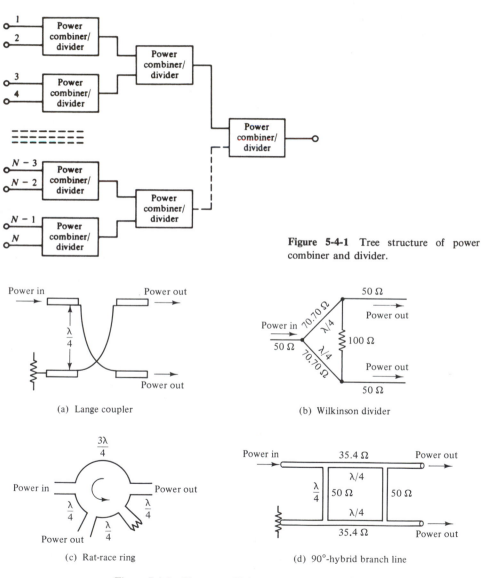

Figure 5-4-1 Tree structure of power combiner and divider.

(a) Lange coupler

(b) Wilkinson divider

(c) Rat-race ring

(d) 90°-hybrid branch line

Figure 5-4-2 Binary combining structures.

Combining efficiency of binary combiner/divider.   The combining efficiency of a binary combiner/divider can be expressed as

$$\eta = 10^{xL/10} \qquad\qquad (5\text{-}4\text{-}1)$$

where $x = \log_2 (N)$
$N = 2^x$ is the number of adders
L = loss per adder, decibels

The two-way power combiners have lossy components and they limit the combining efficiency as shown in Fig. 5-4-3. The combining efficiency of a binary combiner/divider is decreased when the number of the devices combined is increased. They all have two features in common: the input ports are matched to the output port and the input ports are isolated from one another. Lange couplers were described in Section 5-2-1. Here the Wilkinson combiner, rat-race ring, and 90°-hybrid branch line are discussed.

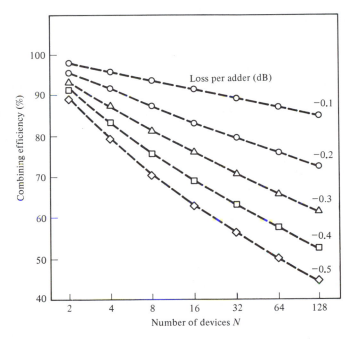

Figure 5-4-3   Combining efficiency for a binary combining structure. (After Kenneth J. Russell [5]. Copyright © 1979 IEEE. Reprinted by permission of the IEEE, Inc.)

Example 5-4-1:   Combining Efficiency of a Binary Combiner

A binary power combiner has 32 adders and each adder has a loss of −0.4 dB. Compute its combining efficiency.

Solution   From Eq. (5-4-1) we have

$$32 = 2^x \qquad \text{and} \qquad x = 5$$

Then the combining efficiency is

$$\eta = 10^{5(-0.4)/10} = 0.63 = 63\%$$

Wilkinson combiner.   The Wilkinson combiner, a popularly used two-way hybrid combiner/divider, is shown in Fig. 5-4-4. When an input power is connected to the input port of the Wilkinson divider with a 50-Ω characteristic impedance, the power is then equally split into two parts in amplitude and phase, and traveling to the two output ports. It appears the input port is connected to two 100-Ω matched lines in parallel. Since the output-port lines are two 50-Ω loads, it is necessary to use two quarter-wave lines to transform the apparent 100-Ω lines to the 50-Ω loads. That is, the characteristic impedance of the quarter-wave line is

$$Z_{(\lambda/4)} = \sqrt{100 \times 50} = 70.70 \ \Omega \tag{5-4-2}$$

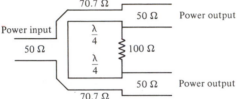

Figure 5-4-4 Wilkinson power divider.

At the two output-load terminals, each 50-Ω line sees two 100-Ω lines in parallel, so the shunt resistance must be

$$R_{\text{shunt}} = 100 \ \Omega \tag{5-4-3}$$

There is no power dissipated in the 100-Ω resistor when equal loads are connected to the two output ports. If a mismatch occurs at either output port, the reflected signals split through the two transmission lines, travel to the input port, split again, and travel back to the output ports. That is, the reflected waves return to the output ports in two equal parts of 180°-phase difference and are perfectly canceled. The 3-dB Wilkinson power combiner/divider circuit is commonly used in the balanced amplifier, as shown in Fig. 5-4-5.

A quarter-wave line is inserted in front of one active device and behind the opposite amplifier. If the input reflection coefficients of the two amplifiers are identical, all reflected input signals appear 180° out of phase across R_2 and are canceled. Similarly, the output is terminated in R_3, and the signals are added in phase at the output port. The S-parameter equations can be expressed as

$$S_{11} = \frac{1}{2}(S_{11b} - S_{11a}) \tag{5-4-4a}$$

$$S_{22} = \frac{1}{2}(S_{22a} - S_{22b}) \tag{5-4-4b}$$

$$\text{power gain} = |S_{21}|^2 = \frac{1}{4}|S_{21a} + S_{21b}|^2 \tag{5-4-5}$$

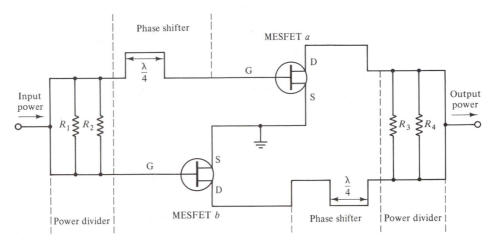

Figure 5-4-5   Balanced amplifier with Wilkinson power dividers.

Example 5-4-2:   Balanced Amplifier with Wilkinson Power Dividers

Two GaAs MESFETs are to be used in a balanced amplifier with Wilkinson power dividers as shown in Fig. 5-4-5. Their S parameters measured at 8 GHz are:
MESFET a:

$$\text{Reflection coefficients} \qquad S_{11a} = 0.78 \; \underline{/-87°}$$

$$S_{22a} = 0.57 \; \underline{/-42°}$$

$$\begin{array}{l}\text{Forward transmission} \\ \text{coefficient}\end{array} \qquad S_{21a} = 2.58 \; \underline{/108°}$$

MESFET b:

$$\text{Reflection coefficients} \qquad S_{11b} = 0.85 \; \underline{/-66°}$$

$$S_{22b} = 0.60 \; \underline{/-43°}$$

$$\begin{array}{l}\text{Forward transmission} \\ \text{coefficient}\end{array} \qquad S_{21b} = 2.48 \; \underline{/123°}$$

Compute (a) the input and output reflection coefficients of the designed balanced amplifier, and (b) the input and output VSWRs. Determine (c) the power gain for the designed balanced amplifier in decibels, and (d) the power loss if one MESFET fails. (e) Calculate the linear output power gain in decibels.

Solution   (a) From Eqs. (5-4-4a) and (5-4-4b) the input and output reflection coefficients are

$$S_{11} = \frac{1}{2} \left(0.85 \; \underline{/-66°} - 0.78 \; \underline{/-87°}\right) = 0.155$$

$$S_{22} = \frac{1}{2} \left(0.57 \; \underline{/-42°} - 0.60 \; \underline{/-43°}\right) = 0.018 \; \underline{/61.8°}$$

(b) The input and output VSWRs are

$$\text{VSWR}_{\text{input}} = \frac{1 + 0.155}{1 - 0.155} = 1.36$$

$$\text{VSWR}_{\text{output}} = \frac{1 + 0.018}{1 - 0.018} = 1.04$$

(c) From Eq. (5-4-5) the power gain is

$$\text{power gain} = \frac{1}{4} |2.58 \ \underline{/108°} + 2.48 \ \underline{/123°}|^2 = 6.28 = 8 \text{ dB}$$

(d) If one MESFET fails, the power loss is

$$\text{power loss} = \frac{1}{4} \ 10 \ \ell og \left(\frac{1}{4}\right) = -6 \text{ dB}$$

(e) The linear output power is increased by

$$P_{\text{out}} = 10 \ \ell og \ (2) = 3 \text{ dB}$$

Rat-race ring.   A rat-race ring (or hybrid ring) circuit consists of an annular line of proper electrical length to sustain standing waves, to which four arms are connected at proper intervals by means of series or parallel junctions. Figure 5-4-6 shows a rat-race ring with series junctions. When a signal wave is fed into port 1, it will not appear at port 3 because the phase difference between the two traveling waves in the clockwise and counterclockwise directions is 180°. Thus the two waves are canceled at port 3. For the same reason, the signal wave fed into port 2 will not emerge at port 4, and so on.

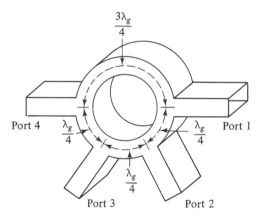

Figure 5-4-6   Rat-race ring circuit.

The **S** matrix for an ideal rat-race ring can be expressed as

$$\mathbf{S} = \begin{bmatrix} 0 & S_{12} & 0 & S_{14} \\ S_{21} & 0 & S_{23} & 0 \\ 0 & S_{32} & 0 & S_{34} \\ S_{41} & 0 & S_{43} & 0 \end{bmatrix} \tag{5-4-6}$$

It should be noted that the phase cancellation occurs only at a designed frequency for an ideal rat-race ring. In actual rat-race rings, there are small leakage couplings, and therefore the zero elements in the matrix equation (5-4-6) are not quite equal to zero.

90°-Hybrid branch line.   The 90°-hybrid branch line is shown in Fig. 5-4-7. The quarter-wave line is used to transform the 50-Ω line to a line with two 50-Ω lines in parallel. That means that the characteristic impedance of the quarter-wave line is

$$Z_{(\lambda/4)} = \sqrt{50 \times \frac{50}{2}} = 35.36 \ \Omega \qquad (5\text{-}4\text{-}7)$$

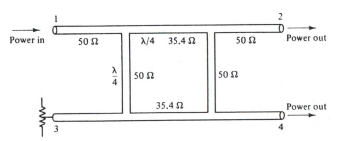

Figure 5-4-7 90°-hybrid branch line.

Thus the signal wave incident at port 1 couples power to ports 2 and 4 but not to port 3. Similarly, the wave incident at port 3 couples to ports 2 and 4 but not to port 1. Therefore, ports 1 and 3 are isolated. The wavelength λ is evaluated to midband frequency.

5-4-2 Nonbinary Combiner/Divider Structures

The nonbinary combining structure, also called a *chain* or *serial combiner*, is shown in Fig. 5-4-8. For an *N*-stage combining structure each successive combiner adds $1/N$ of its output power to the total output power. The number of the stages determines the required coupling coefficient for that stage. For a lossless structure the coupling coefficient is expressed as

$$\text{coupling coefficient} = 10 \ \ell og \ N \qquad (5\text{-}4\text{-}8)$$

where *N* is the number of the couplers.

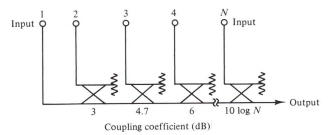

Figure 5-4-8 Nonbinary combining structure.

The advantage of the chain configuration is that an additional stage can easily be added. However, the combining efficiency and attainable bandwidth are very low due to the coupling losses. Figure 5-4-9 shows the combining efficiency for a nonbinary combining structure. The loss in decibels refers to the coupling loss in each power path of each stage. The combining efficiency can be written as

$$\text{combining efficiency} = \frac{1}{N}\left[10^{(N-1)L/10} + \sum_{K=0}^{N-2} 10^{(1+K)L/10} \right] \quad (5\text{-}4\text{-}9)$$

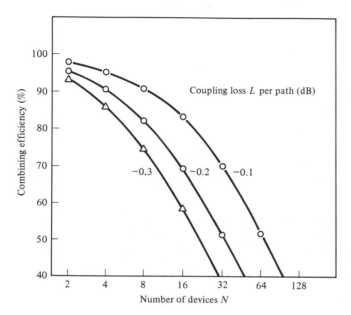

Figure 5-4-9 Combining efficiency of nonbinary structures. (After Kenneth J. Russell [5]. Copyright © 1979 IEEE. Reprinted by permission of the IEEE, Inc.)

where L is the coupling loss in decibels per path. The nonbinary combining structure can be realized in a number of different transmission circuits, such as microstrip line, coaxial line, or waveguide.

Example 5-4-3:   Combining Efficiency of Four Nonbinary Combiners

A microwave power combiner is to be designed to combine four small power sources together in a nonbinary structure as shown in Fig. 5-4-8. Calculate its combining efficiency if each path has a loss of -0.3 dB.

Solution   From Eq. (5-4-9) we have

$$\text{combining efficiency} = \frac{1}{4}\left[10^{(4-1)(-0.3)/10} + 10^{(-0.3)/10} + 10^{2(-0.3)/10} + 10^{3(-0.3)/10} \right]$$

$$= \frac{1}{4}(0.81 + 0.93 + 0.87 + 0.81)$$

$$= 0.86 = 86\%$$

5-4-3 Resonant-Cavity N-Way Combiners/Dividers

The N-way resonant-cavity combining structure sums the power of the N power sources directly in a single stage without having to proceed through several combining steps. This technique may increase the combining efficiency because the generated power does not need to pass through several combining stages. There are two types of resonant cavities for N-way combining structures: waveguide cavity and coaxial cavity.

Waveguide resonant cavity.   The sum of output power from N power sources is obtained by coupling their outputs to a single-cavity resonator. Figure 5-4-10 shows a schematic diagram for a waveguide combiner. In this waveguide combiner, the power from 12 IMPATT-diode oscillators is combined. Each diode is mounted at one end of a stabilized coaxial line which is coupled to the maximum magnetic field at the sidewall of a waveguide cavity. The diode pairs are spaced one-half wavelength apart along the waveguide for resonance. At 9.1 GHz, the combined output power is 10.5 W CW and the operating efficiency is 6.2%.

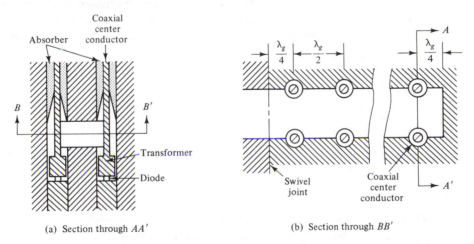

(a) Section through AA' (b) Section through BB'

Figure 5-4-10   Cross section of a waveguide combiner. (After Kenneth J. Russell [5]. Copyright © 1979 IEEE. Reprinted by permission of the IEEE, Inc.)

Coaxial resonant cavity.   A number of coaxial modules are placed around the circular periphery of a TM_{ONO}-mode coaxial resonant cavity as shown in Fig. 5-4-11. Part (a) shows the magnetic coupling to a TM_{010}-mode module in a single coaxial cavity, and part (b) illustrates the top view of the module positions around the periphery in the resonant cavity. In TM_{ONO} modes the maximum magnetic field is located at the cavity wall and it couples to the center conductor of the coaxial modules, inducing a high impedance in series with the coaxial line. The coaxial modules incorporate quarter-wave transformers at the diode end and its center conductors serve as a bias

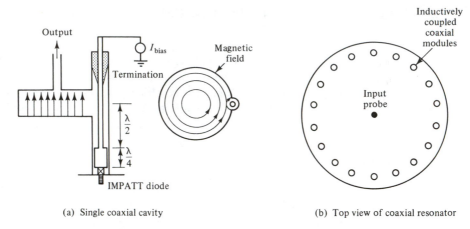

(a) Single coaxial cavity

(b) Top view of coaxial resonator

Figure 5-4-11 Coaxial-cavity resonator. (After R S. Harp and H. L. Stover [7]. Copyright © 1973 IEEE. Reprinted by permission of the IEEE, Inc.)

to the IMPATT diodes. The TM_{ONO} modes have a maximum of electric field at the central axis of the combiner, and the output power is obtained from the combiner by placing a probe at this point. The output power from the N-way combiner is

$$\text{power} = NP_{in}\ 10^{L_2/10} \tag{5-4-10}$$

where P_{in} = power generated by each diode
  L_2 = power losses of each coaxial module, decibels
  N = number of devices combined

The combining efficiency for a given cavity is independent of the number of devices combined and it is given by

$$\text{combining efficiency} = 10^{(L_1 + L_2)/10} \tag{5-4-11}$$

where L_1 = power losses in the cavity and output coaxial line, decibels
  L_2 = power losses for each coaxial module, decibels

The combining efficiency of 7.5% for a 32-diode combiner is achieved in a TM_{020}-mode cavity at the X-band and the peak power of 385 W is recorded.

Example 5-4-4:   Combining Efficiency of a Coaxial-Cavity Combiner

A coaxial-cavity combiner has the following parameters:

Number of devices combined	$N = 18$
Power generated by each device	$P_{in} = 200\text{ mW}$
Power losses in cavity and coaxial line	$L_1 = -0.1\text{ dB}$
Power losses by each coaxial module	$L_2 = -0.2\text{ dB}$

Compute (a) the combining efficiency, and (b) the output power.

Solution   (a) From Eq. (5-4-11) the combining efficiency is

$$\eta = 10^{(L_1 + L_2)/10} = 10^{(-0.1 - 0.2)/10}$$

$$= 0.93 = 93\%$$

(b) From Eq. (5-4-10) the output is

$$P_{\text{out}} = NP_{\text{in}}\, 10^{L_2/10} = 18 \times 0.2 \times 10^{-0.2/10}$$

$$= 3.42\ \text{W}$$

5-4-4 Nonresonant-Cavity N-Way Combiners/ Dividers

The nonresonant combining techniques offer the prospect of broadband operation. There are three versions of nonresonant combiners: the Wilkinson N-way combiner, the Rucker 5-way combiner, and the 12-way micro-strip-radial-line combiner.

Wilkinson N-way combiner.   Figure 5-4-12 shows a Wilkinson N-way combiner. The input ports of impedance Z_0 feed into N-output lines of characteristic impedance $\sqrt{N}\, Z_0$ in parallel which are one quarter-wave long. Isolation between the N ports is achieved by means of the resistive star connected to the N ports. The 2-way Wilkinson combiner described in Section 5-4-1 is the best known version of the Wilkinson circuit. The major problem with the Wilkinson combining technique is that it is generally not feasible to connect the isolation resistors in the manner shown in Fig. 5-4-12 when N is greater than 2. Nor can the resistors be connected as shown when planar circuits are used. However, a six-GaAs MESFET power combiner has yielded a 10% bandwidth and a 78% combining efficiency at the X-band [8].

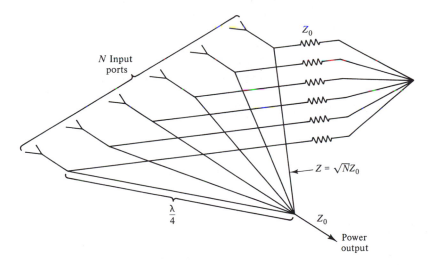

Figure 5-4-12   Wilkinson N-way combiner.

Example 5-4-5: Four-Way Wilkinson Power Divider

A 4-way Wilkinson power divider is to be designed to handle an input power of 2 W. It is assumed to have equal phase and amplitude per divider. The loss per divider is −0.3 dB. Compute (a) the output power per divider, and (b) the dividing efficiency.

Solution   (a) the four-way Wilkinson power divider can be arranged in the manner as shown in Fig. 5-4-13. The output power per divider is

$$P_{out} = \frac{1}{4} P_{in} = \frac{1}{4} \times 2 = 0.5 \, \text{W}$$

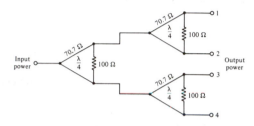

Figure 5-4-13   Four-way Wilkinson power divider.

(b) From Eq. (5-4-1) the dividing efficiency is

$$\eta = 10^{xL/10} = 10^{2(-0.3)/10} = 0.87 = 87\%$$

where $N = 2^x$ (i.e., $4 = 2^2$).

Rucker five-way combiner.   The Rucker five-way power combiner is also commonly used in power-combining techniques, and it is shown in Fig. 5-4-14, where a five-avalanche-diode oscillator has five coaxial lines, each approximately one-quarter-wave long, terminated by a device and arranged radially about a common bias network and a common output network [9]. A resistor R_s is incorporated into each coaxial center conductor for stability. The capacitor C_c between the output coupling disk and each coaxial center conductor provides the necessary coupling to the common load R_ℓ. The bypass capacitor C_b in the bias network is located approximately one-quarter wavelength from the center hub of the oscillator. The five-diode oscillator has about 5% conversion efficiency at 7 to 9 GHz.

12-Way microstrip-radial-line combiner.   Figure 5-4-15 shows a 12-way microstrip-radial-line combiner. The 12-way power combiner is fabricated on a fused silica substrate and uses dielectric-filled sectored radial lines for the power divider/combiner structures. The isolation between ports is accomplished with tantalum thin-film resistors deposited between adjacent ports. The radial transmission line is provided on microstrip with a fused silica substrate. Coaxial transmission lines are abrupt, and stepped transitions to the microstrip radial lines are used for input and output terminals. The combiner has demonstrated a combining efficiency of 87% and a 1-dB bandwidth of 20% at 8.5 GHz operating as an amplifier.

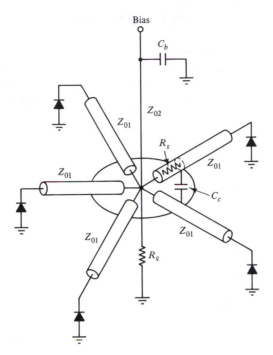

Figure 5-4-14   Rucker five-way combiner.

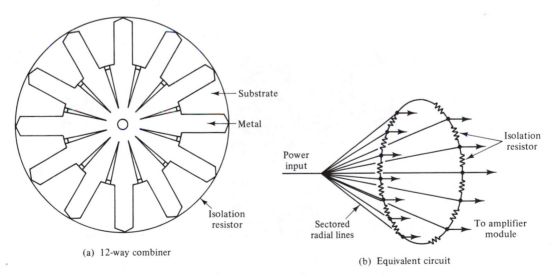

(a) 12-way combiner

(b) Equivalent circuit

Figure 5-4-15   12-way microstrip-radial-line combiner. (After J. M. Schellenberg and M. Cohn [10]. Copyright © 1978 IEEE. Reprinted by permission of the IEEE, Inc.)

5-5 POWER-COMBINING/DIVIDING DESIGN EXAMPLES

Microwave power-combining/dividing techniques are commonly used in microwave power circuits to obtain either a large amount of output power or to reduce the input power to a lower acceptable level. Some examples are the power combining of an IMPATT-diode amplifier or oscillator at the X- and Ku-bands and a GaAs-MESFET balanced amplifier. Here two design examples are described.

5-5-1 Power Combiner with 90°-Hybrid Couplers

A power combiner with 90°-hybrid couplers is shown in Fig. 5-5-1.

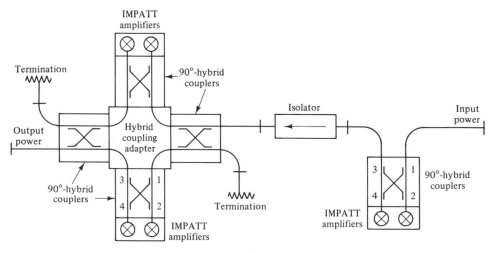

(a) Power-combiner block diagram

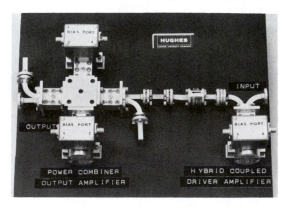

(b) Two-stage power combiner

Figure 5-5-1   Two-stage IMPATT amplifier/combiner. (After H. J. Kuno and David L. English [11]. Copyright © 1976 IEEE. Reprinted by permission of the IEEE, Inc.)

90°-Hybrid couplers.   The power-combining circuits used in this example are the 90°-hybrid couplers. When the input power is applied to port 1, the power is equally split to port 2 and port 4, and port 3 is isolated from port 1. If ports 2 and 4 are terminated by a pair of balanced loads such as matched reflection-type amplifiers, a signal applied at port 1 is amplified and reflected from ports 2 and 4. The reflected waves are then added at port 3 and canceled at port 1, due to the phase relationship between the two reflected waves. Isolation between the input and output ports can be provided by the 90°-hybrid couplers for improving the stability and ensuring gain–bandwith performance.

Power amplifier circuits.   The power amplifier circuit used in this example is the reduced-height waveguide shown in Fig. 5-5-2. The IMPATT diode is mounted in the reduced-height waveguide cavity. The adjustable coaxial section provides a series tuning function and the movable short serves as shunt tuning element. The quarter-wave transformer is used to transform the circuit impedance to the load impedance for output matching purpose. The real and imaginary components of the

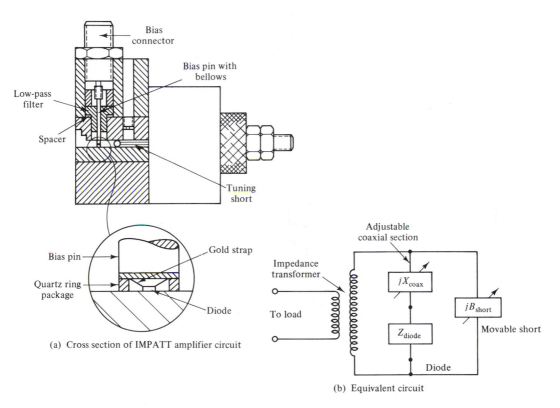

(a) Cross section of IMPATT amplifier circuit

(b) Equivalent circuit

Figure 5-5-2   Reflection-type power amplifier circuit. (After H. J. Kuno and David L. English [11]. Copyright © 1976 IEEE. Reprinted by permission of the IEEE, Inc.)

circuit impedance can be matched to those of the device to achieve optimum performance over a specified frequency range.

Electronic performances.   The four-diode power combiner produces an output power of 1.075 W at 59 GHz and its power gain is 3.6 dB. The 3-dB bandwidth is about 7 GHz and the dc-to-RF power conversion efficiency is about 3%. The two-stage power combiner has an output power of 0.98 W at 59 GHz and its power gain is 9 dB.

5-5-2 16-Way TM$_{010}$-Mode Symmetric Combiner

A 16-way circular-cavity TM$_{010}$-mode symmetric power combiner is shown in Fig. 5-5-3. The symmetric N-way combining can be determined in terms of scattering parameters. However, when N becomes large, there may not be adequate physical space

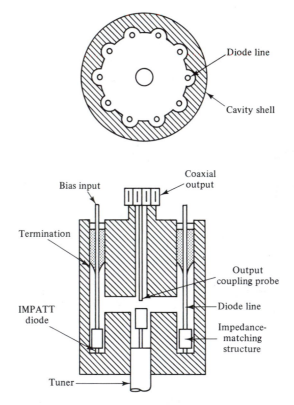

Figure 5-5-3   Schematic diagram of a circular-cavity TM$_{010}$-mode power combiner (After D. M. Kinman et al. [12]. Copyright © 1982 IEEE. Reprinted by permission of the IEEE, Inc.)

available to attach the necessary connectors and the matched loads for S-parameter measurements. An alternative approach is to use only two input ports, in addition to the output or sum port, with the remaining ports shorted. The method is equally applicable at any reference plane, as shown in Fig. 5-5-4. The combiner itself is reference plane A and the combiner plus matching is reference plane B or C; which reference plane is to be used depends on where the shorts are placed. The shorted reference plane must be placed so as not to reflect a short circuit at the summing point.

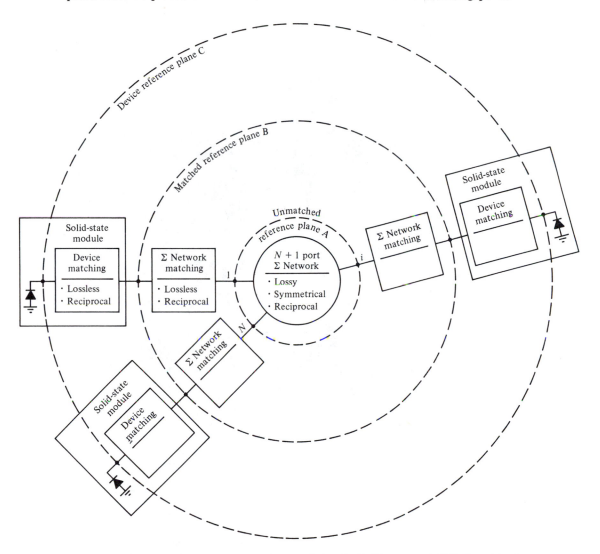

Figure 5-5-4 Power-combiner block diagram. (After D. M. Kinman et al. [12]. Copyright © 1982 IEEE. Reprinted by permission of the IEEE, Inc.)

Combining efficiency and input impedance.   If the adjacent lines are not close together in the TM_{010}-mode cavity, the combiner becomes perfectly symmetric and the transmission S parameters are given by

$$S_{ij} = S_{ik} \qquad (5\text{-}5\text{-}1)$$

where $j = 1, 2, 3,...,N$
$k = 1, 2, 3,...,N$
$j \ncong i$
$k \ncong i$

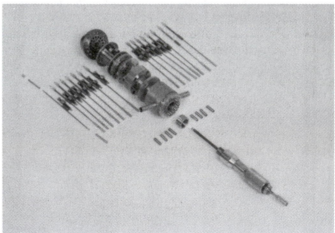

Figure 5-5-5   16-way power combiner. (After D. M. Kinman et al. [12]. Copyright © 1982 IEEE. Reprinted by permission of the IEEE, Inc.)

The combining efficiency and input impedance can be expressed, respectively, as

$$\eta = \frac{N \, |S_{\Sigma i}|^2}{1 - |S_{ii} + (N - 1)S_{ji}|^2} \qquad (5\text{-}5\text{-}2)$$

and

$$Z_{in} = Z_0 \frac{1 + [S_{ii} + (N - 1)S_{ji}]}{1 - [S_{ii} + (N - 1)S_{ji}]} \qquad (5\text{-}5\text{-}3)$$

where S_{ii} = reflection coefficient
S_{ji} = transmission coefficient
$S_{\sigma i}$ = sum port transmission coefficient

16-Way TM$_{010}$-mode combiner.   A 16-way TM$_{010}$-mode symmetric combiner has been built [12] and is shown in Fig. 5-5-5. It has been reported that the 16-way TM$_{010}$-mode symmetric power combiner has a combining efficiency of about 90% and an input impedance of $(4.5 - j0.5)Z_o$ ohms at 10 GHz.

REFERENCES

1. Lange, Julius, Interdigitated stripline quadrature hybrid. *IEEE Trans. Microwave Theory and Techniques*, Vol. MTT-17, No. 12, December 1969, pp. 1150–1151.

2. Ou, Wen Pin, Design equations for an interdigitated directional coupler. *IEEE Trans. Microwave Theory and Techniques*, Vol. MTT-23, No. 2, February 1975, pp. 253–255.

3. Presser, Adolph, Interdigitated microstrip coupler design. *IEEE Trans. Microwave Theory and Techniques*, Vol. MTT-26, No. 10, October 1978, pp. 801–805.

4. Kurokawa, K., Design theory of balanced transistor amplifiers. *Bell System Technical Journal*, October 1965.

5. Russell, Kenneth J., Microwave power combining techniques. *IEEE Trans. Microwave Theory and Techniques*, Vol. MTT-27, No. 5, May 1979, pp. 472–478.

6. Wilkinson, E. J., An *N*-way hybrid power divider. *IEEE Trans. Microwave Theory and Techniques*, Vol. MTT-8, No. 1, January 1960, pp. 116–118.

7. Harp, R. S., and H. L. Stover, Power combining of X-band IMPATT circuit modules. *IEEE Int. Solid-State Circuits Conf.*, 1973.

8. Nagai, N., et al., New *N*-way hybrid power dividers. *IEEE Trans. Microwave Theory and Techniques*, Vol. MTT-25, No. 12, December 1977, p. 1008.

9. Rucker, C. T., A multiple-diode high-average power avalanche-diode oscillator. *IEEE Trans. Microwave Theory and Techniques*, Vol. MTT-17, No. 12, December 1969, p. 1156.

10. Schellenberg, J. M., and M. Cohn, A wideband radial power combiner for FET amplifiers. *1978 IEEE Int. Solid-State Circuits Conf., Dig. Tech. Repts.*, February 1978.

11. Kuno, H. J., and David L. English, Millimeter-wave IMPATT power amplifier combiner.

IEEE Trans. Microwave Theory and Techniques, Vol. MTT-24, No. 11, November 1976, pp. 758–767.

12. Kinman, D. M., et al., Symmetric combiner analysis using S parameters. *IEEE Trans. Microwave Theory and Techniques*, Vol. MTT-30, No. 3, March 1982, pp. 268–277.

PROBLEMS

5-1 Lange Couplers

5-1-1. A four-strip Lange coupler is to be designed for a balanced amplifier and its specifications are as follows:

Coupling coefficient	$K = 3$ dB
Strip length	$\ell = \lambda/4$
Center frequency	$f = 8$ GHz
Relative dielectric constant of substrate	$\epsilon_r = 10$

Determine:

(a) The strip length ℓ.
(b) The impedance ratio R.
(c) The normalized odd-mode impedance z.
(d) The strip width w, spacing between strips S, and substrate height h.
(e) The odd-mode impedance Z_{0o} and the even-mode impedance Z_{0e}.

5-1-2. A four-strip Lange coupler is to be designed for a balanced amplifier and its specifications are as follows:

Coupling coefficient	$K = 3$ dB
Strip length	$\ell = \lambda/4$
Center frequency	$f = 14$ GHz
Relative dielectric constant of substrate	$\epsilon_r = 10$

Determine:

(a) The strip length ℓ.
(b) The impedance ratio R.
(c) The normalized odd-mode impedance z.
(d) The strip width w, spacing between strips S, and the substrate height h.
(e) The odd-mode impedance Z_{0o} and the even-mode impedance Z_{0e}.

5-1-3. A four-strip Lange coupler is to be designed for a balanced amplifier and its specifications are as follows:

Coupling coefficient	$K = 3$ dB
Strip length	$\ell = \lambda/4$
Center frequency	$f = 10$ GHz
Relative dielectric constant of substrate	$\epsilon_r = 10$

Determine:

(a) The strip length ℓ.
(b) The impedance ratio R.

(c) The normalized odd-mode impedance z.

(d) The strip width w, spacing between strips S, and substrate height h.

(e) The odd-mode impedance Z_{0o} and the even-mode impedance Z_{0e}.

5-2 Balanced Amplifier Design

5-2-1. Two 3-dB quadrature Lange couplers are used in a GaAs-MESFET balanced amplifier circuit with the following parameters:

MESFET a:

Reflection coefficients $S_{11a} = 0.7488 \; \underline{/-158.3°}$

$S_{22a} = 0.8521 \; \underline{/-155.7°}$

Forward transmission coefficient $S_{21a} = 1.3500 \; \underline{/-8.5°}$

MESFET b:

Reflection coefficients $S_{11b} = 0.6210 \; \underline{/175.9°}$

$S_{22b} = 0.7727 \; \underline{/-151.4°}$

Forward transmission coefficient $S_{21b} = 1.2200 \; \underline{/-19.1°}$

Compute:

(a) The input and output VSWRs.

(b) The power gain in decibels for the balanced amplifier.

(c) The power loss in decibels if one MESFET fails.

(d) The linear output power gain in decibels.

5-2-2. A GaAs MESFET has the following parameters:

MESFET a:

Reflection coefficients $S_{11a} = 0.81 \; \underline{/53°}$

$S_{22a} = 0.55 \; \underline{/20°}$

Forward transmission coefficient $S_{21a} = 6.22 \; \underline{/180°}$

MESFET b:

Reflection coefficients $S_{11b} = 0.79 \; \underline{/55°}$

$S_{22b} = 0.50 \; \underline{/22°}$

Forward transmission coefficient $S_{21b} = 6.10 \; \underline{/178°}$

(a) Calculate the input and output reflection coefficients of the balanced amplifier.

(b) Compute the input and output VSWRs.

(c) Determine the power gain in decibels for the balanced amplifier.

(d) Find the power loss in decibels if one MESFET fails.

(e) Calculate the linear output power capability in comparison with two MESFETs in series.

5-2-3. A GaAs-MESFET balanced amplifier has the following parameters:
MESFET a:

Reflection coefficients $S_{11a} = 0.57 \; \underline{/178°}$
$S_{22a} = 0.62 \; \underline{/180°}$

Forward transmission
coefficient $S_{21a} = 5.60 \; \underline{/50°}$

MESFET b:

Reflection coefficients $S_{11b} = 0.50 \; \underline{/160°}$
$S_{22b} = 0.64 \; \underline{/170°}$

Forward transmission
coefficient $S_{21b} = 6.00 \; \underline{/60°}$

(a) Calculate the input and output reflection coefficients of the balanced amplifier.
(b) Compute the input and output VSWRs.
(c) Determine the power gain in decibels for the balanced amplifier.
(d) Find the power loss in decibels if one MESFET fails.
(e) Calculate the linear output power capability as compared with two MESFETs in series.

5-2-4. A GaAs-MESFET balanced amplifier has the following parameters:
MESFET a:

Reflection coefficients $S_{11a} = 0.77 \; \underline{/-135°}$
$S_{22a} = 0.85 \; \underline{/-101°}$

Forward transmission
coefficient $S_{21a} = 1.377 \; \underline{/37°}$

MESFET b:

Reflection coefficients $S_{11b} = 0.74 \; \underline{/-135°}$
$S_{22b} = 0.89 \; \underline{/-106°}$

Forward transmission
coefficient $S_{21b} = 1.291 \; \underline{/27°}$

(a) Calculate the input and output reflection coefficients of the balanced amplifier.
(b) Compute the input and output VSWRs.
(c) Determine the power gain in decibels for the balanced amplifier.
(d) Find the power loss in decibels if one MESFET fails.
(e) Calculate the linear output power capability as compared with two MESFETs in series.

5-2-5. A GaAs-MESFET balanced amplifier has the following parameters:
MESFET a:

Reflection coefficients $S_{11a} = 0.79 \; \underline{/-127°}$
$S_{22a} = 0.83 \; \underline{/-102°}$

Forward transmission $S_{21a} = 1.29 \; \underline{/39°}$
coefficient

MESFET b:

Reflection coefficients $S_{11b} = 0.75 \; \underline{/-132°}$
$$S_{22b} = 0.81 \; \underline{/-107°}$$
Forward transmission $S_{21b} = 1.21 \; \underline{/28°}$
coefficients

(a) Calculate the input and output reflection coefficients of the balanced amplifier.
(b) Compute the input and output VSWRs.
(c) Determine the power gain in decibels for the balanced amplifier.
(d) Find the power loss in decibels if one MESFET fails.
(e) Calculate the linear output power capability as compared with two MESFETs in series.

5-4 Power-Combining Techniques

5-4-1. A 90°-hybrid branch line is to be designed to match a line of 75-Ω characteristic impedance as shown in Fig. P5-4-1. Determine the charactristic impedance of the quarter-wave section.

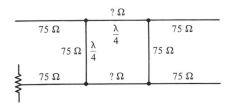

Figure P5-4-1 Branch line.

5-4-2. A Wilkinson combiner uses a two-way hybrid combiner/divider of 75-Ω characteristic impedance as shown in Fig. P5-4-2. Design the quarter-wave section of the combiner and the shunt section.

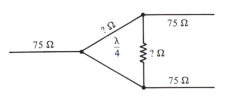

Figure P5-4-2 Wilkinson two-way combiner.

5-4-3. A binary combiner has 16-adders and each has a loss of −0.3 dB. Determine its combining efficiency.

5-4-4. A binary combiner has 64 adders and each adder has a loss of −0.1 dB. Find its combining efficiency.

5-4-5. A microwave power combiner is to be designed to combine six small power sources together in a nonbinary structure. Each path has a loss of −0.2 dB. Determine its coupling efficiency.

5-4-6. A coaxial-cavity combiner has the following parameters:

Number of devices combined	$N = 16$
Power generated by each device	$P_{in} = 100$ mW
Power losses in cavity and coaxial line	$L_1 = -0.2$ dB
Power losses by each coaxial module	$L_2 = -0.1$ dB

Compute:

(a) The combining efficiency.

(b) The output power.

5-4-7. An eight-way Wilkinson power divider is to be designed to handle an input power of 10 W. It is assumed to have equal phase and amplitude per divider. The loss per divider is −0.2 dB. Calculate:

(a) The output power per divider.

(b) The dividing efficiency.

Chapter 6

Microwave Striplines and Stripline-Type Amplifier Design

6-0 INTRODUCTION

Prior to 1965, nearly all microwave equipments utilized coaxial, waveguide, or parallel stripline circuits. In recent years, with the advent of monolithic microwave integrated circuits (MMICs), microstrip lines and coplanar striplines have been extensively used, for they provide one free and accessible surface on which solid-state devices can be placed. In this chapter, microstrip lines, coplanar striplines, parallel striplines, shielded striplines, and slot striplines, are described; they are shown in Fig. 6-0-1. In addition, the applications of microstrip lines to microwave amplifier circuits such as stripline-type amplifier are discussed.

6-1 MICROSTRIP LINES

A microstrip line consists of a strip conductor and a ground plane separated by a dielectric material as shown in Fig. 6-1-1. The electric and magnetic field lines are not contained entirely in the substrate, so the wave propagation in the microstrip line is not a pure transverse electromagnetic (TEM) mode but a quasi-TEM mode. For a quasi-TEM mode, the phase velocity of the propagating wave in a microstrip line is given by

$$v_p = \frac{c}{\sqrt{\epsilon_{re}}} \qquad (6\text{-}1\text{-}1)$$

where $c = 3 \times 10^8$ m/s is the velocity of light in vacuum
ϵ_{re} = effective relative dielectric constant of the substrate board

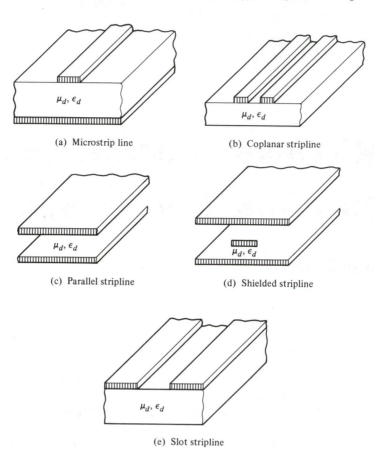

(a) Microstrip line   (b) Coplanar stripline

(c) Parallel stripline   (d) Shielded stripline

(e) Slot stripline

Figure 6-0-1   Microwave striplines.

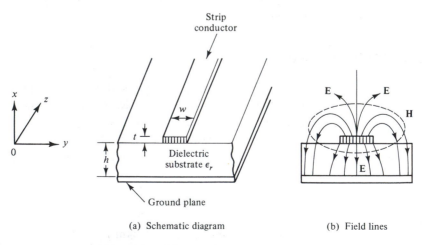

(a) Schematic diagram   (b) Field lines

Figure 6-1-1   Diagrams for a microstrip line.

6-1-1 Dielectric Substrates

The dielectric material sandwiched between the strip conductor and the ground plane serves as a substrate board for the line. Table 6-1-1 lists several dielectric materials which are commonly used as the substrate boards for microstrip lines.

TABLE 6-1-1   DIELECTRIC SUBSTRATE MATERIALS

Material	Relative dielectric constant ϵ_r	Loss tangent $(\tan \theta \times 10^4$ at 10 GHz)	Thermal conductivity K (W/cm²-°K)	Remarks
Alumina (Al₂O₃)	10	2	0.30	Low cost
Beryllia (BeO)	6	1	2.50	
Duroid	2.56	1		
Fused silica (quartz)	3.78	4	0.01	Fragile
GaAs	13.10	16	0.03	Fragile
Sapphire crystal (A1₂O₃)	10	1	0.04	Expensive
Silicon	11.7			
Teflon-fiberglass	2.32			

The effective relative dielectric constant of the microstrip line is a function of the relative dielectric constant ϵ_r, the height h of the dielectric substrate, and the width w of the strip conductor. An experimental equation [1] of the effective relative dielectric constant for $t/h \ll 0.005$ is given by

$$\epsilon_{re} = \frac{\epsilon_r + 1}{2} + \frac{\epsilon_r - 1}{2}\left[\left(1 + \frac{12}{w/h}\right)^{-1/2} + 0.04\left(1 - \frac{w}{h}\right)^2\right] \qquad \text{for } \frac{w}{h} \leq 1 \quad (6\text{-}1\text{-}2)$$

or

$$\epsilon_{re} = \frac{\epsilon_r + 1}{2} + \frac{\epsilon_r - 1}{2}\left(1 + \frac{12}{w/h}\right)^{-1/2} \qquad \text{for } \frac{w}{h} \geq 1 \quad (6\text{-}1\text{-}3)$$

The wavelength in the microstrip line for $t/h \leq 0.005$ is given by the following equation [2]:

$$\lambda = \frac{\lambda_0}{\sqrt{\epsilon_r}}\left[\frac{\epsilon_r}{1 + 0.63(\epsilon_r - 1)(w/h)^{0.1255}}\right]^{1/2} \qquad \text{for } \frac{w}{h} \geq 0.6 \quad (6\text{-}1\text{-}4)$$

or

$$\lambda = \frac{\lambda_0}{\sqrt{\epsilon_r}}\left[\frac{\epsilon_r}{1 + 0.60(\epsilon_r - 1)(w/h)^{0.0297}}\right]^{1/2} \qquad \text{for } \frac{w}{h} < 0.6 \quad (6\text{-}1\text{-}5)$$

where $\lambda_0 = c/f$ is the wavelength in free space

$c = 3 \times 10^8$ m/s is the velocity of light in vacuum

ϵ_r = relative dielectric constant of the substrate board

Figure 6-1-2 shows the normalized wavelength of a microstrip line versus w/h.

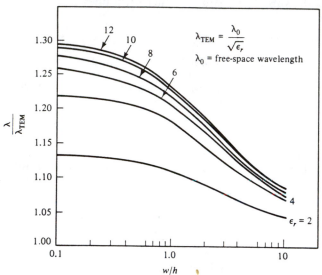

Figure 6-1-2   Normalized wavelength of a microstrip line versus w/h. (After H. Sobol [2]. Copyright © 1971 IEEE. Reprinted by permission from the IEEE, Inc.)

6-1-2 Characteristic Impedance

The characteristic impedance of a microstrip line for $t/h < 0.005$ can be expressed as [1]

$$Z_0 = \frac{60}{\sqrt{\epsilon_{re}}} \ell n \left(\frac{8}{w/h} + 0.25 \frac{w}{h} \right) \qquad \text{for } \frac{w}{h} \leq 1 \qquad (6\text{-}1\text{-}6)$$

or

$$Z_0 = \frac{120\pi / \sqrt{\epsilon_{re}}}{w/h + 1.393 + 0.667 \, \ell n \, (w/h + 1.444)} \qquad \text{for } \frac{w}{h} \geq 1 \qquad (6\text{-}1\text{-}7)$$

Figure 6-1-3 shows the characteristic impedance of a microstrip line versus w/h with ϵ_r as a parameter.

Thus far, the equations described are valid for zero thickness t or $t/h \leq 0.005$. However, if the thickness t of the strip conductor over the substrate board is greater than 0.005 (i.e., $t/h \geq 0.005$), the modified effective width w_{eff} must be replaced for w in all related equations. When $t < h$ and $t < w/2$, w_{eff} becomes

$$w_{eff} = w + \frac{t}{\pi} \left[1 + \ell n \left(\frac{2}{t/h} \right) \right] \qquad \text{for } \frac{w}{h} \geq \frac{1}{2\pi} \qquad (6\text{-}1\text{-}8a)$$

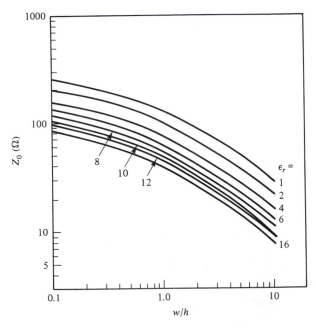

Figure 6-1-3 Characteristic impedance of a microstrip line versus w/h. (After H. Sobol [2]. Copyright © 1971 IEEE. Reprinted by permission from the IEEE, Inc.)

or

$$w_{\text{eff}} = w + \frac{t}{\pi}\left[1 + \ell n\left(\frac{4\pi}{t/w}\right)\right] \qquad \text{for } \frac{w}{h} \le \frac{1}{2\pi} \qquad (6\text{-}1\text{-}8b)$$

Example 6-1-1:   Characteristic Impedance of a Microstrip Line

A microstrip line is to be designed and its specifications are:

Strip thickness	$t \le 0.005h$
Substrate board	Alumina
Substrate relative dielectric constant	$\epsilon_r = 10$
Ratio of w/h	$w/h = 0.95$

Compute (a) the effective relative dielectric constant, (b) the characteristic impedance, (c) the phase velocity, and (d) the wavelength.

Solution   (a) From Eq. (6-1-2) the effective relative dielectric constant is

$$\epsilon_{\text{re}} = \frac{10 + 1}{2} + \frac{10 - 1}{2}\left[\left(1 + \frac{12}{0.95}\right)^{-1/2} + 0.04(1 - 0.95)^2\right]$$

$$= 6.71$$

(b) From Eq. (6-1-6) the characteristic impedance is

$$Z_0 = \frac{60}{\sqrt{6.71}} \, \ell n \left[\frac{8}{0.95} + 0.25(0.95)\right] = 50 \; \Omega$$

(c) The phase velocity is

$$v_{ph} = \frac{3 \times 10^8}{\sqrt{6.71}} = 1.16 \times 10^8 \text{ m/s}$$

(d) The wavelength is

$$\lambda = \frac{\lambda_0}{\sqrt{10}} \left[\frac{10}{1 + 0.63(10 - 1)(0.95)^{0.1255}} \right]^{1/2} = 0.43\lambda_0$$

6-1-3 Losses in Microstrip Lines

The problem of losses in a microstrip line is a serious one and depends on the geometrical factors, the electronic properties of the substrate and conductors, and the frequency. For a nonmagnetic dielectric substrate, there are two types of losses in the dominant microstrip mode: dielectric loss in the substrate and ohmic skin loss in the strip conductor and the ground plane. The sum of these two losses may be expressed as losses per unit length in terms of an attenuation factor α. From ordinary transmission-line theory, the power carried by a wave traveling in the positive z direction is given by

$$P = \frac{1}{2} VI^* = \frac{1}{2} (V_+ e^{-\alpha z} I_+ e^{-\alpha z}) = \frac{1}{2} \frac{|V_+|^2}{Z_0} e^{-2\alpha z} = P_0 e^{-2\alpha z} \qquad (6\text{-}1\text{-}9)$$

where $P_0 = \dfrac{|V_+|^2}{2Z_0}$ is the power at $z = 0$. The attenuation constant α can be expressed as

$$\alpha = -\frac{dP/dz}{2P(z)} = \alpha_d + \alpha_c \qquad (6\text{-}1\text{-}10)$$

where α_d = dielectric attenuation constant
α_c = ohmic attenuation constant

The gradient of power in the z direction in Eq. (6-1-10) can be further expressed in terms of the power loss per unit length dissipated by the resistance and the power loss per unit length in the dielectric. That is

$$-\frac{dP(z)}{dz} = -\frac{d}{dz}\left(\frac{1}{2} VI^*\right) = \frac{1}{2}\left(-\frac{dV}{dz}\right)I^*$$

$$+ \frac{1}{2}\left(-\frac{dI^*}{dz}\right)V = \frac{1}{2}(RI)I^* + \frac{1}{2}\sigma V^* V \qquad (6\text{-}1\text{-}11)$$

$$= \frac{1}{2}|I|^2 R + \frac{1}{2}|V|^2 \sigma = P_c + P_d$$

where σ is the conductivity of the dielectric substrate board. Substitution of Eq. (6-1-11) in Eq. (6-1-10) results in

$$\alpha_d = \frac{P_d}{2P(z)} \quad \text{Np/cm} \tag{6-1-12a}$$

$$\alpha_c = \frac{P_c}{2P(z)} \quad \text{Np/cm} \tag{6-1-12b}$$

Dielectric losses.   From electric field theory, when the conductivity of a dielectric cannot be neglected, the electric and magnetic fields in the dielectric are no longer in time phase. The dielectric attenuation constant is usually expressed as

$$\alpha_d = \frac{\sigma}{2} \sqrt{\frac{\mu}{\epsilon}} \quad \text{Np/cm} \tag{6-1-13}$$

This dielectric constant can be expressed in terms of dielectric loss tangent as

$$\tan \theta = \frac{\sigma}{\omega \epsilon} \tag{6-1-14}$$

Then the dielectric attenuation constant is expressed by

$$\alpha_d = \frac{\omega}{2} \sqrt{\mu \epsilon} \tan \theta \quad \text{Np/cm} \tag{6-1-15}$$

Since the microstrip line is a nonmagnetic mixed dielectric system, the upper dielectric above the microstrip ribbon is air with no loss. Welch and Pratt [3] derived an expression for the attenuation constant of a dielectric substrate. Later, Pucel and his coworkers [4] modified Welch's equation. The result is

$$\alpha_d = 4.34 \frac{q\sigma}{\sqrt{\epsilon_{re}}} \sqrt{\frac{\mu_0}{\epsilon_0}} = 1.634 \times 10^3 \frac{q\sigma}{\sqrt{\epsilon_{re}}} \quad \text{dB/cm} \tag{6-1-16}$$

In Eq. (6-1-16), the conversion factor of 1 Np equal to 8.636 dB has been replaced, ϵ_{re} is the effective dielectric constant of the substrate as given by Eq. (6-1-2), and q denotes the dielectric filling factor defined by Wheeler [5] as

$$q = \frac{\epsilon_{re} - 1}{\epsilon_r - 1} \tag{6-1-17}$$

Usually, the attenuation constant is expressed per wavelength as

$$\alpha_d = 27.3 \left(\frac{q\epsilon_r}{\epsilon_{re}} \right) \frac{\tan \theta}{\lambda_g} \quad \text{dB}/\lambda_g \tag{6-1-18}$$

where $\lambda_g = \lambda_0 / \sqrt{\epsilon_{re}}$ and λ_0 is the wavelength in free space, or
  $\lambda_g = c / (f \sqrt{\epsilon_{re}})$
  c = velocity of light in vacuum

If the loss tangent $\tan \theta$ is frequency independent, the dielectric attenuation per wavelength is also frequency independent. On the other hand, if the substrate conductivity is frequency independent, as for a semiconductor, the dielectric attenuation per unit is also frequency independent. Since q is a function of ϵ_r and w/h, the filling factor for the loss tangent $q\epsilon_r/\epsilon_{re}$ and for the conductivity $q/\sqrt{\epsilon_{re}}$ are also functions of these quantities. Figure 6-1-4 shows the loss tangent filling factor against w/h for a range of dielectric constants suitable for microwave integrated circuits. For most practical purposes, this factor can be approximated by unity. Figure 6-1-5 illustrates the product $\alpha_d p$ against w/h for two semiconducting substrates which are being used for integrated microwave circuits, such as silicon and gallium arsenide. For designing purposes, the conductivity filling factor exhibits only a mild dependence on w/h which probably can be ignored in practice.

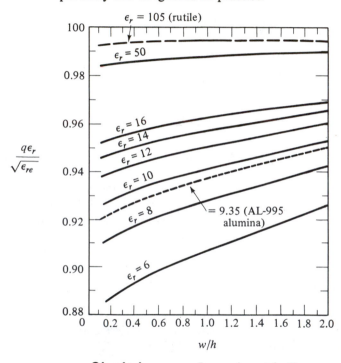

Figure 6-1-4 Filling factor for loss tangent of microstrip substrate as a function of w/h. (After R. A. Pucel et al. [4]. Copyright © 1968 IEEE. Reprinted by permission of the IEEE, Inc.)

Ohmic losses.   In a microstrip line over a low-loss dielectric substrate, the predominant sources of losses at microwave frequencies are the nonperfect conductors. The current density in the conductors of a microstrip line is concentrated in a sheet approximately skin depth deep inside the conductor surface exposed to the electric field. Both the strip conductor thickness t and the ground plane thickness are assumed to be at least three or four skin depths. The current density in the strip conductor and the ground conductor is not uniform in the transverse plane. The microstrip conductor contributes the major part of the ohmic loss. A diagram of the current density J for a microstrip line is shown in Fig. 6-1-6.

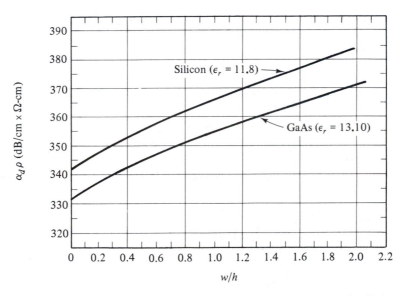

Figure 6-1-5   Dielectric attenuation factor of microstrip as a function of w/h for silicon and gallium arsenide substrates. (After R. A. Pucel et al. [4]. Copyright © 1968 IEEE. Reprinted by permission of the IEEE, Inc.)

Because of the mathematic complexity, exact expressions for the current density of a microstrip line with nonzero thickness have never been derived [4]. Several researchers [6] assumed for simplicity that the current distribution is uniform and equal to I/w in both conductors, and confined to the region $|x| < w/2$. With this assumption the conducting attenuation constant of a wide microstrip line is given by

$$\alpha_c = \frac{8.68R_s}{Z_0 w} \qquad \text{dB/cm} \qquad \text{for } \frac{w}{h} > 1 \qquad (6\text{-}1\text{-}19)$$

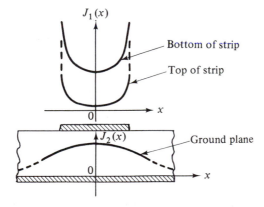

Figure 6-1-6   Current distribution on microstrip conductors. (After R. A. Pucel et al. [4]. Copyright © 1968 IEEE. Reprinted by permission of the IEEE, Inc.)

where $R_s = \sqrt{\pi f \mu / \sigma}$ is the surface skin resistance in ohms or

$$R_s = 1/(\delta\sigma), \text{ ohms}$$

$$\delta = \sqrt{\frac{1}{\pi f \mu \sigma}} \text{ is the skin depth, centimeters.}$$

However, for a narrow microstrip line with $w/h < 1$, Eq. (6-1-19) is not applicable. This is because the current distribution in the conductor is unknown (i.e., not uniform, as assumed).

Purcel and coworkers [4,7] derived the following three formulas from the results of Wheeler's work [5]:

$$\frac{\alpha_c Z_0 h}{R_s} = \frac{8.68}{2\pi}\left[1 - \left(\frac{w'}{4h}\right)^2\right]\left[1 + \frac{h}{w'} + \frac{h}{\pi w'}\left(\ell n \frac{4\pi w}{t} + \frac{t}{w}\right)\right] \tag{6-1-20}$$

$$\text{for } \frac{w}{h} \leq \frac{1}{2\pi}$$

$$\frac{\alpha_c Z_0 h}{R_s} = \frac{8.68}{2\pi}\left[1 - \left(\frac{w'}{4h}\right)^2\right]\left[1 + \frac{h}{w'} + \frac{h}{w'}\left(\ell n \frac{2h}{t} - \frac{t}{h}\right)\right] \tag{6-1-21}$$

$$\text{for } \frac{1}{2\pi} < \frac{w}{h} \leq 2$$

$$\frac{\alpha_c Z_0 h}{R_s} = \frac{8.68}{\left\{\frac{w'}{h} + \frac{2}{\pi}\ell n\left[2\pi e\left(\frac{w'}{2h} + 0.94\right)\right]\right\}^2}\left[\frac{w'}{h} + \frac{w'/(\pi h)}{w'/(2h) + 0.94}\right]$$
$$\left[1 + \frac{h}{w'} + \frac{h}{\pi w'}\left(\ell n \frac{2h}{t} - \frac{t}{h}\right)\right] \tag{6-1-22}$$

$$\text{for } 2 \leq \frac{w}{h}$$

where α_c is in dB/cm, and

$$e = 2.718,$$

$$w' = w + \Delta w, \tag{6-1-23}$$

$$\Delta w = \frac{t}{\pi}\left(\ell n \frac{4\pi w}{t} + 1\right) \quad \text{for } \frac{2t}{h} < \frac{w}{h} \leq \frac{1}{2\pi} \tag{6-1-24}$$

$$\Delta w = \frac{t}{\pi}\left(\ell n \frac{2h}{t} + 1\right) \quad \text{for } \frac{w}{h} \geq \frac{1}{2\pi} \tag{6-1-25}$$

The values of α_c computed from Eqs. (6-1-20) through (6-1-22) are plotted in Fig.

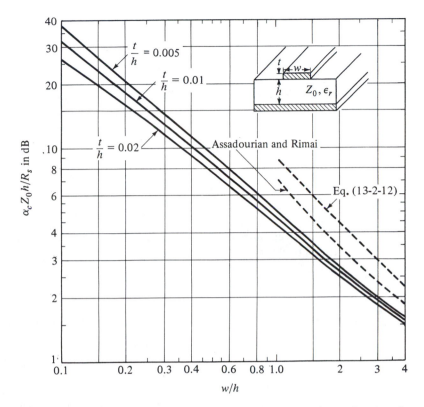

Figure 6-1-7 Theoretical conductor attenuation factor of microstrip as a function of w/h. (After R. A. Pucel et al. [4]. Copyright © 1968 IEEE. Reprinted by permission of the IEEE, Inc.)

6-1-7. For the purposes of comparison, α_c based on Assadourian and Rimai's equation (6-1-19) [6] is also shown in the same diagram.

Radiation losses.   In addition to the conductor and dielectric losses, a microstrip line also has radiation losses. The radiation loss from a microstrip line depends on the substrate's thickness and dielectric constant, as well as the geometry. Lewin [8] has calculated the radiation for several discontinuities with the following approximations:

1. TEM transmission
2. Uniform dielectric in the neighborhood of the strip, equal in magnitude to an effective value
3. Neglect of radiation from the transverse electric (TE) field component parallel to the strip
4. Substrate thickness much less than the free-space wavelength

From the results given by Lewin, the ratio of radiated power to total dissipated power for an open-circuited microstrip line is

$$\frac{P_{rad}}{P_t} = 240\pi^2 \left(\frac{h}{\lambda_0}\right)^2 \frac{F(\epsilon_{re})}{Z_0}$$ (6-1-26)

where $F(\epsilon_{re})$ is a radiation factor given by

$$F(\epsilon_{re}) = \frac{\epsilon_{re} + 1}{\epsilon_{re}} - \frac{\epsilon_{re} - 1}{2\epsilon_{re}\sqrt{\epsilon_{re}}} \ell n \left(\frac{\sqrt{\epsilon_{re}} + 1}{\sqrt{\epsilon_{re}} - 1}\right)$$ (6-1-27)

and

ϵ_{re} = effective dielectric constant

$\lambda_0 = c/f$ is the free-space wavelength

It can be seen that the radiation factor decreases with increasing substrate dielectric constant. Alternatively, Eq. (6-1-26) can be expressed as

$$\frac{P_{rad}}{P_t} = \frac{R_r}{Z_0}$$ (6-1-28)

where R_r is the radiation resistance of an open-circuited microstrip and is given by

$$R_r = 240\pi^2 \left(\frac{h}{\lambda_0}\right)^2 F(\epsilon_{re})$$ (6-1-29)

The ratio of the radiation resistance R_r to the real part of the characteristic impedance Z_0 of the microstrip line is equal to a small fraction of the power radiated from a single open-circuit discontinuity. In view of Eq. (6-1-26), the radiation loss decreases when the characteristic impedance increases. For lower-dielectric-constant substrates, radiation is significant at higher impedance levels. For higher-dielectric-constant substrates, radiation becomes significant until very low impedance levels are reached.

6-1-4 Quality Factor Q

Many microwave integrated circuits require very high Q resonant circuits. The quality factor Q of a microstrip line is very high, but it is limited by the radiation losses of the substrates and with low dielectric constant. For a uniform current distribution in the microstrip line, the ohmic attenuation constant of a wide microstrip line is given by Eq. (6-1-19) as

$$\alpha_c = \frac{8.68 R_s}{Z_0 w} \quad \text{dB/cm}$$ (6-1-30)

The charateristic impedance of a wide microstrip line can be expressed as

$$Z_0 = \frac{h}{w}\sqrt{\frac{\mu}{\epsilon}} = \frac{377}{\sqrt{\epsilon_r}}\frac{h}{w} \quad \text{ohms}$$ (6-1-31)

The wavelength in the microstrip line is given by

$$\lambda_g = \frac{30}{f\sqrt{\epsilon_r}} \quad \text{cm} \tag{6-1-32}$$

where f is frequency in gigahertz. Since Q_c is related to the conductor attenuation constant by

$$Q_c = \frac{27.3}{\alpha_c} \tag{6-1-33}$$

where α_c is in dB/λ_g, the Q_c of the wide microstrip line is expressed as

$$Q_c = 39.5\left(\frac{h}{R_s}\right)f_{\text{GHz}} \tag{6-1-34}$$

where h is in centimeters and R_s is given by

$$R_s = \sqrt{\frac{\pi f \mu}{\sigma}} = 20\pi\sqrt{\frac{f_{\text{GHz}}}{\sigma}} \quad \text{(MKS units)} \tag{6-1-35}$$

Finally, the quality factor Q_c of a wide microstrip line is

$$Q_c = 0.63h\sqrt{\sigma f_{\text{GHz}}} \tag{6-1-36}$$

where σ is the conductivity of dielectric substrate board per ohm-meter. For a copper strip, the conductivity σ is 5.8×10^7, whence Q_c becomes

$$Q_{\text{cop}} = 4780h\sqrt{f_{\text{GHz}}} \tag{6-1-37}$$

For a 25-mil alumina at 10 GHz, the maximum Q_c achievable from wide microstrip lines is 954 [9].

Similarly, there is a qualify factor Q_d which is related to the dielectric attenuation constant by

$$Q_d = \frac{27.3}{\alpha_d} \tag{6-1-38}$$

where α_d is in dB/λ_g. Substitution of Eq. (6-1-18) in Eq. (6-1-38) yields

$$Q_d = \frac{\lambda_0}{\sqrt{\epsilon_{\text{re}}}\tan\theta} = \frac{1}{\tan\theta} \tag{6-1-39}$$

where λ_0 is the free-space wavelength in centimeters. It should be noted that the Q_d which is due to the dielectric attenuation constant of a microstrip line is approximately the reciprocal of the dielectric loss $\tan\theta$ and relatively constant with frequency.

6-1-5 Microstrip-Line Realization

From transmission-line theory in Chapter 2, an open-circuited termination is equivalent to a capacitance, and and a short-circuited termination is equal to an inductance when the line is very short. In microwave integrated circuits, the lumped capacitor and

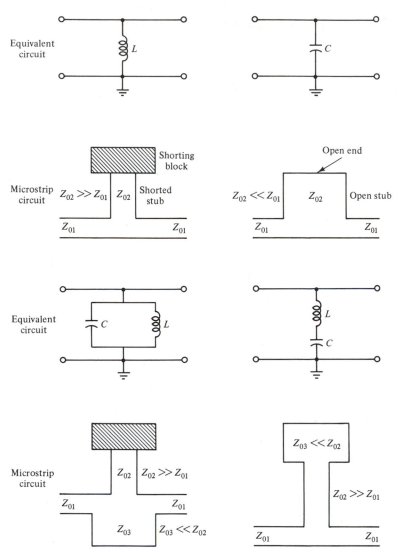

Figure 6-1-8 Realization of microstrip-line circuits.

inductor can easily be realized by using microstrip lines in different forms. Figure 6-1-8 shows the realization of several lumped circuits by microstrip lines. In Fig. 6-1-8, only the conductor shapes of microstrip lines are shown; their characteristic impedances must be calculated from the standard equation.

Example 6-1-2: Realization of Microstrip Elements

An equivalent circuit of a complicated electronic network is shown in Fig. 6-1-9. It is necessary to realize the equivalent circuit by using microstrip circuit and draw the circuit.

Solution The microstrip circuit is designed and shown in Fig. 6-1-10.

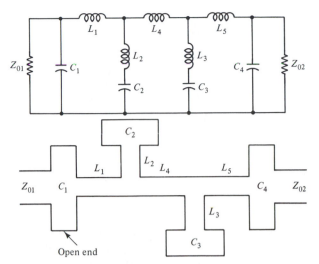

Figure 6-1-9   Equivalent circuit.

Figure 6-1-10   Microstrip circuit.

Open end

6-2 STRIPLINE-TYPE AMPLIFIER DESIGN

At microwave frequencies above 4 GHz, it is difficult to obtain purely lumped elements such as inductors with low loss. Therefore, it is necessary to approximate lumped elements whenever appropriate in microwave integrated circuits (MICs) with striplines. This is because the striplines can be fabricated on dielectric substrates with low loss and low cost. Those semilumped elements, such as series inductors and shunt capacitors, are short lengths of stripline that can be made to approximate the characteristics of the lumped elements by proper choice of their length and characteristic impedance. Two examples are illustrated for stripline-type amplifier design.

6-2-1 Design of a Narrowband Amplifier with Stripline Matching Networks

A narrowband amplifier design for maximum power gain using three-eighths-wavelength stubs and quarter-wave transformers is discussed in Section 4-5. This design example uses stripline matching networks.

Design Example 6-2-1:   Narrowband Amplifier Design for Maximum Power Gain

A GaAs MESFET has the following parameters measured at 6 GHz with a 50-Ω reference:

Source reflection coefficient for maximum power gain $\Gamma_{sm} = 0.761 \;/177°$

Load reflection coefficient for maximum power gain $\Gamma_{\ell m} = 0.719 \;/104°$

Maximum available power gain $G_{a\max} = 12$ dB

Design the input and output matching networks by using 50-Ω stripline elements as shown in Fig. 6-2-1.

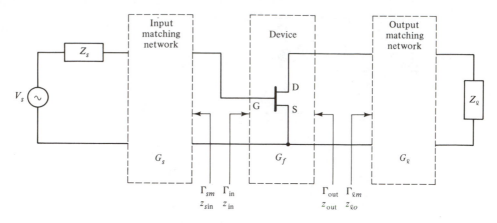

Figure 6-2-1   Stripline matching network to be designed.

Design Steps

Step 1. Enter Γ_{sm} and $\Gamma_{\ell m}$ on the Smith chart as shown in Fig. 6-2-2. Read

$$z_{sin} = 0.14 + j0.03$$
$$z_{\ell o} = 0.26 + j0.75$$

Step 2. Read the normalized admittances from the same chart.

$$y_{sin} = 7.50 - j1.60$$
$$y_{\ell o} = 0.40 - j1.20$$

Step 3. Use the distance from the origin of the Smith chart to y_{sin} and $y_{\ell o}$ as a radius to make two circular arcs from y_{sin} and $y_{\ell o}$ in the counterclockwise direction (or toward the load) and stop at point A or point B on the unity resistance circle. Read

$$y_A = 1 + j2.40 \qquad \text{at point } A$$
$$y_B = 1 - j2.10 \qquad \text{at point } B$$

Step 4. Realize two series elements. If the alumina is used for the substrate board, its effective relative dielectric constant ϵ_{re} is 6.71 for $w/h = 0.95$ (see Example 6-1-1). The characteristic impedance of a microstrip line with $w/h = 0.95$ is 50 Ω and the wavelength is

$$\lambda = 0.43\lambda_o = 0.43 \times 5 = 2.15 \text{ cm} \qquad \text{at 6 GHz}$$

The lengths of the input and output series 50-Ω striplines are

$$\ell_A = 0.055\lambda = 0.055 \times 2.15 = 0.12 \text{ cm}$$
$$\ell_B = 0.046\lambda = 0.046 \times 2.15 = 0.10 \text{ cm}$$

Step 5. Realize two shunt elements. Two shunt open stubs are needed to contribute $+j2.40$

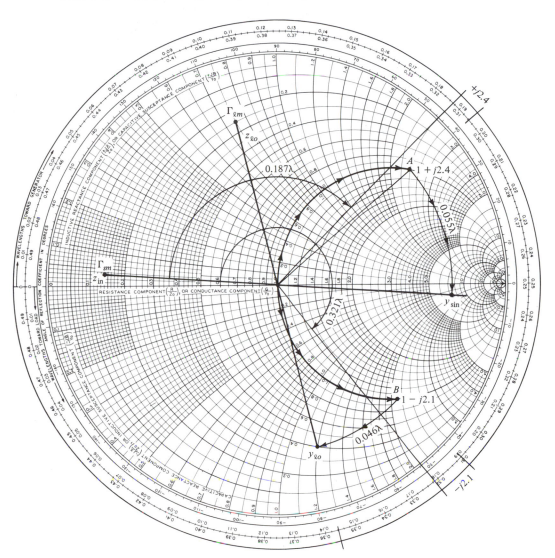

Figure 6-2-2 Graphical solution of a stripline-type amplifier for Example 6-2-1.

and $-j2.10$ for matching. The length of the input shunt open stub for $+j2.40$ is

$$\ell_s = 0.187\lambda = 0.187 \times 2.15 = 0.40 \text{ cm}$$

The length of the output shunt open stub for $-j2.10$ is

$$\ell_o = 0.321\lambda = 0.321 \times 2.15 = 0.69 \text{ cm}$$

Step 6. Sketch the complete input and output stripline matching networks (Fig. 6-2-3).

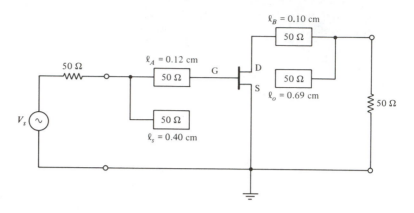

Figure 6-2-3   Stripline matching network.

6-2-2 Design of a Minimum-Noise Amplifier with Stripline Matching Networks

A narrowband amplifier design for a minimum-noise figure by using quarter-wave transformers and three-eighths-wavelength stubs is discussed in Section 4-5. This design example uses stripline matching networks.

Design Example 6-2-2:   Minimum-Noise Amplifier

A GaAs MESFET has the following parameters measured at 6 GHz for minimum-noise figure with a 50-Ω reference:

Minimum-noise figure	$F_{\min}$	= 3 dB
Optimum source reflection coefficient for $F_{\min}$	Γ_o	= 0.541 $\underline{/140°}$
Load reflection coefficient for $F_{\min}$	Γ_ℓ	= 0.543 $\underline{/105°}$
Equivalent noise resistance	R_n	= 10Ω

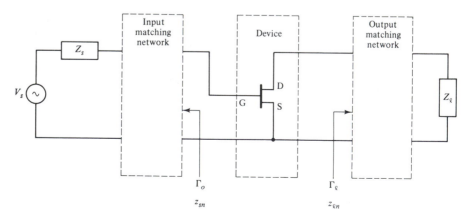

Figure 6-2-4   A stripline matching network to be designed.

Design the input and output matching networks by using 50-Ω stripline elements as shown in Fig. 6-2-4.

Design Steps

Step 1. Enter Γ_o and Γ_ℓ on the Smith chart as shown in Fig. 6-2-5.
Read

$$z_{sn} = 0.34 + j0.33$$

$$z_{\ell n} = 0.45 + j0.67$$

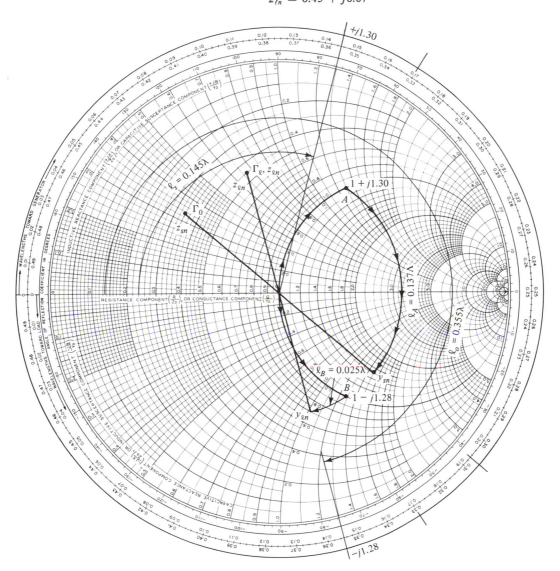

Figure 6-2-5 Graphical solution of a stripline-type amplifier for Example 6-2-2.

Step 2. Read the normalized admittances from the same chart.

$$y_{sn} = 1.50 - j1.50$$

$$y_{\ell n} = 0.70 - j1.05$$

Step 3. Use the distance from the origin of the Smith chart to y_{sn} or $y_{\ell n}$ as a radius to make two circular arcs from y_{sn} and $y_{\ell n}$ in the counterclockwise direction (or toward the load) and stop at points A and B on the unity resistance circle. Read

$$y_A = 1 + j1.30 \qquad \text{at point } A$$

$$y_B = 1 - j1.28 \qquad \text{at point } B$$

Step 4. Realize two series elements. It is assumed that alumina is used for the substrate board. Its effective relative dielectric constant ϵ_{re} is 6.71 for $w/h = 0.95$ (see Example 6-1-1). The characteristic impedance of a microstrip line with $w/h = 0.95$ is 50 Ω and the wavelength is

$$\lambda = 0.43\lambda_o = 0.43 \times 5 = 2.15 \text{ cm} \qquad \text{for 6 GHz}$$

The lengths of the input and output series 50-Ω striplines are

$$\ell_A = 0.137\lambda = 0.137 \times 2.15 = 0.29 \text{ cm}$$

$$\ell_B = 0.025\lambda = 0.025 \times 2.15 = 0.05 \text{ cm}$$

Step 5. Realize two shunt elements. Two shunt open stubs are needed to contribute $+ j1.30$ and $- j1.28$ for stripline matching. The lengths of the input and output shunt open stubs are

$$\ell_s = 0.145\lambda = 0.145 \times 2.15 = 0.31 \text{ cm} \qquad \text{for } +j1.30$$

$$\ell_o = 0.355\lambda = 0.355 \times 2.15 = 0.76 \text{ cm} \qquad \text{for } - j1.28$$

Step 6. Sketch the complete input and output stripline matching networks (Fig. 6-2-6).

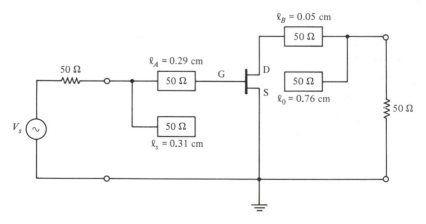

Figure 6-2-6   Stripline matching networks.

6-3 COPLANAR STRIPLINES

A coplanar stripline consists of two conducting strips on one substrate surface with one strip grounded as shown in Fig. 6-3-1. The coplanar stripline has advantages over the conventional parallel stripline as described in Section 6-4 because the former has its two strips on the same substrate surface for convenient connections. In microwave integrated circuits (MICs) the wire bonds have always been a serious factor in reliability and reproducibility. The coplanar striplines eliminate the difficulties in connecting the shunt elements between the hot and ground strips. As a result, the reliability is increased, the reproducibility is enhanced, and the production cost is decreased.

The characteristic impedance of a coplanar stripline can be expressed as

$$Z_0 = \frac{2P_{avg}}{I_0^2} \tag{6-3-1}$$

where I_0 = total peak current in one strip
  P_{avg} = average power flowing in the positive z direction
The average flowing power can be expressed as

$$P_{avg} = 1/2 \operatorname{Re} \iint (\overline{E} \times \overline{H}^*) \cdot \overline{u}_z \, dx \, dy \tag{6-3-2}$$

where E = electric field in the positive x direction
  H = magnetic field in the positive y direction
  $*$ = conjugate

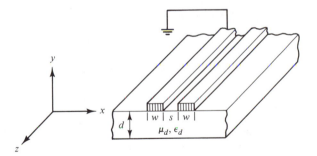

Figure 6-3-1 Coplanar stripline.

Example 6-3-1: Characteristic Impedance of a Coplanar Stripline

A coplanar stripline carries an average power of 250 mW with a peak current of 100 mA. Determine the characteristic impedance of the coplanar stripline.

Solution From Eq. (6-3-1) the characteristic impedance of the coplanar stripline is found to be

$$Z_0 = \frac{2 \times 250 \times 10^{-3}}{(100 \times 10^{-3})^2} = 50 \ \Omega$$

6-4 PARALLEL STRIPLINES

A parallel stripline consists of two perfectly parallel strips separated by a perfect dielectric slab of uniform thickness as shown in Fig. 6-4-1. The plate width is w, the separation distance is d, and the relative dielectric constant of the slab is ϵ_{rd}.

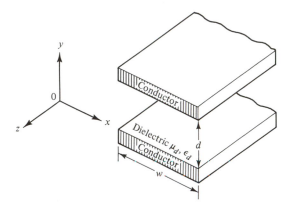

Figure 6-4-1   Schematic diagram of a parallel stripline.

6-4-1 Distributed Parameters

In a microwave integrated circuit a stripline can easily be fabricated on a dielectric substrate by using printed-circuit techniques. A parallel stripline is similar to a two-conductor transmission line, so it can support the TEM mode. Let us consider a TEM mode propagating in the positive z direction in a lossless stripline ($R = G = 0$). The electric field is in the y direction and the magnetic field is in the x direction. If the width w is much larger than the separation distance d, the fringing capacitance is negligible. The inductance along the two conducting strips can be written as

$$L = \frac{\mu_c d}{w} \qquad \text{H/m} \tag{6-4-1}$$

where μ_c is the permeability of the conductor.

The capacitance between the two conducting strips can be expressed as

$$C = \frac{\epsilon_d w}{d} \qquad \text{F/m} \tag{6-4-2}$$

where ϵ_d is the permittivity of the dielectric slab. However, if the two parallel strips have some surface resistance and the dielectric substrate has some shunt conductance, the parallel stripline would have some losses. The series resistance for both strips is

$$R = \frac{2R_s}{w} = \frac{2}{w}\sqrt{\frac{\pi f \mu_c}{\sigma_c}} \qquad \Omega/\text{m} \tag{6-4-3}$$

where $R_s = \sqrt{\dfrac{\pi f \mu_c}{\sigma_c}}$ = conductor surface resistance, mhos

σ = conductor conductivity, mhos per meter

The shunt conductance of the stripline is given by

$$G = \frac{\sigma_d w}{d} \qquad \text{mhos/m} \tag{6-4-4}$$

where σ_d is the conductivity of the dielectric substrate.

6-4-2 Characteristic Impedance

Lossless striplines.   The characteristic impedance of a lossless parallel strip-line is

$$Z_0 = \sqrt{\frac{L}{C}} = \frac{d}{w}\sqrt{\frac{\mu_d}{\epsilon_d}} = \frac{377}{\sqrt{\epsilon_{rd}}}\frac{d}{w} \qquad \text{ohms} \quad \text{for } w \gg d \tag{6-4-5}$$

The phase velocity along a parallel stripline is

$$v_p = \frac{\omega}{\beta} = \frac{1}{\sqrt{LC}} = \frac{1}{\sqrt{\mu_d \epsilon_d}} = \frac{c}{\sqrt{\epsilon_{rd}}} \qquad \text{m/s} \quad \text{for } \mu_c = \mu_0 \tag{6-4-6}$$

Lossy striplines.   The characteristic impedance of a lossy parallel stripline at microwave frequencies ($R \ll \omega L$ and $G \ll \omega C$) can be approximated as

$$Z_0 \simeq \sqrt{\frac{L}{C}} = \frac{377}{\sqrt{\epsilon_{rd}}}\frac{d}{w} \qquad \text{ohms} \quad \text{for } w \gg d \tag{6-4-7}$$

6-4-3 Attenuation Losses

The propagation constant of a parallel stripline at microwave frequencies can be expressed as

$$\gamma = \sqrt{(R + j\omega L)(G + j\omega C)} \qquad \text{for } R \ll \omega L \text{ and } G \ll \omega C$$

$$\simeq \frac{1}{2}\left(R\sqrt{\frac{C}{L}} + G\sqrt{\frac{L}{C}}\right) + j\omega\sqrt{LC} \tag{6-4-8}$$

Thus the attenuation and phase constants are given by

$$\alpha = \frac{1}{2}\left(R\sqrt{\frac{C}{L}} + G\sqrt{\frac{L}{C}}\right) \qquad \text{Np/m} \tag{6-4-9}$$

$$\beta = \omega\sqrt{LC} \qquad \text{rad/m} \tag{6-4-10}$$

Substitution of the distributed parameters of a parallel stripline in Eq. (6-4-9) yields the attenuation constants for the conductor and dielectric losses as

$$\alpha_c = \frac{1}{2} R \sqrt{\frac{C}{L}} = \frac{1}{d} \sqrt{\frac{\pi f \epsilon_d}{\sigma_c}} \qquad \text{Np/m} \qquad (6\text{-}4\text{-}11)$$

$$\alpha_d = \frac{1}{2} G \sqrt{\frac{L}{C}} = \frac{188 \sigma_d}{\sqrt{\epsilon_{rd}}} \qquad \text{Np/m} \qquad (6\text{-}4\text{-}12)$$

Example 6-4-1:   Characteristics of a Parallel Stripline

A lossless parallel stripline has a conducting strip width w. The substrate dielectric that separates the two conducting strips has a relative dielectric constant ϵ_{rd} of 6 (beryllia, BeO) and a thickness d of 4 mm. Calculate (a) the required conducting strip width w in order to have a characteristic impedance of 50 Ω, (b) the stripline capacitance, (c) the stripline inductance, and (d) the phase velocity of the wave in the parallel stripline.

Solution   (a) From Eq. (6-4-5) we have the conducting strip w as

$$w = \frac{377}{\sqrt{\epsilon_{rd}}} \frac{d}{Z_0} = \frac{377}{\sqrt{6}} \frac{4 \times 10^{-3}}{50} = 12.31 \times 10^{-3} \qquad \text{m}$$

(b) The stripline capacitance is

$$C = \frac{\epsilon_d w}{d} = \frac{8.854 \times 10^{-12} \times 6 \times 12.31 \times 10^{-3}}{4 \times 10^{-3}}$$
$$= 163.50 \quad \text{pF/m}$$

(c) The stripline inductance is

$$L = \frac{\mu_c d}{w} = \frac{4\pi \times 10^{-7} \times 4 \times 10^{-3}}{12.31 \times 10^{-3}}$$
$$= 0.41 \quad \mu\text{H/m}$$

(d) The phase velocity is

$$v_p = \frac{c}{\sqrt{\epsilon_{rd}}} = \frac{3 \times 10^8}{\sqrt{6}} = 1.22 \times 10^8 \text{ m/s}$$

6-5 SHIELDED STRIPLINES

A partially shielded stripline is one whose strip conductor is embedded in a dielectric medium and whose top and bottom ground planes have no connection, as shown in Fig. 6-5-1. The characteristic impedance for a wide strip ($w/d \geq 0.35$) is given by [10]

$$Z_0 = \frac{94.15}{\sqrt{\epsilon_r}} \left(\frac{w}{d} K + \frac{C_f}{8.854 \epsilon_r} \right)^{-1} \qquad \text{ohms} \qquad (6\text{-}5\text{-}1)$$

where $K = \dfrac{1}{1 - t/d}$

t = strip thickness

d = distance between the two ground planes

$C_f = \dfrac{8.854\epsilon_r}{\pi} [2K \ln (K + 1) - (K - 1) \ln (K^2 - 1)]$ is the fringe

capacitance in pF/m

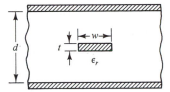

Figure 6-5-1 Partially shielded stripline.

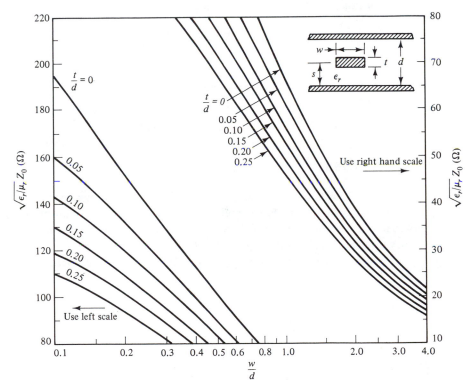

Figure 6-5-2 Characteristic impedance Z_0 of a partially shielded stripline with t/d ratio as parameter. (After S. Cohn [11]. Copyright © 1954 IRE [now IEEE]. Reprinted by permission of the IEEE, Inc.)

Figure 6-5-2 shows the characteristic impedance Z_0 for a partially shielded stripline with t/d ratio as a parameter.

Example 6-5-1:   Characteristic Impedance of a Shielded Stripline

A shielded stripline has the following parameters:

Relative dielectric constant of the insulator ϵ_r = 2.56
  (polystyrene)
Strip width w = 25 mils
Strip thickness t = 14 mils
Shield depth d = 70 mils

Calculate (a) the K factor, (b) the fringe capacitance, and (c) the characteristic impedance of the line.

Solution   (a) From Eq. (6-5-1) the K factor is

$$K = \left(1 - \frac{t}{d}\right)^{-1} = \left(1 - \frac{14}{70}\right)^{-1} = 1.25$$

(b) From Eq. (6-5-1) the fringe capacitance is

$$C_f = \frac{8.854 \times 2.56}{3.1416}\left[2 \times 1.25 \ \ell n \ (1.25 + 1) - (1.25 - 1) \ \ell n \ (1.25^2 - 1)\right]$$

$$= 15.61 \text{ pF/m}$$

(c) From Eq. (6-5-1) the characteristic impedance is

$$Z_0 = \frac{94.15}{\sqrt{2.56}}\left[\frac{25}{70}(1.25) + \frac{15.61}{8.854 \times 2.56}\right]^{-1}$$

$$= 50.29 \ \Omega$$

6-6 SLOT STRIPLINES

Slot stripline consists of a narrow slot or gap in a thin conductive layer on one side of a dielectric substrate of fairly high permittivity, the other side of the substrate being bare, as shown in Fig. 6-6-1. Slot line is actually an inverse form or complementary configuration of microstrip line.

In connection with increased applications in monolithic microwave integrated

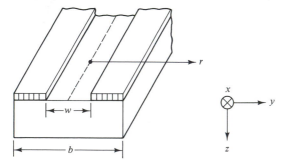

Figure 6-6-1   Schematic diagram of a slot stripline.

circuits (MMICs) without the need to penetrate the dielectric substrate for shunt-element connections as in the case of microstrip lines, coplanar slot stripline can be used either alone or with microstrip line on the opposite side of the substrate for filters, couplers, and ferrite devices. However, slot stripline is different from waveguide because it has no cutoff frequency.

6-6-1 Slot-Mode Wavelength

When the permittivity of the substrate board is sufficient high, such as $\epsilon_r = 10$ to 30, the slot-mode wavelength will be much shorter than the free-space wavelength, and the electromagnetic fields will be closely confined near the slot with negligible radiation loss. When a voltage difference exists between the slot edges, the electric field extends across the slot and the magnetic field is perpendicular to the slot. The slot-mode wavelength can be expressed [1] as

$$\lambda = \frac{\lambda_0}{\sqrt{\epsilon_{re}}} = \lambda_0 \sqrt{\frac{2}{\epsilon_r + 1}} \qquad (6\text{-}6\text{-}1)$$

where $\lambda_0 = c/f$ is the free-space wavelength
$c = 3 \times 10^8$ m/s is the velocity of light in vacuum
$\epsilon_{re} = (\epsilon_r + 1)/2$ is the effective relative dielectric constant
$\epsilon_r =$ relative dielectric constant of the substrate.

6-6-2 Field Intensities

The magnetic field component in the air region of the slot at a distance r from the slot center line can be written as

$$H_z = H_{0z} H_n(\chi r) \qquad \text{for } r \gg w \qquad (6\text{-}6\text{-}2)$$

where $H_n(\chi r) =$ Hankel function of the first kind
$n =$ order of Hankel function
$\chi r =$ argument of Hankel function
$\chi = \sqrt{k^2 - \beta_z^2}$ is the propagation constant
$k = \omega\sqrt{\mu_0 \epsilon_0} = 2\pi/\lambda_0$ is the wave number in free space
$\lambda_0 =$ wavelength in free space
$\beta_z = \omega/v_\epsilon = 2\pi/\lambda$ is the phase constant in substrate
$\lambda =$ wavelength in dielectric substrate
The propagation constant χ can be expressed as

$$\chi = \sqrt{\left(\frac{2\pi}{\lambda_0}\right)^2 - \left(\frac{2\pi}{\lambda}\right)^2}$$

$$= j\frac{2\pi}{\lambda_0}\sqrt{\left(\frac{\lambda_0}{\lambda}\right)^2 - 1} \qquad (6\text{-}6\text{-}3)$$

$$= j\frac{2\pi}{\lambda}\sqrt{1 - \left(\frac{\lambda}{\lambda_0}\right)^2}$$

At zero order, the value of χ is

$$\chi = j \frac{2\pi}{\lambda_0} \sqrt{\frac{\epsilon_r - 1}{2}} \qquad (6\text{-}6\text{-}4)$$

The other field componenets are

$$H_r = \frac{-j\beta_z}{\chi^2} \frac{\partial H_z}{\partial r} = H_{0z} \left[1 - \left(\frac{\lambda}{\lambda_0}\right)^2 \right]^{-1/2} H_1^{(1)}(\chi r) \qquad (6\text{-}6\text{-}5)$$

$$E_\phi = \frac{j\omega\mu}{\chi^2} \frac{\partial H_z}{\partial r} = -\eta \left(\frac{\lambda}{\lambda_0}\right) H_{0z} \left[1 - \left(\frac{\lambda}{\lambda_0}\right)^2 \right]^{-1/2} H_1^{(1)}(\chi r) \qquad (6\text{-}6\text{-}6)$$

where $\partial H_z / \partial r = -H_1^{(1)}(\chi r)$ is used.

The Hankel function having an imaginary argument approaches zero. Equation (6-6-3) indicates that the argument (χr) is imaginary for $\lambda/\lambda_0 < 1$. Hence a relative wavelength ratio of λ/λ_0 less than unity is a sufficient condition to ensure decay of the slot-mode field with radial distance. As λ/λ_0 is decreased, the decay becomes sharper and the field becomes more tightly bound to the slot.

Example 6-6-1:   Characteristics of a Slot Stripline

A slot stripline is fabricated on a GaAs substrate board with the following specifications:

Relative dielectric constant of GaAs substrate	ϵ_r =	13.10
Operating frequency	f =	10 GHz
Distance r from the centerline of the slot	r =	1 cm

Compute (a) the wavelength ratio λ/λ_0, (b) the argument χr, and (c) the value of Hankel function.

Solution   (a) The wavelength ratio is

$$\lambda/\lambda_0 = \sqrt{\frac{2}{13.1 + 1}} = 0.38$$

(b) The argument is

$$\chi r = j \frac{2\pi}{3} \sqrt{\frac{13.1 - 1}{2}} \times 1 = j5.15$$

(c) The value of Hankel function is

$$H_0(j5.15) \rightarrow 0$$

6-6-3 Characteristic Impedance

The slot-stripline characteristic impedance for the approximate TEM mode is given by [11]

$$Z_0 = \frac{591.7}{\sqrt{\epsilon_{re}} \, \ell n \left(\frac{8b}{\pi w}\right)} = \frac{591.7(\lambda/\lambda_0)}{\ell n \left(\frac{8b}{\pi w}\right)} \quad \text{ohms} \quad (6\text{-}6\text{-}7)$$

where ϵ_{re} = $(\epsilon_r + 1)/2$ is the effective relative dielectric constant of the sub-strate

ϵ_r = relative dielectric constant of the substrate

λ = slot-mode wavelength

λ_0 = free-space wavelength

b = substrate width

w = slot width

Example 6-6-2:   Characteristic Impedance of a Slot Stripline

A slot stripline of 50-Ω characteristic impedance is to be designed on a GaAs substrate. Determine the ratio of b over w.

Solution   The effective relative dielectric constant is

$$\epsilon_{re} = \frac{13.1 + 1}{2} = 7.05$$

From Eq. (6-6-7) we have

$$\ell n \left(\frac{8b}{\pi w}\right) = \frac{591.7}{\sqrt{7.05} \times 50} = 4.45$$

$$\frac{b}{w} = 33.77$$

6-7 PLANAR LUMPED ELEMENTS

The choice of lumped or distributed elements for amplifier matching networks depends on the operating frequency. When the frequency is up to X-band, its wavelength is very short and a smaller lumped element exhibits a negligible phase shift. Because of the advanced thin-film technology, the size of lumped elements can be reduced considerably and their operating frequencies can reach up to 20 GHz. Beyond that, distributed elements are preferred. In monolithic microwave integrated circuits (MMICs) lumped resistors are very useful in thin-film resistive terminations for couplers, lumped capacitors are absolutely essential for bias bypass applications, and planar inductors are extremely useful for matching purposes, especially at lower microwave frequencies where stub inductors are physically too large [12].

6-7-1 Planar Resistors

A planar resistor consists of a thin resistive film deposited on an insulating substrate. Thin-film resistor materials are aluminum, copper, gold, nichrome, titanium, tantalum, and so on, with resistivity ranging from 30 to 1000 Ω per square. The planar resistors are essential for terminations for hybrid couplers, power combiners/dividers, and bias-voltage circuits. Some design considerations:

1. Sheet resistivity available
2. Thermal stability or temperature coefficient of the resistive material
3. Thermal resistance of the load
4. Frequency bandwidth

Planar resistors can be grouped into three categories:

1. Semiconductor films
2. Deposited metal films
3. Cermets

Planar resistors based on semiconductors can be fabricated by forming an isolated land of conducting epitaxial film on the substrate by mesa etching or by isolation implant of the surrounding conducting film. Another way is by implanting a high-resistivity region within the semi-insulating substrate. Metal-film resistors are formed by evaporating a metal layer over the substrate and forming the desired pattern by photolithography. Cermet resistors are formed from films consisting of a mixture of metal and a dielectric. Figure 6-7-1 shows several examples of planar resistor design.

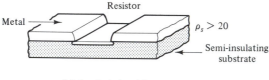

(a) Implanted resistor

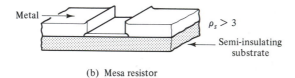

(b) Mesa resistor

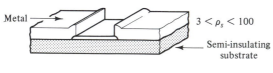

(c) Deposited resistor

Figure 6-7-1 Configurations of planar resistors.

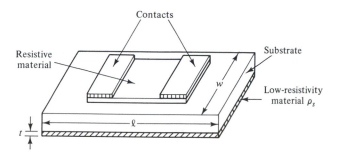

Figure 6-7-2   Thin-film resistor.

The resistance of a planar resistor as shown in Fig. 6-7-2 can be expressed as

$$R = \frac{\ell \rho_s}{wt} \quad \text{ohms} \qquad (6\text{-}7\text{-}1)$$

where ℓ = length of the resistive film
   w = width of the film
   ρ_s = sheet resistivity of the film, ohm-meters
   t = film thickness, meters

   When units of length ℓ and width w are chosen to have equal magnitude, the result is a square, Therefore, the resistance R in ohms per square is independent of the dimension of the square.

Example 6-7-1:   Resistance of a Planar Resistor

   A planar resistor has the following parameters:

Resistive film thickness	t = 0.1 μm
Resistive film length	ℓ = 10 mm
Resistive film width	w = 10 mm
Sheet resistivity of gold film	ρ_s = 2.44 $\times$ 10^{-8} Ω-m

   Calculate the planar resistance.

Solution   The planar resistance is

$$R = \frac{10 \times 2.44 \times 10^{-8}}{10 \times 1 \times 10^{-7}} = 0.244 \ \Omega/\text{square}$$

6-7-2 Planar Inductors

Planar inductors for monolithic circuits can be realized in a number of configurations, such as square spiral and circular spiral, as shown in Fig. 6-7-3. Typical inductance values for monolithic circuits range from 0.5 to 10 nH. The equations of inductance are different for different configurations.

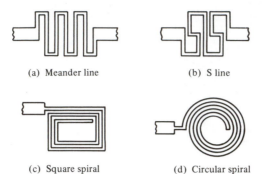

(a) Meander line   (b) S line

(c) Square spiral   (d) Circular spiral

Figure 6-7-3 Configurations of planar inductors.

Ribbon inductor.   The inductance of a ribbon inductor can be expressed as [13]

$$L = 5.08 \times 10^{-3} \ell \left[\ell n \left(\frac{\ell}{w + t} \right) + 1.19 + 0.022 \left(\frac{w + t}{\ell} \right) \right] \quad \text{nH/mil} \quad (6\text{-}7\text{-}2)$$

where ℓ = ribbon length, mils
    t = ribbon thickness, mils
    w = ribbon width, mils

Round-wire inductor.   The inductance of a round-wire inductor is given by

$$L = 5.08 \times 10^{-3} \ell \left[\ell n \left(\ell/d + 0.386 \right] \right. \quad \text{nH/mil} \quad (6\text{-}7\text{-}3)$$

where d = wire diameter, mils
    ℓ = wire length, mils

Circular spiral inductor.   The inductance of a circular spiral inductor is expressed as

$$L = 0.03125 n^2 d_o \quad \text{nH/mil} \quad (6\text{-}7\text{-}4)$$

where d_o = $5d_i$ = $2.5n(w + s)$, mils
    n = number of turns
    s = separation, mils
    w = film width, mils
Figure 6-7-4 shows the schematic diagram of a circular spiral inductor.

Circular loop inductor.   The inductance of a single-turn flat circular loop inductor is given by

$$L = 5.08 \times 10^{-3} \ell \left[\ell n \left(\frac{t}{w + t} \right) - 1.76 \right] \quad \text{nH/mil} \quad (6\text{-}7\text{-}5)$$

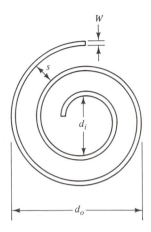

Figure 6-7-4 Circular spiral inductor.

Square spiral inductor.   The inductance of a square spiral inductor can be written as

$$L = 8.5 A^{1/2} n^{5/3} \qquad \text{nH} \tag{6-7-6}$$

where A = surface area, cm^2

$\quad\;\; n$ = number of turns

Example 6-7-2: Calculation of a Planar Circular Spiral Inductor

A circular spiral inductor has the following parameters:

Number of turns $n = 5$
Separation $s = 100$ mils
Film width $w = 50$ mils

Compute the inductance.

Solution   The inductance is

$$L = 0.03125(5)^2 \times 2.5(5)(50 + 100)$$
$$= 1464.84 \text{ nH/mil}$$

6-7-3 Planar Capacitors

There are two types of planar capacitors commonly used for MMICs: the metal–oxide–metal capacitor and the interdigitated capacitor, shown in Fig. 6-7-5.

Metal–oxide–metal capacitor.   The metal–oxide–metal capacitor has three layers; the middle dielectric layer is sandwiched by the top and bottom electrode layers, as shown in Fig. 6-7-5(a). The capacitance can be expressed as

$$C = \epsilon_0 \epsilon_r \frac{\ell w}{h} \qquad \text{farads} \tag{6-7-7}$$

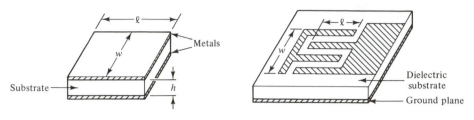

(a) Metal–oxide–metal capacitor   (b) Interdigitated capacitor

Figure 6-7-5   Schematic diagrams of planar capacitors.

where ϵ_0 = 8.84 $\times$ 10^{-12} F/m
   ϵ_r = relative dielectric constant of the dielectric material
   ℓ = metal length
   w = metal width
   h = height of the dielectric material

Interdigitated capacitor.   The interdigitated capacitor consists of a single-layer structure and it can be fabricated easily on substrates as microstrip lines with values between 0.1 and 15 pF, as shown in Fig. 6-7-5(b). The capacitance can be approximated as [13]

$$C = \frac{\epsilon_r + 1}{w} \ell[(N - 3)A_1 + A_2] \qquad \text{pF/cm} \qquad (6\text{-}7\text{-}8)$$

where N = number of fingers
   A_1 = 0.089 pF/cm is the contribution of the interior finger for $h > w$
   A_2 = 0.10 pF/cm is the contribution of the two external fingers for
       $h > w$
   ℓ = finger length, centimeters
   w = finger-base width, centimeters

Example 6-7-3:   Computations of a Planar Capacitor

An interdigitated capacitor fabricated on a GaAs substrate has the following parameters:

Number of fingers	$N = 8$
Relative dielectric constant of GaAs	$\epsilon_r = 13.10$
Substrate height	$h = 0.254$ cm
Finger length	$\ell = 0.00254$ cm
Finger-base width	$w = 0.051$ cm

Compute the capacitance.

Solution   The capacitance is

$$C = \frac{13.10 + 1}{0.051} \times 0.00254 \times [(8 - 3) \times 0.089 + 0.10]$$

$$= 0.380 \quad \text{pF/cm}$$

REFERENCES

1. Bahl, I. J., and D. K. Triveli, A designer's guide to microstrip line. *Microwaves,* May 1977.

2. Sobol, H., Applications of interdigitated circuit technology to microwave frequencies. *Proc. IEEE,* August 1971.

3. Welch, J. D., and H. J. Pratt, Losses in microstrip transmission systems for integrated microwave circuits. *NEREM Rec.,* 8, 1966, pp. 100–101.

4. Pucel, Robert A., et al., Losses in microstrip. *IEEE Trans. Microwave Theory and Techniques,* Vol. MTT-16, No. 6, June 1968, pp. 342–350.

5. Wheeler, H. A., Transmission-line properties of parallel strips separated by a dielectric sheet. *IEEE Trans. Microwave Theory and Techniques,* Vol. MTT-3, No. 3, March 1965, pp. 172–185.

6. Assadourian, H., and E. Rimai, Simplified theory of microwave transmission systems. *Proc. IRE,* 40, December 1952, pp. 1651–1657.

7. Pucel, Robert A., et al., Correction to "Losses in microstrip." *IEEE Trans. Microwave Theory and Techniques,* Vol. MTT-16, No. 12, December 1968, p. 1064.

8. Lewin, L., Radiation from discontinuities in stripline. *IEEE Monograph No. 358E,* February 1960.

9. Vendeline, George D., Limitations on stripline Q. *Microwave Journal,* May 1970, pp. 63–69.

10. Cohn, S., Characteristic impedance of the shielded-strip transmission line. *IRE Trans. Microwave Theory and Techniques,* Vol. MTT-2, No. 7, July 1954, p. 52.

11. Cohn, S., Slot line on a dielectric substrate. *IEEE Trans. Microwave Theory and Techniques,* Vol. MTT-17, No. 10, October 1969, pp. 768–778.

12. Pucel, Robert A., Design considerations for monolithic microwave circuits. *IEEE Trans. Microwave Theory and Techniques,* Vol. MTT-29, No. 6, June 1981, pp. 513–534.

13. Young, L., *Advances in Microwaves.* New York: Academic Press, Inc., 1974, pp. 148–158.

PROBLEMS

6-1 Microstrip Lines

6-1-1. A certain microstrip line has the following parameters:

ϵ_r = 5.23 is the relative dielectric constant of the fiberglass–epoxy board material
h = 8 mils
t = 2.8 mils
w = 10 mils

Write a FORTRAN program to compute the characteristic impedance Z_0 of the line. Use a READ statement to read in the input values, the F10.5 format for numerical outputs, and the Hollerith format for character outputs.

6-1-2. Since modes on microstrip lines are only quasi-transverse electric and magnetic (TEM), the theory of TEM coupled lines applies only approximately. Derive from the basic theory of a lossless line that the inductance and capacitance of a microstrip line are given by

$$L = \frac{Z_0}{v} = \frac{Z_0 \sqrt{\epsilon_r}}{c}$$

$$C = \frac{1}{Z_0 v} = \frac{\sqrt{\epsilon_r}}{Z_0 c}$$

where Z_0 = characteristic impedance of the microstrip line
v = wave velocity in the microstrip line
c = 3×10^8 m/s is the velocity of light in vacuum
ϵ_r = relative dielectric constant of the board material

6-1-3. A microstrip line is constructed of a perfect conductor and a lossless dielectic board. The relative dielectric constant of the fiberglass–epoxy board is 5.23 and the line characteristic impedance is 50 Ω. Calculate the line inductance and the line capacitance.

6-1-4. A certain microstrip line is constructed of a copper conductor and nylon phenolic board. The relative dielectric constant of the board material is 4.19 measured at 25 GHz and its thickness is 0.4836 mm (19 mils). The line width is 0.635 mm (25 mils) and the line thickness is 0.071 mm (2.8 mils).
(a) Compute the characteristic impedance Z_0 of the microstrip line.
(b) Calculate the dielectric filling factor q.
(c) Compute the dielectric attenuation constant α_d.
(d) Find the surface skin resistivity R_s of the copper conductor at 25 GHz.
(e) Determine the conductor attenuation constant α_c.

6-1-5. A certain microstrip line is made of a copper conductor 0.254 mm (10 mils) wide on a G-10 fiberglass–epoxy board 0.20 mm (8 mils) in height. The relative dielectric constant ϵ_r of the board material is 4.8 measured at 25 GHz. The microstrip line of 0.035 mm (1.4 mils) thick is to be used for 10 GHz.
(a) Calculate the characteristic impedance Z_0 of the microstrip line.
(b) Compute the surface resistivity R_s of the copper conductor.
(c) Calculate the conductor attenuation constant α_c.
(d) Determine the dielectric attenuation constant α_d.
(e) Find the quality factors Q_c and Q_d.

6-2 Stripline-Type Amplifier Design

6-2-1. A stripline-type amplifier for maximum power gain is to be designed by using three-eighths-wavelength stubs and quarter-wave transformers. The parameters of a certain GaAs MESFET are as follows:

Maximum available power gain $\qquad$ G_{amax} = 10 dB
Source reflection coefficient for maxi- $\qquad$ Γ_{sm} = 0.72 $\underline{/165°}$
mum power gain
Load reflection coefficient for maxi- $\qquad$ $\Gamma_{\ell m}$ = 0.67 $\underline{/110°}$
mum power gain

Design the input and output matching networks for the amplifier by using 50-Ω microstrip line at 8 GHz.

6-2-2. A stripline-type amplifier for minimum noise is to be designed by using three-eighths-wavelength stubs and quarter-wave transformers. The parameters of a certain GaAs MESFET are as follows:

Minimum-noise figure	$F_{min} = 2.5$ dB
Optimum source reflection coefficient for F_{min}	$\Gamma_o = 0.532 \; \underline{/115°}$
Load reflection coefficient for F_{min}	$\Gamma_\ell = 0.594 \; \underline{/108°}$
Equivalent noise resistance	$R_n = 8 \; \Omega$

Design the input and output matching networks for the amplifier by using 50-Ω microstrip line at 6 GHz.

6-3 Coplanar Striplines

6-3-1. A 50-Ω coplanar stripline has the following parameters:

Relative dielectric constant of alumina	$\epsilon_{rd} = 10$
Strip width	$w = 4$ mm
Strip thickness	$t = 1$ mm
TEM-mode field intensities	

$$E_y = 3.16 \times 10^3 \sin\left(\frac{\pi x}{w}\right) e^{-j\beta z}$$

$$H_x = 63.20 \sin\left(\frac{\pi x}{w}\right) e^{-j\beta z}$$

Compute:
(a) The average power flow.
(b) The peak current in one strip.

6-3-2. Design a coplanar stripline for a characteristic impedance of 50 Ω. If the average power is 200 mW, what is its peak current?

6-4 Parallel Striplines

6-4-1. A gold parallel stripline has the following parameters:

Relative dielectic constant ofTeflon	$\epsilon_{rd} = 2.1$
Strip width	$w = 26$ mm
Separation distance	$d = 5$ mm
Conductivity of gold	$\sigma_c = 4.1 \times 10^7$ mhos/m
Frequency	$f = 10$ GHz

Compute:
(a) The surface resistance of the gold strip.
(b) The characteristic impedance of the stripline.
(c) The phase velocity.

6-4-2. A gold parallel stripline has the following parameters:

Relative dielectric constant of poly-ethylene	$\epsilon_{rd} = 2.25$

Strip width	$w = 25$ mm
Separation distance	$d = 5$ mm

Calculate:

(a) The characteristic impedance of the stripline.

(b) The stripline capacitance.

(c) The stripline inductance.

(d) The phase velocity.

6-5 Shielded Striplines

6-5-1. A shielded stripline has the following parameters:

Relative dielectric constant of the insulator polyethylene	$\epsilon_{rd} = 2.25$
Strip width	$w = 2$ mm
Strip thickness	$t = 0.5$ mm
Shield depth	$d = 4$ mm

Calculate:

(a) The K factor.

(b) The fringe capacitance.

(c) The characteristic impedance.

6-5-2. A shielded stripline is made of a gold strip in a polystyrene dielectric insulator and its parameters are:

Relative dielectric constant of polystyrene	$\epsilon_{rd} = 2.56$
Strip thickness	$t = 0.7$ mm
Strip width	$w = 1.4$ mm
Shield depth	$d = 3.5$ mm

Determine:

(a) The K factor.

(b) The fringe capacitance.

(c) The characteristic impedance.

6-6 Slot Striplines

6-6-1. A slot stripline is fabricated on an alumina substrate board with the following parameters:

Relative dielectric constant of alumina substrate	$\epsilon_r = 10$
Operating frequency	$f = 9$ GHz
Distance r from the center line of the slot	$r = 1.5$ cm

Calculate:

(a) The wavelength ratio, λ/λ_0.

(b) The argument, χr.

(c) The value of Hankel function.

6-6-2. A slot stripline of 50-Ω characteristic impedance is to be designed on a beryllia (ϵ_r = 6) substrate. Determine the ratio of substrate width b over slot width w.

6-6-3. A slot stripline is to be designed with a characteristic impedance of 50 Ω on an alumina substrate. The operating frequency is 10 GHz and the slot width w is 1 mm. Determine the substrate width b in millimeters.

6-7 Planar Lumped Elements

6-7-1. A planar film resistor has the following parameters:

Resistive film thickness	$t = 0.1 \ \mu m$
Resistive film length	$\ell = 2$ mm
Resistive film width	$w = 2$ mm
Sheet resistivity of gold film	$\rho_s = 2.44 \times 10^{-8} \ \Omega\text{-m}$

Determine the planar film resistance in ohms per square.

6-7-2. A planar film resistor has the following parameters:

Resistive film thickness	$t = 0.01 \ \mu m$
Resistive film length	$\ell = 5$ mm
Resistive film width	$w = 1$ mm
Sheet resistivity of gold film	$\rho_s = 2.44 \times 10^{-8} \ \Omega\text{-m}$

Determine the planar film resistance in ohms.

6-7-3. A square spiral inductor has a surface area of 2 cm² and two turns. Calculate its inductance.

6-7-4. A circular spiral inductor has the following parameters:

Number of turns	$n = 4$
Separation	$s = 120$ mils
Film width	$w = 20$ mils

Find the inductance.

6-7-5. A metal–oxide–metal capacitor fabricated on a duroid substrate has the following parameters:

Relative dielectric constant of duroid	$\epsilon_r = 2.23$
Metal length	$\ell = 2$ cm
Metal width	$w = 2$ cm
Dielectric height	$h = 0.1$ mm

Calculate the capacitance.

6-7-6. An interdigitated capacitor fabricated on an alumina substrate has the following parameters:

Number of fingers	$N = 6$
Relative dielectric constant of alumina	$\epsilon_r = 10$
Substrate height	$h = 0.254$ cm
Finger length	$\ell = 0.051$ cm
Finger-base width	$w = 0.051$ cm

Compute the capacitance.

Chapter 7

Large-Signal and Broadband Amplifier Design

7-0 INTRODUCTION

In Chapter 4, small-signal and narrowband amplifier designs were presented. However, in many microwave applications, such as communications or surveillance electronic systems, large-signal and broadband amplifiers are often required. The purpose of this chapter is to discuss various large-signal and broadband amplifier design techniques. At microwave frequencies, the reverse transmission S parameter S_{12} of microwave transistor or GaAs MESFET is very small and can be negligible for a simple unilateral case. Figure 7-0-1 shows a large-signal operation of a typical microwave amplifier.

7-1 LARGE-SIGNAL AMPLIFIER DESIGN

As described in Chapter 4, a small-signal amplifier is based on the small-signal S parameters of the device. These S parameters are easily measured and usually supplied by the device manufacturers. For linear operation in the Class A mode, a large-signal amplifier can be designed by using small-signal S parameters. However, for Class C nonlinear operation, the small-signal S parameters are not useful for design purposes. On the other hand, the measurements of large-signal S parameters are sometimes not possible because the test equipment does not have the capability to handle high-power signals. For example, the power levels of the Hewlett-Packard automatic network analyzer HP8409 series are under the power levels required by the large-signal power (say 400 mW).

236

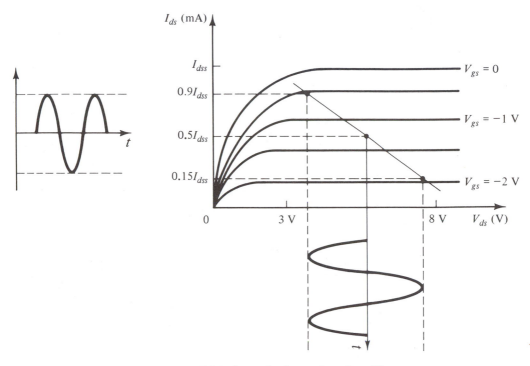

Figure 7-0-1   Large-signal operation of amplifier.

7-1-1 Large-Signal Measurements

In practical design, a large signal can be approximated by using small-signal S parameters. This is because all of the S parameters except the magnitude of S_{21} are constant under large signals. The large-signal magnitude of S_{21} is reduced as the power levels and frequency are increased. If large-signal S parameters are very desirable for design processes, the test setup shown in Fig. 7-1-1 can be used for measurements.

The magnitude of S_{12} and S_{21} can be determined by measuring the power loss or gain through the device using calibrated directional couplers as shown in Fig. 7-1-1(a). For phase measurements, a phase shifter is placed in the line, and the input and output waves are summed in a mixer. The phase shifter is adjusted to give a minimum signal from the mixer corresponding to 180° phase cancellation. When the device is inserted, the phase difference is measured using a calibrated precision phase shifter. The difference in phase is the phase of S_{12} and S_{21}, depending on which direction the device is inserted. S_{11} and S_{22} can be measured similarly using the test setup shown in Fig. 7-1-1(b).

However, in this section, we describe several design examples by using small-signal devices with their S parameters operating in the linear region for high power output.

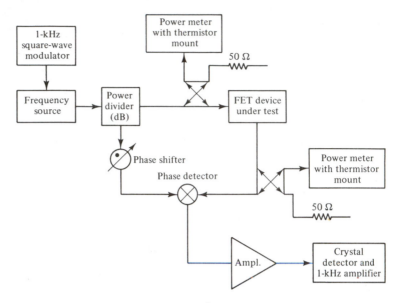

(a) For S_{12} and S_{21} measurements

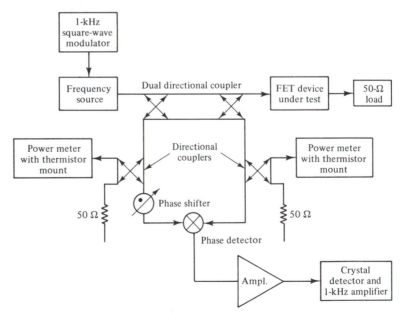

(b) For S_{11} and S_{22} measurements

Figure 7-1-1 Large-signal S-parameter measurement setups. (After K. M. Johnson [1]. Copyright © 1979 IEEE. Reprinted by permission of the IEEE, Inc.)

7-1-2 Design Example

When a signal becomes larger and larger, the magnitude of the forward transmission coefficient S_{21} is increased accordingly, but the magnitudes of the coefficients S_{11}, S_{12}, and S_{22} remain almost constant. The design approach for a large-signal amplifier in a simple case is to use the S parameters of a small-signal device with a large magnitude of the forward transmission coefficient S_{21}. Because the signal is large, the device may operate in its nonlinear region. Therefore, it is very possible that the device is potentially unstable and that its stability factor K may be less than 1. Let's design a large-signal amplifier using a potential unstable device.

Design Procedure.   The design steps can be listed as follows:

1. Check the delta factor Δ and stability factor K for stability.
2. Determine the maximum stable gain.
3. Plot the load stability circles.
4. Plot the desired power-gain circle.
5. Plot the source stability circle.
6. Select the load reflection coefficient and determine the source reflection coefficient.
7. Realize the input and output matching networks.
8. Balance the shunt stubs to minimize the transition interactions.

Design Example 7-1-1:   Large-Signal Amplifier Design Using a Potentially Unstable Device

At 3 GHz a certain GaAs MESFET biased for a large signal has the following S parameters:

$$S_{11} = 0.876 \;\underline{/-50°} \quad f = 3 \text{ GHz}$$

$$S_{12} = 0.050 \;\underline{/-118.9°}$$

$$S_{21} = 3.114 \;\underline{/130°}$$

$$S_{22} = 0.754 \;\underline{/-30°}$$

Design the matching networks of the amplifier for 20-dB power gain shown in Fig. 7-1-2.

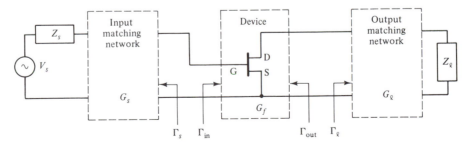

Figure 7-1-2   Desired matching networks for Example 7-1-2.

Design Steps

Step 1. Check the delta factor Δ and the stability factor K. The delta factor is

$$\Delta = S_{11}S_{22} - S_{12}S_{21}$$

$$= 0.876 \,\underline{/-50^\circ} \times 0.754 \,\underline{/-30^\circ} - 0.050 \,\underline{/-118.9^\circ} \times 3.114 \,\underline{/130^\circ}$$

$$= 0.68 \,\underline{/-93.2^\circ}$$

$$|\Delta| = 0.68 < 1$$

$$|\Delta|^2 = 0.46$$

The stability factor is

$$K = \frac{1 + |\Delta|^2 - |S_{11}|^2 - |S_{22}|^2}{2|S_{12}S_{21}|}$$

$$= \frac{1 + |0.68|^2 - |0.876|^2 - |0.754|^2}{2|0.050 \times 3.114|}$$

$$= 0.40 < 1$$

The device is potentially unstable.

Step 2. Determine the maximum stable gain. From Eq. (3-4-14b) the maximum stable gain is

$$G_{msg} = \frac{|S_{21}|}{|S_{12}|} = \frac{3.114}{0.050} = 62.28 = 18 \text{ dB}$$

Step 3. Plot the stability circles. From Eq. (3-5-9) the center of the load stability circle is

$$c_\ell = \frac{C_\ell^*}{|S_{22}|^2 - |\Delta|^2} = \frac{S_{22}^* - \Delta^* S_{11}}{|S_{22}|^2 - |\Delta|^2}$$

$$= \frac{0.754 \,\underline{/30^\circ} - 0.68 \,\underline{/93.2^\circ} \times 0.876 \,\underline{/-50^\circ}}{|0.754|^2 - |0.68|^2} = \frac{0.222 \,\underline{/-7.77^\circ}}{0.109}$$

$$= 2.00 \,\underline{/-7.77^\circ}$$

From Eq. (3-5-8) the radius of the load stability circle is

$$r_\ell = \frac{|S_{12}S_{21}|}{\left||S_{22}|^2 - |\Delta|^2\right|} = \frac{|0.050 \times 3.114|}{\left||0.754|^2 - |0.68|^2\right|}$$

$$= 1.43$$

As $|S_{11}| < 1$, the region of the Smith chart outside the load stability circle is the load stable region.

Step 4. Plot the desired 18-dB power-gain circle. The normalized power gain from Eq. (3-7-1) is

$$g_p = \frac{G_p}{|S_{21}|^2} = \frac{62.28}{|3.114|^2} = 6.42$$

The distance of the circle center is

$$d_p = \frac{g_p C_t^*}{|1 + g_p(|S_{22}|^2 - |\Delta|^2)|}$$

$$= \frac{6.42 \times 0.222 \; \underline{/-7.77°}}{|1 + 6.42 \,(|0.754|^2 - |0.68|^2)|}$$

$$= 0.84 \; \underline{/-7.77°}$$

The radius of the 20-dB power-gain circle is

$$r_p = \frac{(1 - 2K|S_{12}S_{21}|g_p + |S_{12}S_{21}|^2 g_p^2)^{1/2}}{|1 + g_p(|S_{22}|^2 - |\Delta|^2)|}$$

$$= \frac{(1 - 2 \times 0.40|0.050 \times 3.114|\, 6.42 + |0.050 \times 3.114|^2 \,(6.42)^2)^{1/2}}{|1 + 6.42\,(|0.754|^2 - |0.68|^2)|}$$

$$= 0.64$$

For $|S_{11}| < 1$ and $|S_{22}| < 1$, the stable region of the Smith chart is the region outside the load stability circle. Any load in this stable region and on the 18-dB constant-power-gain circle will make the amplifier stable and yield 18-dB gain.

Step 5. Select the load reflection coefficient:

$$\Gamma_t = 0.29 \, \underline{/-42°}$$

Read

$$z_t = 1.4 - j0.60 \qquad y_t = 0.60 + j0.26$$

then

$$Z_t = 70 - j30 \qquad Y_t = 12 \times 10^{-3} + j5.2 \times 10^{-3}$$

Step 6. Plot the source stability circle from Eqs. (3-5-6) and (3-5-7): The center of the source stability circle is

$$c_s = \frac{C_s^*}{|S_{11}|^2 - |\Delta|^2} = \frac{S_{11}^* - \Delta^* S_{22}}{|S_{11}|^2 - |\Delta|^2}$$

$$= \frac{0.876 \; \underline{/50°} - 0.68 \; \underline{/93.2°} \times 0.754 \; \underline{/-30°}}{|0.876|^2 - |0.68|^2}$$

$$= 1.29 \; \underline{/32.97°}$$

The radius of the source stability circle is

$$r_s = \frac{|S_{12}S_{21}|}{\left||S_{11}|^2 - |\Delta|^2\right|} = \frac{|0.031 \times 3.114|}{\left||0.876|^2 - |0.68|^2\right|}$$

$$= 0.51$$

Plot the source stability circle on the Smith chart.

Step 7. Determine the source reflection coefficient to yield a 18-dB power gain with the selected load reflection coefficient or load impedance from Eq. (3-4-8):

$$\Gamma_s = \left(\frac{S_{11} - \Delta\Gamma_\ell}{1 - S_{22}\Gamma_\ell}\right)^*$$

$$= \left(\frac{0.876\ \underline{/-50°} - 0.68\ \underline{/-93.2°} \times 0.29\ \underline{/-42°}}{1 - 0.754\ \underline{/-30°} \times 0.29\ \underline{/-42°}}\right)^*$$

$$= 0.50\ \underline{/21.83°}$$

Read

$$z_s = 2.50 + j1.50 \qquad y_s = 0.30 - j0.17$$

then

$$Z_s = 125 + j75 \qquad Y_s = 6 \times 10^{-3} - j3.4 \times 10^{-3}$$

Step 8. Realize the input and output matching networks by using microstrip lines. The microstrip lines are made on a beryllia substrate with a relative dielectric constant ϵ_r of 6 and its w/h ratio is 0.8. The effective relative dielectric constant is computed from Eq. (6-1-2) as

$$\epsilon_{re} = \frac{6 + 1}{2} + \frac{6 - 1}{2}\left[\left(1 + \frac{12}{0.8}\right)^{-1/2} + 0.04\,(1 - 0.8)^2\right] = 4.13$$

The wavelength in the substrate is

$$\lambda = \frac{\lambda_0}{\sqrt{\epsilon_{re}}} = \frac{\lambda_0}{\sqrt{4.13}} = 0.492\lambda_0 = 4.92\ \text{cm}$$

where λ_0 = wavelength in free space
          = 10 cm for 3 GHz

   Output matching network: The shortest length of microstrip line plus stub for the output matching network is obtained by using a short-circuited shunt stub of length 0.159λ (0.78 cm) to move from the center of the Smith chart to point B, then using a transmission-line length of 0.131λ (0.645 cm) to move from B to y_ℓ.

   Input matching network: The shortest length of microstrip line plus stub for the input matching network is determined by using a short-circuited shunt stub of 0.10λ (0.492 cm) to move from the center of the Smith chart to point A, then using a transmission-line length of 0.121λ (0.595 cm) to move from A to y_s, as shown in Fig. 7-1-3.

Step 9. Draw the input and output matching networks. The characteristic impedance of all microstrip lines is 50 Ω. The complete input and output matching networks are shown in Fig. 7-1-4.

Step 10. Balance the shunt stubs to minimize the transition interactions along the lines. To minimize the transition interactions between the shunt stubs and the series microstrip lines, the shunt stubs are often balanced along the series line. Figure 7-1-5 shows a schematic diagram of the designed amplifier with two parallel shunt

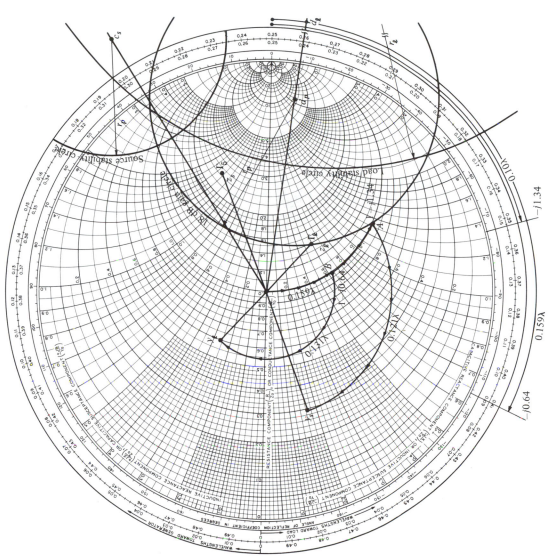

Figure 7-1-3 Graphical solution for Example 7-1-1.

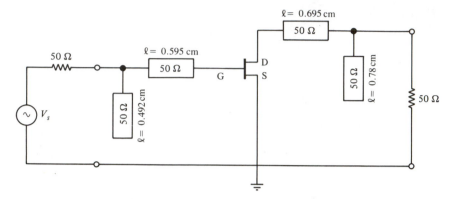

Figure 7-1-4 Complete input and output matching networks for the designed amplifier.

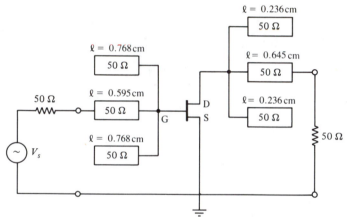

Figure 7-1-5 Matching networks with balanced stubs.

stubs. Admittance of each side of the balanced stubs must be equal to a half of the total admittance. For the input matching network, the total susceptance is $-j1.34$, and then each balanced stub should have $-j0.67$. Thus the length of each side stub must be 0.156λ (0.768 cm). Similarly, for the output matching network, each balanced stub has $-j0.32$ and its length is 0.048λ (0.236 cm).

7-2 HIGH-POWER AMPLIFIER DESIGN

High-power amplifiers are commonly used in electronic systems such as transmitters. For design purposes, a set of large-signal S parameters is usually needed to characterize the device for power applications. However, the measurements of large-signal S parameters are not well defined. An alternative method is to obtain the source and load reflection coefficients in terms of output power and gain at its 1-dB gain compression point.

7-2-1 Gain Compression Point

The 1-dB gain compression point, which is often called G_{1dB} or G_{1dBgcp}, is defined as the power gain at the nonlinear region of the microwave devices reduces 1-dB power gain over the small-signal linear power gain. That is,

$$G_{1dB} = G_{st}(dB) - 1(dB) \tag{7-2-1}$$

where $G_{st}(dB)$ is the small-signal linear power gain in decibels.

In ordinary power measurement, the small-signal power gain is defined as the ratio of the output power over input power. That is,

$$
\begin{aligned}
G_p &= \log \frac{P_{out}}{P_{in}} \qquad dB \\
&= P_{out}(dBm) - P_{in}(dBm) \qquad dB
\end{aligned}
\tag{7-2-2}
$$

Then

$$P_{out}(dBm) = P_{in}(dBm) + G_p(dB) \tag{7-2-3}$$

At the 1-dB gain compression point, the output power can be written as

$$P_{1dB}(dBm) = P_{in}(dBm) + G_p(dB) - 1(dB) \tag{7-2-4}$$

Figure 7-2-1 shows power-gain compression by 1 dB. The linear region represents the power levels between the minimum detectable signal output power $P_{o,mds}$ and P_{1dB}. This linear region is also called the *dynamic range* (DR), where the amplifier has a linear power gain. The low power level of the dynamic range is limited by the noise power level. A minimum detectable input signal $P_{i,mds}$ can be detectable only if its

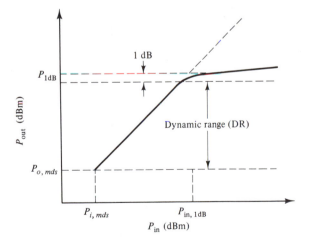

where $P_{o,mds}$ = minimum detectable signal output power
$P_{i,mds}$ = minimum detectable signal input power

Figure 7-2-1 Power-gain compression point.

output power level $P_{o,mds}$ is above the noise power level. The dynamic range is then given by

$$DR = P_{1dB} - P_{o,mds} \tag{7-2-5}$$

7-2-2 Design Example

The design of a high-power amplifier is not easy because the large-signal S parameters are sometime unavailable or unmeasurable. The alternative design method is to utilize the data of source and load reflection coefficients plus the 1-dB gain compression point.

Design Procedure. The design steps for a high-power amplifier can be stated as follows:

1. Check the delta factor Δ and the stability factor K for stability condition.
2. Determine the maximum available power gain.
3. Compute the large-signal source and load matching reflection coefficients.
4. Calculate the actual transducer power gain.
5. Find the required input power.
6. Realize the input matching network.
7. Realize the output matching network.
8. Draw the complete matching networks for the designed high-power amplifier.

Design Example 7-2-1: High-Power Amplifier Design

A power microwave transistor has the following scattering and power parameters measured with a 50-Ω resistance at 3 GHz.

$$S_{11} = 0.62 \; \underline{/140°} \qquad P_{1dB} = 30 \text{ dBm}$$

$$S_{12} = 0.06 \; \underline{/-10°} \qquad G_{1dB} = 12 \text{ dB}$$

$$S_{21} = 2.58 \; \underline{/20°}$$

$$S_{22} = 0.53 \; \underline{/-120°}$$

Design the input and output matching networks of the amplifier for maximum output power as shown in Fig. 7-2-2.

Design Steps

Step 1. Check the delta factor Δ and the stability factor K for stability condition.

$$\Delta = S_{11}S_{22} - S_{12}S_{21}$$

$$= 0.62 \; \underline{/140°} \times 0.53 \; \underline{/-120°} - 0.06 \; \underline{/-10°} \times 2.58 \; \underline{/20°}$$

$$= 0.20 \; \underline{/26.57°}$$

$$|\Delta| = 0.20 < 1$$

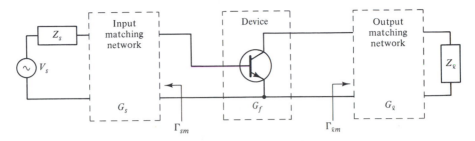

Figure 7-2-2 Desired matching networks for Example 7-2-1.

$$K = \frac{1 + |\Delta|^2 - |S_{11}|^2 - |S_{22}|^2}{2|S_{12}S_{21}|}$$

$$= \frac{1 + |0.20|^2 - |0.62|^2 - |0.53|^2}{2|0.06 \times 2.58|}$$

$$= 1.23 > 1$$

The device is unconditionally stable.

Step 2. Determine the maximum available power gain. From Eq. (3-4-14a), we have

$$G_{a\max} = \frac{|S_{21}|}{|S_{12}|}\left|K - (K^2 - 1)^{1/2}\right|$$

$$= \frac{|2.58|}{|0.06|}\left|1.23 - [(1.23)^2 - 1]^{1/2}\right| = 22.36 = 13.50 \text{ dB}$$

Step 3. Compute the large-signal source and load-matching reflection coefficients Γ_{sm} and $\Gamma_{\ell m}$ from Eqs. (3-5-13) and (3-5-15) for output power at the 1-dB gain compression point. From Eq. (3-5-13), we have

$$\Gamma_{sm} = C_s^*\left[\frac{B_s \pm \sqrt{B_s^2 - 4|C_s|^2}}{2|C_s|^2}\right]$$

where $\begin{aligned} C_s^* &= S_{11}^* - \Delta^* S_{22} \\ &= 0.62\ \underline{/-140°} - 0.20\ \underline{/-26.57°} \times 0.53\ \underline{/-120°} \\ &= 0.51\ \underline{/-138.18°} \end{aligned}$

$\begin{aligned} B_s &= 1 + |S_{11}|^2 - |S_{22}|^2 - |\Delta|^2 \\ &= 1 + |0.62|^2 - |0.53|^2 - |0.20|^2 \\ &= 1.06 \quad \text{positive} \end{aligned}$

$$\sqrt{B_s^2 - 4|C_s|^2} = \sqrt{(1.06)^2 - 4|0.51|^2}$$

$$= 0.28$$

Then

$$\Gamma_{\ell m} = 0.51\ \underline{/-138.18°}\left[\frac{1.06 - 0.28}{2|0.51|^2}\right]$$

$$= 0.77\ \underline{/-138.18°}$$

From Eq. (3-5-15), we have

$$\Gamma_{\ell m} = C_\ell^* \left[\frac{B_\ell \pm \sqrt{B_\ell^2 - 4\,|C_\ell|^2}}{2|C_\ell|^2} \right]$$

where
$$\begin{aligned}
C_\ell^* &= S_{22}^* - \Delta^* S_{11} \\
&= 0.53\,\underline{/120°} - 0.20\,\underline{/-26.57°} \times 0.62\,\underline{/140°} \\
&= 0.40\,\underline{/121°} \\
B_\ell &= 1 + |S_{22}|^2 - |S_{11}|^2 - |\Delta|^2 \\
&= 1 + |0.53|^2 - |0.62|^2 - |0.20|^2 \\
&= 0.86 \qquad \text{positive}
\end{aligned}$$

Then
$$\Gamma_{\ell m} = 0.40\,\underline{/121°} \left[\frac{0.86 - \sqrt{(0.86)^2 - 4|0.40|^2}}{2|0.40|^2} \right]$$

$$= 0.68\,\underline{/121°}$$

Step 4. Calculate the actual transducer power gain under the optimum linear output power condition from Eq. (3-4-5).

$$\begin{aligned}
G_t &= \frac{(1 - |\Gamma_{sm}|^2)\,|S_{21}|^2\,(1 - |\Gamma_{\ell m}|^2)}{|(1 - S_{11}\Gamma_{sm})(1 - S_{22}\Gamma_{\ell m}) - S_{12}S_{21}\Gamma_{sm}\Gamma_{\ell m}|^2} \\[2mm]
&= \frac{(1 - |0.77|^2)|2.58|^2(1 - |0.68|^2)}{\left| \begin{array}{l} (1 - 0.62\underline{/140°} \times 0.77\underline{/-138.18°})(1 - 0.53\underline{/-120°} \times 0.68\underline{/121°} \\ - 0.06\underline{/-10°} \times 2.58\underline{/20°} \times 0.77\underline{/-138.18°} \times 0.68\underline{/121°} \end{array} \right|^2} \\[2mm]
&= 24.50 = 13.9\ \text{dB}
\end{aligned}$$

Step 5. Find the required input power. To obtain an output power if $P_{1dB} = 30$ dBm at the 1-dB gain compression point, the input power from Eq. (7-2-4) must be

$$P_{in}(\text{dBm}) = P_{1dB}(\text{dBm}) - G_{amax}(\text{dB}) + 1\ \text{dB}$$

$$= 30\ \text{dBm} - 13.50 + 1\ \text{dB}$$

$$= 17.50\ \text{dBm}$$

Step 6. Realize the input matching network. Enter $\Gamma_{sm} = 0.77\,\underline{/-138.18°}$ on the Smith chart as shown in Fig. 7-2-3 and read

$$z_{sm} = 0.14 - j0.37 \qquad y_{sm} = 0.90 + j2.35$$

Then

$$Z_{sm} = 7.00 - j18.50 \qquad Y_{sm} = 0.018 + 0.047$$

An open-circuited stub one-eighth-wavelength long looks like a shunt element with admittance $+jY_o$. Here $Y = jY_o \tan(\beta\ell) = jY_o = j0.047$. Two parallel stubs are used to minimize the transition interactions, so $Y = j0.0235$. Each stub has a characteristic impedance

$$Z_{o1} = \frac{1}{0.0235} = 42.55\ \Omega \qquad \text{and} \qquad \ell_1 = 1.25\ \text{cm}$$

An alumina substrate with $\epsilon_r = 10$ is used to support the microstrip lines and

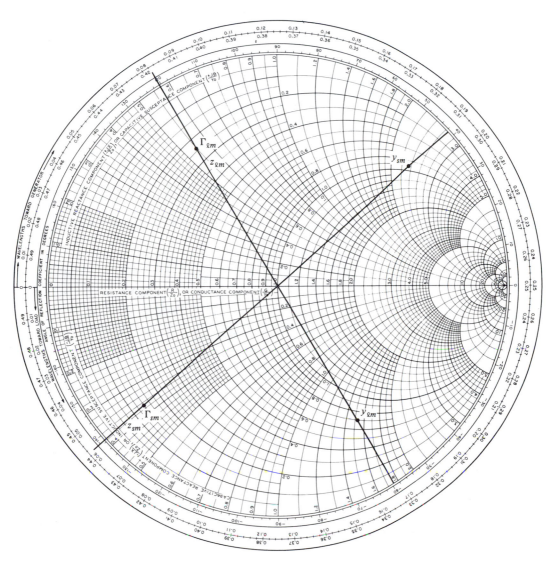

Figure 7-2-3 Graphical solution for Example 7-2-2.

the wavelength is 10 cm at 3 GHz. The length of the open stub is then 1.25 cm.

A quarter-wave transformer that is used to transform the conductance of 0.018 mho to a 50-Ω line must have a characteristic impedance

$$Z_{o2} = \sqrt{\frac{50}{0.018}} = 52.70 \ \Omega \quad \text{and} \quad \ell_2 = 2.50 \text{ cm}$$

Step 7. Realize the output matching network. Enter $\Gamma_{\ell m} = 0.68 \ \underline{/121°}$ on the Smith chart and read

$$z_{\ell m} = 0.24 + j0.58 \qquad y_{\ell m} = 0.70 - j1.55$$

Then

$$Z_{tm} = 12.00 + j29.0 \qquad Y_{tm} = 0.014 - j0.031$$

An open-circuited stub that is three-eighths wavelength long looks like a shunt inductor of a susceptance $-jY_o$. For two stubs,

$$Z_{o1} = \frac{1}{0.0155} = 64.52 \ \Omega \qquad \text{and} \qquad \ell_3 = 3.75 \ \text{cm}$$

A quarter-wave transformer is needed to transform the conductance of 0.014 mho to a 50-Ω line and its characteristic impedance must be

$$Z_{o2} = \sqrt{\frac{50}{0.014}} = 59.76 \ \Omega \qquad \text{and} \qquad \ell_4 = 2.50 \ \text{cm}$$

Step 8. Draw the complete input and output matching networks (Fig. 7-2-4).

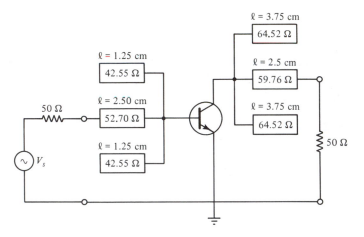

Figure 7-2-4   Complete matching networks for amplifier.

This design example is a tedious work. If a computer-aided design program is available, the design processes should be much easier and more accurate. The contours of the source and load matching reflection coefficients can be plotted on the Smith chart. However, we electronics engineers must understand the basic design processes.

7-3 LOW-NOISE AMPLIFIER DESIGN

As described previously, one important consideration for a high-power amplifier design is the dynamic range, which is defined as the range of the input or output power with linear gain. At a low power level, the dynamic range is limited by the noise figure or the minimum detectable signal P_{mds}; and at a high power level, it is limited by the power level where small-signal gain has been compressed by 1 dB. This was described in Fig. 7-2-1.

7-3-1 Minimum Detectable Power

The input and output minimum detectable signal powers are related to the noise power and they can be expressed as

$$P_{i,mds} = KTB(\text{dBm}) + M(\text{dB}) + B(\text{dB}) + F(\text{dB}) \tag{7-3-1}$$

and

$$P_{o,mds} = KTB(\text{dBm}) + M(\text{dB}) + B(\text{dB}) + F(\text{dB}) + G_p(\text{dB}) \tag{7-3-2}$$

where KTB = $-114\,\text{dBm/MHz}$ (or $-174\,\text{dBm/Hz}$ or $-84\,\text{dBm/GHz}$) is the thermal noise power at 300°K

K = 1.38×10^{-23} J/°K is the Boltzmann constant

T = absolute temperature, degrees Kelvin

B = bandwidth, megahertz

M = 3 dB is the minimum detectable signal power above the thermal noise power

F = noise figure

G_p = power gain

The dynamic range is then given by Eq. (7-2-5) as

$$\text{DR} = P_{1\text{dB}} - P_{o,mds} \tag{7-3-3}$$

Example 7-3-1:   Dynamic Range of High-Power Amplifier

A high-power amplifier has the following parameters:

Available power gain	G_a	= 20 dB
Bandwidth	B	= 8 GHz
Minimum detectable signal power	M	= 3 dB
Noise figure	F	= 7 dB
Operating temperature	T	= 300°K
Power at 1-dB gain compression point	$P_{1\text{dB}}$	= 30 dBm

Compute: (a) the input minimum detectable signal power, (b) the output minimum detectable signal power, and (c) the dynamic range.

Solution   (a) The input minimum detectable signal power is

$$P_{i,mds} = -114\ \text{dBm} + 3\ \text{dB} + 39\ \text{dB} + 7\ \text{dB}$$

$$= -65\ \text{dBm}$$

(b) The output minimum detectable signal power is

$$P_{o,mds} = -114\ \text{dBm} + 3\ \text{dB} + 39\ \text{dB} + 7\ \text{dB} + 20\ \text{dB}$$

$$= -45\ \text{dBm}$$

(c) The dynamic range is

$$\text{DR} = 30\ \text{dBm} + 45\ \text{dBm}$$

$$= 75\ \text{dBm}$$

7-3-2 Design Example

The low-noise amplifier design can be illustrated by the following example.

Design Procedure.   The design of a high-power amplifier with low noise is similar to the low-power amplifier design. The differences are the source reflection coefficient Γ_{sp}, the load reflection coefficient $\Gamma_{\ell p}$, and the noise figure F are measured at the 1-dB gain compression point. Normally, the noise figure is increased when the power level is increased. Figure 7-3-1 shows a circuit diagram to be designed for a low-noise amplifier.

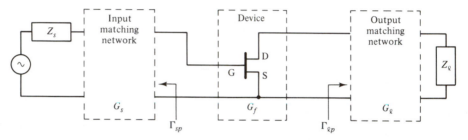

Figure 7-3-1   Matching networks of a low-noise amplifier to be designed.

Design Example 7-3-2:   Low-Noise Amplifier Design

A silicon microwave transistor has the following parameters measured with a 50-Ω resistance at 3 GHz:

Power at 1-dB compression point	$P_{1dB} = 25$ dBm
Source reflection coefficient at P_{1dB}	$\Gamma_{sp} = 0.715 \; \underline{/60°}$
Load reflection coefficient at P_{1dB}	$\Gamma_{\ell p} = 0.612 \; \underline{/-60°}$
Minimum-noise figure at P_{1dB}	$F_{min} = 2$ dB
Optimum source reflection coefficient at F_{min}	$\Gamma_o = 0.485 \; \underline{/155°}$
Noise resistance	$R_n = 4 \; \Omega$

Design the input and output matching networks for the transistor amplifier with a noise figure of 5 dB.

Design Steps:

Step 1. Plot the 4-dB and 5-dB noise circles on the Smith chart as shown in Fig. 7-3-2. Their centers are at

$$c_n = 0.24 \; \underline{/155°} \qquad \text{for 4 dB}$$

$$c_n = 0.18 \; \underline{/155°} \qquad \text{for 5 dB}$$

and radii are at

$$r_n = 0.66 \qquad \text{for 4 dB}$$

$$r_n = 0.76 \qquad \text{for 5 dB}$$

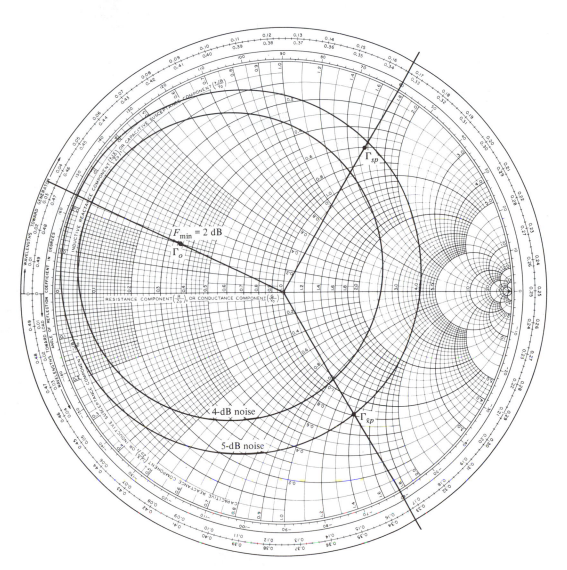

Figure 7-3-2 Graphical solution for Example 7-3-1.

Step 2. Plot the source and load reflection coefficients Γ_{sp}, $\Gamma_{\ell p}$ on the same Smith chart.

Step 3. Check the noise requirement: Since Γ_{sp} and $\Gamma_{\ell p}$ are located around the 5-dB noise circle, the 5-dB noise requirement can be achieved.

Step 4. Determine the input matching network: Calculate the source equivalent impedance from Eq. (4-5-4),

$$Z_{in} = \frac{50(1 - |0.715|^2) + j2 \times 50 \times |0.715| \sin (60°)}{1 + |0.715|^2 - 2|0.715| \cos (60°)}$$

$$= 84.29 \ \underline{/68.41°} = 31.02 + j78.38 \ \Omega$$

$$Y_{in} = \frac{1}{Z_{in}} = 11.86 \times 10^{-3} \ \underline{/-68.41;°}$$

$$= 4.36 \times 10^{-3} - j11.03 \times 10^{-3} \ \text{mho}$$

Use two shunt open-circuited stubs three-eighths-wavelength long to provide $-jY_o$. The characteristic impedance of each stub is

$$Z_{o1} = \frac{1}{5.51 \times 10^{-3}} = 181.32 \ \Omega \qquad \text{and} \qquad \ell_1 = 3.75 \ \text{cm}$$

Use a quarter-wave transformer to transform the 50 Ω of the line to the source equivalent conductance of 4.36×10^{-3} mho. The characteristic impedance of the quarter-wave transformer is

$$Z_{o2} = \sqrt{\frac{50}{4.36 \times 10^{-3}}} = 1\,07.10 \ \Omega \qquad \text{and} \qquad \ell_2 = 2.50 \ \text{cm}$$

Step 5. Find the output matching network: Compute the load equivalent impedance from Eq. (4-5-5),

$$Z_{to} = \frac{50(1 - |0.612|^2) + j2 \times 50 \times |0.612| \sin (-60°)}{1 + |0.612|^2 - 2|0.612| \cos (-60°)}$$

$$= 812 \ \underline{/-69.28°} = 28.70 - j75.87 \ \Omega$$

$$Y_{to} = \frac{1}{Z_{to}} = 12.33 \times 10^{-3}$$

$$= 4.36 \times 10^{-3} + j11.53 \times 10^{-3} \ \text{mho}$$

Similarly, two parallel shunt stubs are used to minimize the transition interactions and the characteristic impedance of each open-circuited stub one-eighth-wavelength long is

$$Z_{o1} = \frac{1}{5.77 \times 10^{-3}} = 173.31 \ \Omega \qquad \text{and} \qquad \ell_3 = 1.25 \ \text{cm}$$

The characteristic impedance of the quarter-wave transformer is

$$Z_{o2} = \sqrt{\frac{50}{4.36 \times 10^{-3}}} = 107.10 \ \Omega \qquad \text{and} \qquad \ell_4 = 2.50 \ \text{cm}$$

It is assumed that microstrip lines on alumina substrate are used. The relative dielectric constant of alumina is 10. For the design of microstrip lines, refer to Chapter 6.

Step 6. Draw the complete matching networks for the low-noise amplifier designed (Fig. 7-3-3).

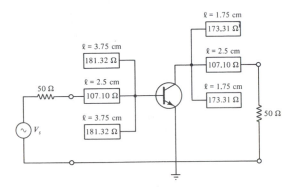

Figure 7-3-3 Matching networks for the amplifier designed.

7-4 BROADBAND AMPLIFIER DESIGN

The design of a broadband microwave amplifier is a problem of the power-gain roll-off between the device and frequency. In broadband amplifier design, *broadband* means that the frequency bandwidth varies with approximate 6 dB per octave. The word *octave* is used in music to denote an interval consisting of eight diatonic degrees. Two notes, one twice the frequency of the other, are an octave apart. For broadband, the octave is associated with two times the frequency. For example, the frequencies are separated by 2, 4, 8, and 16; the bandwidths on the octave scale are 6, 12, 18, and 24 dB apart; and they are separated by 6 dB/octave.

For a broadband amplifier, the input and output impedances of a microwave transistor can usually be represented by a resistor and reactance model. For example, the GaAs-MESFET chip can be represented by a high-pass circuit in the input port and a low-pass circuit in the output port as shown in Fig. 7-4-1(a); and the silicon microwave transistor chip can usually be represented by a low-pass circuit in the input port and a high-pass circuit in the output port, as shown in Fig. 7-4-1(b). In matching circuit design, the parasitic elements of the device must also be taken into account.

In general, there are four types of broadband amplifier design techniques:

1. *Compensated matching techniques*. The compensated matching technique allows a small mismatch of the input and output matching networks to compensate for the variations of the forward transmission parameter of S_{21} with frequency. This method can be achieved by using the Smith chart and will be demonstrated later. If a computer-aided design (CAD) program is available, the design processes should be much easier.

2. *Network synthesis techniques*. The intrinsic input and output impedances (say R, L, and C elements) of a microwave device chip are measurable and available. To match the R, L, and C elements, a well-known network synthesis method, such as Butterworth response or Chebyshev response, can be used analytically. The disadvantage of this method is to use lumped elements for matching pur-

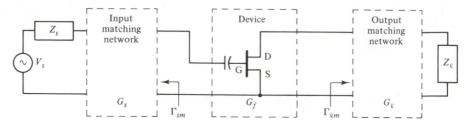

(a) Input and output impedances for a GaAs MESFET chip

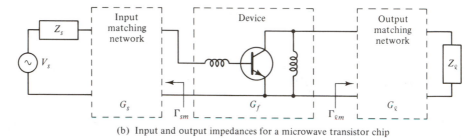

(b) Input and output impedances for a microwave transistor chip

Figure 7-4-1   Input and output impedances of microwave devices.

poses. A few CAD programs such as **AMPSYN** are commercially available. However, this type of matching technique is beyond the scope of this book.

3. *Feedback network techniques.* Feedback network is often used in a broad-band amplifier to provide a flat gain response and to reduce the input and output voltage-standing-wave ratio (VSWR).

4. *Balanced-amplifier techniques.* A balanced amplifier is also commonly used to obtain a broadband amplifier with a flat gain and good input and output VSWR. This type of broadband amplifier was discussed in Chapter 5.

7-4-1 Bandwidth and Quality Factor

Bandwidth BW is a measure of the frequency width between either side of the resonant frequency at the half-power points. Figure 7-4-2 shows the equivalent circuits for a conjugately matched input port of an amplifier.

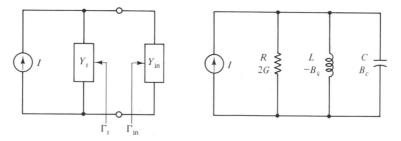

Figure 7-4-2   Equivalent circuits of a matched input port.

Under conjugate matched conditions, $\Gamma_s = \Gamma_{in}^*$ or $Y_s = Y_{in}^*$,

$$Y_s = G - jB_t \qquad (7\text{-}4\text{-}1)$$

$$Y_{in} = G + jB_c \qquad (7\text{-}4\text{-}2)$$

The total admittance of the parallel circuit is

$$Y = Y_s + Y_{in} = 2G + j(B_c - B_t) \qquad (7\text{-}4\text{-}3)$$

At resonant frequency

$$\omega_0 C - \frac{1}{\omega_0 L} = 0$$

Then the resonant frequency is

$$f_0 = \frac{1}{2\pi \sqrt{LC}} \qquad (7\text{-}4\text{-}4)$$

At the lower side, the frequency $f_1 < f_0$,

$$\frac{1}{\omega_1 L} - \omega_1 C = G \qquad \text{and} \qquad \omega_1 = -\frac{G}{2C} + \sqrt{\left(\frac{G}{2C}\right)^2 + \frac{1}{LC}} \qquad (7\text{-}4\text{-}5)$$

At the upper side, the frequency $f_2 > f_0$,

$$\omega_2 C - \frac{1}{\omega_2 L} = G \qquad \text{and} \qquad \omega_2 = \frac{G}{2C} + \sqrt{\left(\frac{G}{2C}\right)^2 + \frac{1}{LC}} \qquad (7\text{-}4\text{-}6)$$

The bandwidth BW is defined as

$$BW = f_2 - f_1 = \frac{\omega_2 - \omega_1}{2\pi} = \frac{G}{2\pi C} \quad \text{Hz} \qquad (7\text{-}4\text{-}7)$$

The resonant frequency is given by

$$f_0 = \sqrt{f_1 f_2} \quad \text{Hz} \qquad (7\text{-}4\text{-}8)$$

The circuit or loaded quality factor at the resonant frequency for a parallel circuit is defined as

$$Q = \frac{\omega_0 C}{G} = \frac{1}{\omega_0 LG} = \omega_0 RC = \frac{R}{\omega_0 L} \qquad (7\text{-}4\text{-}9)$$

Then the bandwidth and the quality factor are related by

$$BW = \frac{G}{2\pi C} = \frac{f_0}{Q} \qquad (7\text{-}4\text{-}10)$$

Applying the optimum conditions to the circuit and substituting Eq. (7-4-9) into Eq. (7-4-10) results in the inherent bandwidth at the input port as

$$BW_{in} = \frac{2f_0 G}{|B|} \qquad (7\text{-}4\text{-}11)$$

where $|B| = \omega_0 C = 1/(\omega_0 L)$

$R = 1/(2G)$ is the resistance at optimum condition

Similarly, the output inherent bandwidth for $Y_\ell = Y_{out}^*$ or $\Gamma_\ell = \Gamma_{out}^*$ is expressed by

$$BW_{out} = \frac{2f_0 G}{|B|} \qquad (7\text{-}4\text{-}12)$$

Example 7-4-1: Bandwidth and Quality Factor

A microwave transistor has the following parameters measured with a 50-Ω resistance at 3 GHz:

Source-matching reflection coefficient $\Gamma_{sm} = 0.77 \,\underline{/-138.2°}$
at P_{1dB}

Load-matching reflection coefficient at $\Gamma_{\ell m} = 0.68 \,\underline{/121°}$
P_{1dB}

Resonant frequency $f_0 = 3$ GHz

Compute the bandwidths at the input and output ports, and the quality factors of the input and output circuits.

Solution

1. From $\Gamma_{sm} = 0.77 \,\underline{/-138.2°}$ on the Smith chart as shown in Fig. 7-4-3, we have

$$y_{sm} = 0.94 + j2.35$$

Then

$$Y_{sm} = 0.018 + j0.047$$

The inherent bandwidth at the input port is

$$BW_{in} = \frac{2 \times 3 \times 10^9 \times 0.018}{|0.047|} = 2.30 \text{ GHz}$$

2. Similarly, from $\Gamma_{\ell m} = 0.68 \,\underline{/121°}$, we have

$$y_{\ell m} = 0.70 - j1.55$$

Then

$$Y_{\ell m} = 0.014 - j0.031$$

The inherent bandwidth at the output port is

$$BW_{out} = \frac{2 \times 3 \times 10^9 \times 0.014}{|0.031|} = 2.71 \text{ Ghz}$$

3. The input capacitance is

$$C_{in} = \frac{B}{\omega_0} = \frac{0.047}{2\pi \times 3 \times 10^9} = 2.49 \text{ pF}$$

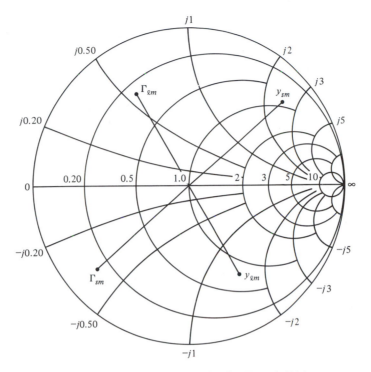

Figure 7-4-3   Graphical solution for Example 7-4-1.

Then the quality factor of the input network is

$$Q_{in} = \frac{\omega_0 C}{G} = \frac{2\pi \times 3 \times 10^9 \times 2.49 \times 10^{-12}}{0.018}$$

$$= 2.61$$

4. The output inductance is

$$L_{out} = \frac{1}{\omega_0 B} = \frac{1}{2\pi \times 3 \times 10^9 \times 0.031} = 1.71 \text{ nH}$$

The quality factor of the output network is

$$Q_{out} = \frac{1}{\omega_0 LG} = \frac{1}{2\pi \times 3 \times 10^9 \times 1.71 \times 19^{-9} \times 0.014}$$

$$= 2.22$$

7-4-2 Design Example

A broadband microwave amplifier can be designed by using the Smith chart. The design procedures are listed below.

Design Procedures

1. Compute the maximum transducer power gain available for the input and output ports at different frequencies.
2. Compensate the power gain for the center frequency.
3. Plot the desired constant output power-gain circles for the output port.
4. Determine the output-matching network.
5. Plot the constant-power-gain circles for the input port.
6. Find the input matching network.
7. Draw the complete matching networks for the broadband amplifier designed.

Design Example 7-4-2:   Broadband Amplifier Design

A GaAs-MESFET NEC70083 has the following S parameters measured at $V_{ds} = 3$ V and $I_{ds} = 30$ mA with a 50-Ω resistance:

4 GHz	6 GHz	8 GHz
$S_{11} = 0.89 \,/\underline{-120°}$	$S_{11} = 0.81 \,/\underline{-103°}$	$S_{11} = 0.73 \,/\underline{-127°}$
$S_{12} = 0.04 \,/\underline{38°}$	$S_{12} = 0.05 \,/\underline{25°}$	$S_{12} = 0.06 \,/\underline{16°}$
$S_{21} = 3.17 \,/\underline{105°}$	$S_{21} = 2.83 \,/\underline{78°}$	$S_{21} = 2.26 \,/\underline{51°}$
$S_{22} = 0.62 \,/\underline{-55°}$	$S_{22} = 0.63 \,/\underline{-73°}$	$S_{22} = 0.70 \,/\underline{-84°}$

Design a broadband amplifier with a bandwidth of 4 GHz from 4 to 8 GHz at the center frequency at 6 GHZ. The power gain must be 9 dB.

Design Steps

Step 1. Compute the maximum transducer power gains.
      For 4 GHz:

$$|S_{12}|^2 = |0.40|^2 \qquad |S_{21}|^2 = |3.17|^2$$
$$= -28 \text{ dB} \qquad\qquad = 10 \text{ dB}$$

$$G_{smax} = \frac{1}{1 - |0.89|^2} \qquad G_{\ell max} = \frac{1}{1 - |0.62|^2}$$
$$= 4.76 = 6.78 \text{ dB} \qquad = 1.61 = 2 \text{ dB}$$

$$G_{umax} = 4.76 \times 10 \times 1.61$$
$$= 76.64 = 18.84 \text{ dB}$$

      For 6 GHz:

$$|S_{12}|^2 = |0.05|^2 \qquad |S_{21}|^2 = |2.83|^2$$
$$\qquad\qquad\qquad\qquad = 9 \text{ dB}$$
$$= -26 \text{ dB}$$

$$G_{smax} = \frac{1}{1 - |0.81|^2} \qquad G_{\ell max} = \frac{1}{1 - |0.63|^2}$$
$$= 2.94 = 4.69 \text{ dB} \qquad = 1.67 = 2.22 \text{ dB}$$

$$G_{u\max} = 2.94 \times 7.95 \times 1.67$$

$$= 33.68 = 15.91 \text{ db}$$

For 8 GHz:

$$|S_{12}|^2 = |0.06|^2 \qquad\qquad |S_{21}|^2 = |2.26|^2$$

$$= -24.4 \text{ dB} \qquad\qquad = 7 \text{ dB}$$

$$G_{s\max} = \frac{1}{1 - |0.73|^2} \qquad\qquad G_{\ell\max} = \frac{1}{1 - |0.70|^2}$$

$$= 2.13 = 3.28 \text{ dB} \qquad\qquad = 1.96 = 3 \text{ dB}$$

$$G_{u\max} = 2.13 \times 5.11 \times 1.96$$

$$= 21.33 = 13.29 \text{ dB}$$

All values are tabulated in Table 7-4-1.

Table 7-4-1   POWER GAINS (dB)

4 GHz	6 GHz	8 GHz
$|S_{21}|^2 = 10$	$|S_{21}|^2 = 9$	$|S_{21}|^2 = 7$
$G_{s\max} = 6.78$	$G_{s\max} = 4.69$	$G_{s\max} = 3.28$
$G_{\ell\max} = 2$	$G_{\ell\max} = 2.22$	$G_{\ell\max} = 3$
$G_{u\max} = 18.78$	$G_{u\max} = 15.91$	$G_{u\max} = 13.28$

Step 2. Determine the constant power gain of 9 dB from 4 to 8 GHz. For 9-dB power gain, the output network should make $G_\ell = -1$ dB at 4 GHz (i.e., to reduce $|S_{21}|^2 = 10$ dB to 9 dB), 0 dB at 6 GHz (no change in $|S_{21}|^2 = 9$ dB), and 2 dB at 8 GHz (to increase $|S_{21}|^2 = 7$ to 9 dB).

	4 GHz	6 GHz	8 GHz
$|S_{21}|^2 =$	10 dB	9 dB	7 dB
	−1 dB	0 dB	+2 dB
Desired:	9 dB	9 dB	9 dB

Step 3. Plot the desired constant-gain circles for the output port on the Smith chart as shown in Fig. 7-4-4.

For 4 GHz:

$$G_\ell = -1 \text{ dB} = 0.79 \qquad\qquad G_{n\ell} = 0.79(1 - |0.62|^2)$$

$$= 0.49$$

$$d_\ell = \frac{0.49 \times 0.62}{1 - 0.38(1 - 0.49)} \qquad\qquad r_\ell = \frac{\sqrt{1 - 0.49} \times 0.62}{0.81}$$

$$= 0.38 \qquad\qquad\qquad = 0.55$$

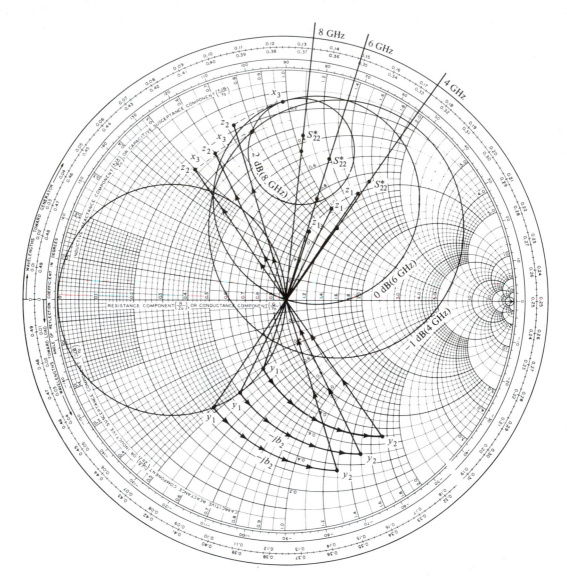

Figure 7-4-4 Graphical solutions of the output matching network.

For 6 GHz

$$G_\ell = 0 \text{ dB} = 1 \qquad g_{n\ell} = 1(1 - |0.63|^2)$$
$$= 0.60$$
$$d_\ell = \frac{0.60 \times 0.63}{1 - 0.4(1 - 0.60)} \qquad r_\ell = \frac{\sqrt{1 - 0.60} \times 0.60}{0.84}$$
$$= 0.45 \qquad\qquad = 0.45$$

For 8 GHz:

$$G_\ell = 2 \text{ dB} = 1.59 \qquad g_{n\ell} = 1.59(1 - |0.70|^2)$$
$$= 0.81$$

$$d_\ell = \frac{0.81 \times 0.70}{1 - 0.49(1 - 0.81)} \qquad r_\ell = \frac{\sqrt{1 - 0.81} \times 0.51}{1 - 0.49(1 - 0.81)}$$
$$= 0.63 \qquad\qquad = 0.24$$

Step 4. Determine the output matching network. By using trial-and-error method, we select

$$z_1 = 1 + jx_1 = 1 + j1.00 \qquad \text{for 6 GHz from the Smith chart}$$

as shown in Fig. 7-4-4. Then the first element from the load is a series inductor. That is,

$$jx_1 = j1.00 \qquad \text{series inductor}$$

Then read $y_1 = 0.50 - j0.50$. To intercept the gain circle of 0 dB at 6 GHz, we add $-jb_2 = -j1.00$ to y_1 and move y_1 to y_2. Read $y_2 = 0.50 - j1.50$. Then the second element is a shunt inductor. That is,

$$-jb_2 = -j1.00 \qquad \text{shunt inductor}$$

Then read $z_2 = 0.20 + j0.58$ from the Smith chart. We add $jx_3 = j0.20$ to jx_2 and move z_2 to the gain circle of 0 dB at 6 GHz. The third element is a series inductor.

$$jx_3 = j0.20 \qquad \text{series inductor}$$

Similarly, the elements for 4 GHz and 8 GHz are also found. Because there are three gain conditions, there should be three elements in the output matching network. The values of the desired elements are tabulated in Table 7-4-2.

TABLE 7-4-2 VALUES OF DESIRED ELEMENTS FOR THE OUTPUT MATCHING NETWORK

4 GHz	6 GHz	8 GHz
$+ jx_1 = + j0.67$	$+jx_1 = + j1.00$	$+jx_1 = +j1.33$
$-jb_2 = -j1.50$	$-jb_2 = -j1.00$	$-jb_2 = -j0.75$
$+jx_3 = +j0.14$	$+jx_3 = +j0.20$	$+jx_3 = +j0.27$

The values of the inductor elements for the output-matching network are calculated as follows:

At 4 GHz:

$$L_1 = \frac{0.67 \times 50}{2\pi \times 4 \times 10^9} = 1.33 \text{ nH} \qquad \text{series inductor}$$

$$L_2 = \frac{50}{2\pi \times 4 \times 10^9 \times 1.50} = 1.33 \text{ nH} \qquad \text{shunt inductor}$$

$$L_3 = \frac{0.14 \times 50}{2\pi \times 4 \times 10^9} = 0.27 \text{ nH} \qquad \text{series inductor}$$

At 6 GHz:

$$L_1 = \frac{1.00 \times 50}{2\pi \times 6 \times 10^9} = 1.33 \text{ nH} \qquad \text{series inductor}$$

$$L_2 = \frac{50}{2\pi \times 6 \times 10^9 \times 1.00} = 1.33 \text{ nH} \qquad \text{shunt inductor}$$

$$L_3 = \frac{0.20 \times 50}{2\pi \times 6 \times 10^9} = 0.27 \text{ nH} \qquad \text{series inductor}$$

At 8 GHz:

$$L_1 = \frac{1.33 \times 50}{2\pi \times 8 \times 10^9} = 1.33 \text{ nH} \qquad \text{series inductor}$$

$$L_2 = \frac{50}{2\pi \times 8 \times 10^9 \times 0.75} = 1.33 \text{ nH} \qquad \text{shunt inductor}$$

$$L_3 = \frac{0.27 \times 50}{2\pi \times 8 \times 10^9} = 0.27 \text{ nH} \qquad \text{series inductor}$$

Step 5. Plot the desired constant-gain circles for the input port on the Smith chart as shown in Fig. 7-4-5.

For 4 GHz:

$$G_s = -1 \text{ dB} = 0.79 \qquad\qquad g_{ns} = 0.79(1 - |0.89|^2)$$
$$= 0.17$$

$$d_s = \frac{0.17|0.89|}{1 - |0.89|^2(1 - 0.17)} \qquad r_s = \frac{\sqrt{1 - 0.17}\,(1 - |0.89|^2)}{1 - |0.89|^2(1 - 0.17)}$$
$$= 0.43 \qquad\qquad\qquad = 0.56$$

For 6 GHz:

$$G_s = 0 \text{ dB} = 1 \qquad\qquad g_{ns} = 1(1 - |0.81|^2)$$
$$= 0.34$$

$$d_s = \frac{0.34|0.81|}{1 - |0.81|^2(1 - 0.34)} \qquad r_s = \frac{\sqrt{1 - 0.34}\,(1 - |0.81|^2)}{1 - |0.81|^2\,(1 - 0.34)}$$
$$= 0.49 \qquad\qquad\qquad = 0.49$$

For 8 GHz:

$$G_s = 2 \text{ dB} = 1.59 \qquad\qquad g_{ns} = 1.59(1 - |0.73|^2)$$
$$= 0.75$$

$$d_s = \frac{0.75|0.73|}{1 - |0.73|^2(1 - 0.75)} \qquad r_s = \frac{\sqrt{1 - 0.75}\,(1 - 0.73|^2)}{1 - |0.73|^2(1 - 0.75)}$$
$$= 0.63 \qquad\qquad\qquad = 0.27$$

Step 6. Find the input matching network. We plot the power-gain circles for 4, 6, and 8 GHz on the Smith impedance chart as shown in Fig. 7-4-5, and then transform them to the Smith admittance chart as shown in Fig. 7-4-6. This transformation is necessary because the input network is out of phase with the output network by 180°. By using the cut-and-try method, we choose $y_1 = 1 - j1.55$ for 6 GHz from the admittance chart. The first element from the source is a shunt inductor.

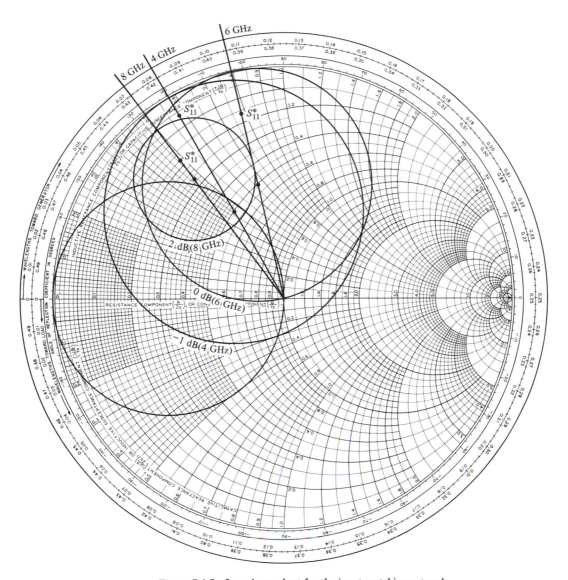

Figure 7-4-5 Impedance chart for the input matching network.

That is,

$$-jb_1 = -j1.55 \qquad \text{shunt inductor}$$

Then read $z_1 = 0.30 + j0.46$. To intersect the 0-dB gain circle of 6 GHz, we add $jx_2 = j0.24$ to z_1 and move z_1 to z_2. Then read $z_2 = 0.30 + j0.70$. Hence the second element is a series inductor.

$$jx_2 = j0.24 \qquad \text{series inductor}$$

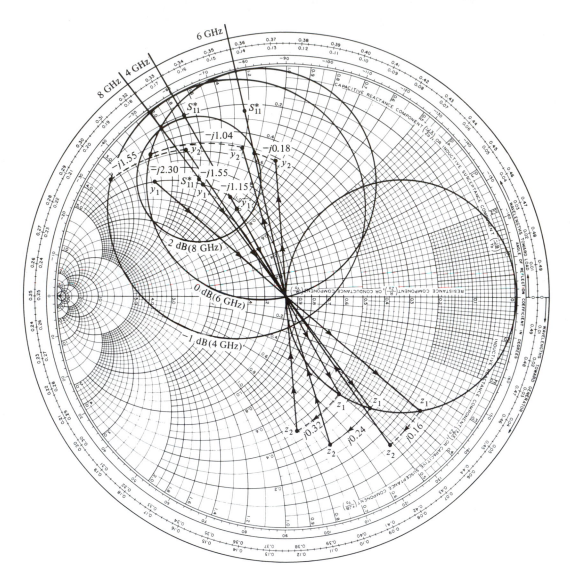

Figure 7-4-6 Admittance chart for the input matching network.

Read $y_2 = 0.53 - j1.22$ from the admittance chart. We add $-jb_3 = -j1.04$ to y_2 and move y_2 to the 0-dB gain circle of 6 GHz. The third element is a shunt inductor. That is,

$$-jb_3 = -j1.04 \qquad \text{shunt inductor}$$

Similarly, the elements for 4 GHz and 8 GHz are also found. Table 7-4-3 tabulates the values of the elements.

TABLE 7-4-3 VALUES OF DESIRED ELEMENTS FOR THE INPUT
MATCHING NETWORK

4 GHz	6 GHz	8 GHz
$-jb_1 = -j2.30$	$-jb_1 = -j1.55$	$-jb_1 = -j1.15$
$+jx_2 = +j0.16$	$+jx_2 = +j0.24$	$+jx_2 = +j0.32$
$-jb_3 = -j1.55$	$-jb_3 = -j1.04$	$-jb_3 = -j0.78$

The values of the inductor elements for the input matching network are calculated as follows:

At 4 GHz:

$$L_1 = \frac{50}{2\pi \times 4 \times 10^9 \times 2.30} = 0.86 \text{ nH} \qquad \text{shunt inductor}$$

$$L_2 = \frac{0.16 \times 50}{2\pi \times 4 \times 10^9} = 0.32 \text{ nH} \qquad \text{series inductor}$$

$$L_3 = \frac{50}{2\pi \times 4 \times 10^9 \times 1.55} = 1.28 \text{ nH} \qquad \text{shunt inductor}$$

At 6 GHz:

$$L_1 = \frac{50}{2\pi \times 6 \times 10^9 \times 1.55} = 0.86 \text{ nH} \qquad \text{shunt inductor}$$

$$L_2 = \frac{0.24 \times 50}{2\pi \times 6 \times 10^9} = 0.32 \text{ nH} \qquad \text{series inductor}$$

$$L_3 = \frac{50}{2\pi \times 6 \times 10^9 \times 1.04} = 1.28 \text{ nH} \qquad \text{shunt inductor}$$

At 8 GHz:

$$L_1 = \frac{50}{2\pi \times 8 \times 10^9 \times 1.15} = 0.86 \text{ nH} \qquad \text{shunt inductor}$$

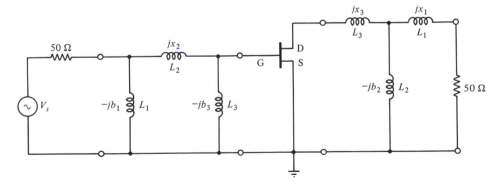

Figure 7-4-7 Complete matching networks.

$$L_2 = \frac{0.32 \times 50}{2\pi \times 8 \times 10^9} = 0.32 \text{ nH} \qquad \text{series inductor}$$

$$L_3 = \frac{50}{2\pi \times 8 \times 10^9 \times 0.78} = 1.28 \text{ nH} \qquad \text{shunt inductor}$$

Step 7. Draw the complete matching networks for the broadband amplifier designed (Fig. 7-4-7).

7-5 FEEDBACK TECHNIQUES

The feedback network is often used in microwave circuits to compensate for the variations of $|S_{21}|$ with frequency. There are two types of feedback network: series feedback and shunt feedback. The series feedback can improve the input match S_{11} over frequency, and the shunt feedback can flatten the power gain S_{21} over frequency. Figure 7-5-1 shows the series feedback and shunt (or parallel) feedbacks of GaAs MESFET and microwave transistor, respectively.

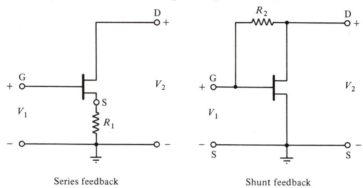

(a) GaAs MESFET with feedbacks

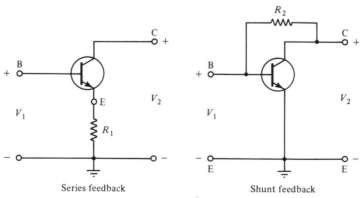

(b) Microwave transistor with feedbacks

Figure 7-5-1 Broadband amplifier with feedback network.

The intrinsic impedances and extrinsic feedback resistors of microwave solid-state devices can be represented by their equivalent circuits as shown in Fig. 7-5-2. Both the series feedback and shunt feedback can yield negative resistance for oscillation.

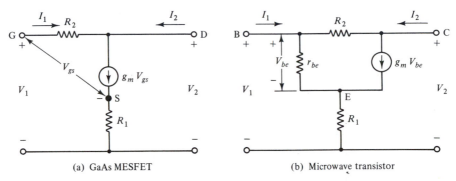

(a) GaAs MESFET   (b) Microwave transistor

Figure 7-5-2   Feedback equivalent circuits for solid-state devices.

From y-parameter theory, the admittance matrix for the network as shown in Fig. 7-5-2(a) can be written as

$$[I] = [y][V] \tag{7-5-1}$$

or

$$\begin{bmatrix} I_1 \\ I_2 \end{bmatrix} = \begin{bmatrix} y_{11} & y_{12} \\ y_{21} & y_{22} \end{bmatrix} \begin{bmatrix} V_1 \\ V_2 \end{bmatrix} \tag{7-5-2}$$

where

$$y_{11} = \frac{1}{R_2} \tag{7-5-3}$$

$$y_{12} = -\frac{1}{R_2} \tag{7-5-4}$$

$$y_{21} = \frac{g_m}{1 + g_m R_1} - \frac{1}{R_2} \tag{7-5-5}$$

$$y_{22} = \frac{1}{R_2} \tag{7-5-6}$$

The y parameters can be converted to the S parameters for the design purpose. Then the S parameters can be expressed in terms of the resistances and transconductance as follows [1]:

$$S_{11} = S_{22} = \frac{1}{D} \left[1 - \frac{g_m Z_0^2}{R_2(1 + g_m R_1)} \right] \tag{7-5-7}$$

$$S_{12} = \frac{2Z_0}{DR_2} \tag{7-5-8}$$

$$S_{21} = \frac{1}{D}\left(\frac{-2g_mZ_0}{1 + g_mR_1} + \frac{2Z_0}{R_2}\right) \tag{7-5-9}$$

where

$$D = 1 + \frac{2Z_0}{R_2} + \frac{g_mZ_0^2}{R_2(1 + g_mR_1)} \tag{7-5-10}$$

For a perfect voltage-standing-wave ratio in the input and output port, VSWR = 1, the required condition is $S_{11} = S_{22} = 0$. Then

$$1 + g_mR_1 = \frac{g_mZ_0^2}{R_2} \tag{7-5-11}$$

or

$$R_1 = \frac{Z_0^2}{R_2} - \frac{1}{g_m} \tag{7-5-12}$$

When $g_m \gg 1$, the feedback resistors become

$$Z_0 = \sqrt{R_1R_2} \tag{7-5-13}$$

Substitution of Eq. (7-5-12) into Eqs. (7-5-8) and (7-5-9) yields

$$S_{12} = \frac{Z_0}{Z_0 + R_2} \tag{7-5-14}$$

and

$$S_{21} = \frac{Z_0 - R_2}{Z_0} \tag{7-5-15}$$

If $|R_2| > Z_0$, $|S_{21}|$ is negative. Equation (7-5-15) indicates that the S_{21} parameter depends only on R_2, not on the other S parameters. Therefore, the gain flattening can be accomplished by using parallel feedback over a broadband frequency. Then

$$R_2 = Z_0(1 + |S_{21}|) \qquad \text{for } |R_2| > Z_0 \tag{7-5-16}$$

From Eq. (7-5-12), there is some minimum transconductance g_m to make R_1 to be zero for a fixed R_2. That is,

$$g_{m,\text{min}} = \frac{R_2}{Z_0^2} \tag{7-5-17}$$

$$= \frac{1 + |S_{21}|}{Z_0} \qquad \text{for } |R_2| > Z_0 \tag{7-5-18}$$

Example 7-5-1:   Feedback Network for Flat Gain over Broadband Frequency

A broadband microwave transistor amplifier is to be designed to have a power gain of 12 dB with a 50-Ω resistance. Compute (a) the required minimum transconductance $g_{m,\text{min}}$, (b) the required shunt feedback resistor R_2 for a flat power gain of 12 dB, and (c) the transducer power gain.

Solution   (a) The required minimum transconductance is

$$g_{m,\text{min}} = \frac{1 + |S_{21}|}{Z_0} = \frac{1 + |3.95|}{50} = 0.099 \text{ mho}$$

(b) The required shunt feedback resistor is

$$R_2 = Z_0(1 + |S_{21}|) = 50(1 + |3.95|)$$
$$= 247.50 \ \Omega$$

(c) The transducer power gain is

$$|S_{21}|^2 = \left| \frac{Z_0 - R_2}{Z_0} \right|^2 = \left| \frac{50 - 247.50}{50} \right|^2$$
$$= 12 \text{ dB}$$

REFERENCES

1. Johnson, K. M., Large signal GaAs MESFET oscillator design. *IEEE Trans. Microwave Theory and Techniques*, Vol. MTT-27, No. 3, March 1979, pp. 217–227.

PROBLEMS

7-1 Large-Signal Amplifier Design

7-1-1. The DXL-3608A X-band power GaAs FET (chip) biased for large signal at $V_{ds} = 5$ V and $I_{ds} = 115$ mA at 10 GHz has the following S parameters with a 50-Ω line:

$$S_{11} = 0.83 \ \underline{/-154°}$$

$$S_{12} = 0.05 \ \underline{/102°}$$

$$S_{21} = 1.51 \ \underline{/53°}$$

$$S_{22} = 0.61 \ \underline{/-75°}$$

Design the input and output matching networks of the amplifier by using the said FET for a power gain of 15 dB.
(a) Check the delta factor Δ and the stability factor K.
(b) Calculate the maximum stable gain.
(c) Plot the load stability circle.
(d) Plot the desired 15-dB power-gain circle.

(e) Select the load reflection coefficient.

(f) Plot the source stability circle.

(g) Determine the input and output matching networks.

(h) Draw the complete matching networks of the amplifier designed.

7-1-2. A certain GaAs MESFET is to be used for a large-signal amplifier and its S parameters measured with a large signal at $V_{ds} = 9$ V and $I_{ds} = 500$ mA at 8 GHz with a 50-Ω line are as follows:

$$S_{11} = 0.754 \ \underline{/-51°}$$

$$S_{12} = 0.025 \ \underline{/65°}$$

$$S_{21} = 2.574 \ \underline{/120°}$$

$$S_{2} = 0.732 \ \underline{/-32°}$$

Design the input and output matching networks of the amplifier for a 20-dB power gain. (See Problem 7-1-1 for the design steps.)

7-2 High-Power Amplifier Design

7-2-1. A certain GaAs MESFET has the following scattering and power parameters measured with a 50-Ω resistance at 4 GHz:

$$S_{11} = 0.65 \ \underline{/135°} \qquad P_{1dB} = 35 \text{ dBm}$$

$$S_{12} = 0.05 \ \underline{/-12°} \qquad G_{1dB} = 14 \text{ dB}$$

$$S_{21} = 2.65 \ \underline{/25°}$$

$$S_{22} = 0.54 \ \underline{/-115°}$$

Design the input and output matching networks of the amplifier by using the said MESFET.

(a) Check the delta factor Δ and the stability factor K.

(b) Calculate the maximum available power gain.

(c) Compute the large-signal source and load matching reflection coefficients.

(d) Find the transducer power gain.

(e) Determine the required input power.

(f) Find the input and output matching networks.

(g) Draw the complete circuits for the designed amplifier.

7-2-2. A microwave power transistor has the following parameters measured with a large signal at 3 GHz with a 50-Ω line:

S parameters	Power parameters
$S_{11} = 0.67 \ \underline{/140°}$	$P_{1dB} = 30$ dBm
$S_{12} = 0.04 \ \underline{/-15°}$	$G_{1dB} = 10$ dB
$S_{21} = 2.32 \ \underline{/30°}$	
$S_{22} = 0.46 \ \underline{/-125°}$	

Design the input and output matching networks of the amplifier by using the said microwave transistor (refer to Problem 7-2-1 for the design steps.)

7-3 Low-Noise Amplifier Design

7-3-1. A high-power amplifier has the following parameters:

Available power gain	$G_a = 18$ dB
Bandwidth	$B = 6$ GHz
Minimum detectable signal power	$M = 3$ dB
Noise figure	$F = 6$ dB
Operating temperature	$T = 300\,°$K
Power at 1-dB gain compression point	$P_{1dB} = 25$ dBm

Compute:

(a) The input minimum detectable signal.
(b) The output minimum detectable signal power.
(c) The dynamic range.

7-3-2. Design a low-noise amplifier by using a GaAs MESFET; the FET has the following parameters measured with a 50-Ω resistance at 8 GHz:

Power at 1-dB compression point	$P_{1dB} = 20$ dbm
Source reflection coefficient at P_{1dB}	$\Gamma_{sp} = 0.732\ \underline{/50°}$
Load reflection coefficient at P_{1dB}	$\Gamma_{\ell p} = 0.625\ \underline{/-50°}$
Minimum-noise figure at P_{1dB}	$F_{min} = 2$ dB
Optimum source reflection coefficient at F_{min}	$\Gamma_o = 0.512\ \underline{/160°}$
Noise resistance	$R_n = 5\ \Omega$

Design the input and output matching networks for the amplifier with a low-noise figure of about 5 dB.

(a) Plot the 4-dB and 5-dB constant-noise circles on a Smith chart.
(b) Plot the source and load reflection coefficients on the same Smith chart.
(c) Determine the input and output matching networks.
(d) Draw the complete circuit diagam for the amplifier designed.

7-3-3. A microwave power transistor has the following parameters measured with a 50-Ω resistance at 4 GHz:

Power at 1-dB compression point	$P_{1dB} = 30$ dBm
Source reflection coefficient at P_{1dB}	$\Gamma_{sp} = 0.593\ \underline{/70°}$
Load reflection coefficient at P_{1dB}	$\Gamma_{\ell p} = 0.655\ \underline{/-70°}$
Minimum-noise figure at P_{1dB}	$F_{min} = 3$ dB
Optimum source reflection coefficient at F_{min}	$\Gamma_o = 0.553\ \underline{/170°}$
Noise resistance	$R_n = 10\ \Omega$

Design the input and output matching networks for the transistor amplifier with a low-noise figure of about 8 dB. (Refer to Problem 7-3-2 for the design steps.)

7-4 Broadband Amplifier Design

7-4-1. A GaAs MESFET has the following parameters measured with a 50-Ω resistance at 10 GHz:

Source-matching reflection coefficient at P_{1dB}	$\Gamma_{sm} = 0.694\ \underline{/-140°}$
Load-matching reflection coefficient at P_{1dB}	$\Gamma_{\ell m} = 0.732\ \underline{/131°}$
Resonant frequency	$f_0 = 10$ GHz

Calculate

(a) The input and output bandwidths.

(b) The input-port and output-port quality factors.

7-4-2. The NE868898 broadband power GaAs MESFET has the following S parameters measured at $V_{ds} = 9$ V and $I_{ds} = 1.2$ A with a 50-Ω reference:

5.4 GHz	7 GHz	8.4 GHz
$S_{11} = 0.32 \underline{/-110°}$	$S_{11} = 0.42 \underline{/-100°}$	$S_{11} = 0.75 \underline{/-130°}$
$S_{12} = 0.11 \underline{/25°}$	$S_{12} = 0.13 \underline{/-10°}$	$S_{12} = 0.14 \underline{/40°}$
$S_{21} = 2.86 \underline{/95°}$	$S_{21} = 2.51 \underline{/25°}$	$S_{21} = 2.00 \underline{/-60°}$
$S_{22} = 0.68 \underline{/-30°}$	$S_{22} = 0.51 \underline{/-70°}$	$S_{22} = 0.67 \underline{/-150°}$

Design a broadband amplifier with a bandwidth of 3 GHz from 5.4 to 8.4 GHz at the center frequency of 7 GHz. The power gain should be 8 dB.

(a) Compute the maximum transducer power gains.

(b) Determine the constant power gain of 8 dB from 5.4 to 8.4 GHz.

(c) Plot the desired constant-gain circles for the output port and determine the output matching network.

(d) Plot the desired constant-gain circles for the input port and determine the input matching network.

(e) Draw the complete matching networks for the broadband amplifier designed.

7-4-3. The DXL-3501A-P100F broadband power GaAs MESFET has the following S parameters measured at $V_{ds} = 8$ V and $I_{ds} = 0.5 I_{dss}$ with a 50-Ω reference:

8 GHz	10 GHz	12 GHz
$S_{11} = 0.69 \underline{/-100°}$	$S_{11} = 0.67 \underline{/-105°}$	$S_{11} = 0.68 \underline{/-140°}$
$S_{12} = 0.07 \underline{/-17°}$	$S_{12} = 0.06 \underline{/-15°}$	$S_{12} = 0.06 \underline{/-9°}$
$S_{21} = 2.23 \underline{/12°}$	$S_{21} = 2.00 \underline{/-3°}$	$S_{21} = 1.59 \underline{/-16°}$
$S_{22} = 0.76 \underline{/-107°}$	$S_{22} = 0.81 \underline{/-115°}$	$S_{22} = 0.81 \underline{/-119°}$

Design a broadband amplifier with a bandwidth of 4 GHz from 8 to 12 GHz at the center frequency of 10 GHz. The power gain should be about 6 dB. (Refer to Problem 7-4-2 for the design steps.)

7-5 Feedback Techniques

7-5-1. Design a broadband microwave amplifier by using feedback technique with a power gain of 10 dB.

Calculate:

(a) The required minimum transconductance $g_{m,min}$.

(b) The required shunt feedback resistance R_2 for a flat power gain of 10 dB.

(c) The transducer power gain.

7-5-2. Verify Eq. (7-5-15).

Chapter 8

Microwave Waveguides and Reflection Amplifier Design

8-0 INTRODUCTION

In general, a waveguide consists of a hollow metallic tube of a rectangular or circular shape used to guide an electromagnetic wave. Waveguides are used principally at microwave frequencies; inconveniently large guides would be required to transmit signal power at longer wavelengths. In waveguides the electric and magnetic fields are confined to the space within the guides. So no power is lost through radiation, and even the dielectric loss is negligible, since the guides are normally air filled. However, there is some power loss as heat in the walls of the guides, but the loss is very small.

It is possible to propagate several modes of electromagnetic waves within a waveguide. These modes correspond to solutions of Maxwell's equations for the particular waveguides. A given waveguide has a definite cutoff frequency for each allowed mode. If the frequency of the impressed signal is above the cutoff frequency for a given mode, the electromagnetic energy can be transmitted down the guide for that particular mode without attenuation. Otherwise, the electromagnetic energy with a frequency below the cutoff frequency for that particular mode will be attenuated to a negligible value in a relative short distance. The dominant mode in a particular guide is the mode having the lowest cutoff frequency. It is advisable to choose the dimensions of a guide in such a way that for a given input signal, only the energy of the dominant mode can be transmitted through the guide.

In the past decade, due to the availability of negative resistance from microwave devices such as Gunn diodes and IMPATT diodes, reflection amplifiers can be designed and built by use of negative-resistance devices incorporated with microwave waveguides, coaxial lines, or hybrid couplers. The purpose of this chapter is to deal with the microwave waveguides and reflection amplifier design.

8-1 RECTANGULAR WAVEGUIDES

A rectangular waveguide is a hollow metallic tube with a rectangular cross section. The conducting walls of the guide serve to confine the fields and, thereby, guide the electromagnetic wave. A number of distinct field configurations or modes can exist in waveguides. When the waves travel longitudinally down the guide, the plane waves are reflected from wall to wall. This results in a component of either electric or magnetic field in the direction of propagation of the resultant wave; therefore, the wave is no longer a TEM wave. Figure 8-1-1 shows that any uniform plane wave in a lossless guide may be resolved into TE and TM waves.

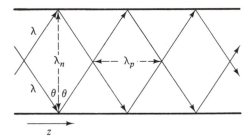

Figure 8-1-1 Plane wave reflected in a waveguide.

It is clear that when the wavelength λ is in the direction of propagation of the incident wave, there will be one component λ_n in the direction normal to the reflecting plane and another λ_p parallel to the plane. These components are

$$\lambda_n = \frac{\lambda}{\cos \theta} \tag{8-1-1a}$$

$$\lambda_p = \frac{\lambda}{\sin \theta} \tag{8-1-1b}$$

where θ = angle of incidence

λ = wavelength of the impressed signal in unbounded medium

This is that the plane wave resolves into two components: one standing wave in the direction normal to the reflecting walls of the guide, and one traveling wave in the direction parallel to the reflecting walls. In lossless waveguides, the modes may be classified as either transverse electric (TE) or transverse magnetic (TM) modes. In rectangular guides, the modes are designated TE_{mn} or TM_{mn}. The integer m denotes the number of half waves of electric or magnetic intensity in the x direction, while n is the number of half waves in the y direction, if the direction of propagation of the wave is assumed in the positive z direction.

The purpose of this chapter is to study the wave modes, power transmission, and power losses in rectangular waveguides.

8-1-1 Solutions of Wave Equations

From wave-propagating theory, there are time-domain and frequency-domain solutions for each wave equation. However, for the simplicity of the solution to the wave equation in three dimensions plus a time-varying variable, only the sinusoidal steady-state or the frequency-do-main solution will be solved. A rectangular coordinate system is shown in Fig. 8-1-2.

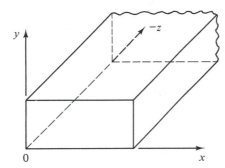

Figure 8-1-2   Rectangular coordinates.

The electric and magnetic wave equations in the frequency domain are given by

$$\nabla^2 \mathbf{E} = \gamma^2 \mathbf{E} \tag{8-1-2a}$$

$$\nabla^2 \mathbf{H} = \gamma^2 \mathbf{H} \tag{8-1-2b}$$

where $\gamma = \sqrt{j\omega\mu(\sigma + j\omega\epsilon)} = \alpha + j\beta$. These are called the *vector wave equations.*
    Rectangular coordinates are the usual right-hand system. The rectangular components of $\mathbf{E}$ or $\mathbf{H}$ satisfy the complex scalar wave equation or Helmholtz equation:

$$\nabla^2 \psi = \gamma^2 \psi \tag{8-1-3}$$

The Helmholtz equation in rectangular coordinates is

$$\frac{\partial^2 \psi}{\partial x^2} + \frac{\partial^2 \psi}{\partial y^2} + \frac{\partial^2 \psi}{\partial z^2} = \gamma^2 \psi \tag{8-1-4}$$

This is a linear and inhomogeneous partial differential equation in three dimensions. By the method of separation of variables, the solution is assumed in the form of

$$\psi = X(x)Y(y)Z(z) \tag{8-1-5}$$

where $X(x)$ = function of the x coordinate only
    $Y(y)$ = function of the y coordinate only
    $Z(z)$ = function of the z coordinate only

Substitution of Eq. (8-1-5) in Eq. (8-1-4) and division of the resultant by Eq. (8-1-5) yields

$$\frac{1}{X}\frac{\partial^2 X}{\partial x^2} + \frac{1}{Y}\frac{\partial^2 Y}{\partial y^2} + \frac{1}{Z}\frac{\partial^2 Z}{\partial z^2} = \gamma^2 \tag{8-1-6}$$

Since the sum of the three terms on the left-hand side is a constant and each term is independently variable, it follows that each term must be equal to a constant. Let the three terms be k_x^2, k_y^2, and k_z^2, respectively; then the separation equation is given by

$$-k_x^2 - k_y^2 - k_z^2 = \gamma^2 \tag{8-1-7}$$

The general solution of each differential equation

$$\frac{\partial^2 X}{\partial x^2} = -k_x^2 X \tag{8-1-8}$$

$$\frac{\partial^2 Y}{\partial y^2} = -k_y^2 Y \tag{8-1-9}$$

$$\frac{\partial^2 Z}{\partial z^2} = -k_z^2 Z \tag{8-1-10}$$

will be in the form of

$$X = A \sin(k_x x) + B \cos(k_x x) \tag{8-1-11}$$

$$Y = C \sin(k_y y) + D \cos(k_y y) \tag{8-1-12}$$

$$Z = E \sin(k_z z) + F \cos(k_z z) \tag{8-1-13}$$

The total solution of the Holmholtz equation in rectangular coordinates is

$$\psi = [A \sin(k_x x) + B \cos(k_x x)][C \sin(k_y y) + D \cos(k_y y)]$$
$$[E \sin(k_z z) + F \cos(k_z z)] \tag{8-1-14}$$

The direction of propagation of the wave in the guide is conventionally assumed in the positive z direction. It should be noted that the propagation constant γ_g in the guide differs from the intrinsic propagation constant γ of the dielectric. Let

$$\gamma_g^2 = \gamma^2 + k_x^2 + k_y^2 = \gamma^2 + k_c^2 \tag{8-1-15}$$

where $k_c = \sqrt{k_x^2 + k_y^2}$ is usually called the *cutoff wave number*. For a lossless dielectric, $\gamma^2 = -\omega^2\mu\epsilon$; then

$$\gamma_g = \pm j\sqrt{\omega^2\mu\epsilon - k_c^2} \tag{8-1-16}$$

There are three cases for the propagation constant, γ_g, in the waveguide.

Case 1. There will be no wave propagation (evanescence) in the guide if $\omega_c^2 \mu\epsilon = k_c^2$ and $\gamma_g = 0$. This is the critical condition for cutoff propagation. The cutoff frequency is expressed as

$$f_c = \frac{1}{2\pi\sqrt{\mu\epsilon}}\sqrt{k_x^2 + k_y^2} \qquad (8\text{-}1\text{-}17)$$

Case 2. The wave will be propagating in the guide if $\omega^2\mu\epsilon > k_c^2$ and

$$\gamma_g = \pm j\beta_g = \pm j\omega\sqrt{\mu\epsilon}\sqrt{1 - \left(\frac{f_c}{f}\right)^2} \qquad (8\text{-}1\text{-}18a)$$

This means that the operating frequency must be above the cutoff frequency for a wave to propagate in the guide.

Case 3. The wave will be attenuated if $\omega^2\mu\epsilon < k_c^2$ and

$$\gamma_g = \pm \alpha_g = \pm\omega\sqrt{\mu\epsilon}\sqrt{\left(\frac{f_c}{f}\right)^2 - 1} \qquad (8\text{-}1\text{-}18b)$$

This means that if the operating frequency is below the cutoff frequency, the wave will be decayed exponentially with respect to a factor of $-\alpha_g z$, and there will be no wave propagation because the propagation constant is a real quantity. Therefore, the traveling-wave solution of the Holmholtz equation in rectangular coordinates is given by

$$\psi = [A \sin(k_x x) + B \cos(k_x x)][C \sin(k_y y) + D \cos(k_y y)]e^{-j\beta_g z}$$

$$(8\text{-}1\text{-}19)$$

8-1-2 TE Modes in Rectangular Waveguides

It has previously been assumed that the waves are propagating in the positive z direction in the waveguide. Figure 8-1-3 shows the coordinates of a rectangular waveguide for TE modes. The TE_{mn} modes in a rectangular guide are characterized by

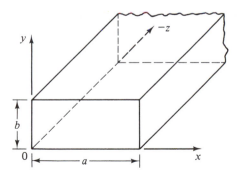

Figure 8-1-3 Coordinates of rectangular guide.

$E_z = 0$. In other words, the z component of the magnetic field, H_z, must exist for energy transmission in the guide. Consequently, from a given Helmholtz equation,

$$\nabla^2 H_z = \gamma^2 H_z \qquad (8\text{-}1\text{-}20)$$

a solution in the form

$$H_z = \left[A_m \sin\left(\frac{m\pi x}{a}\right) + B_m \cos\left(\frac{m\pi x}{a}\right) \right]\left[C_n \sin\left(\frac{n\pi y}{b}\right) + D_n \cos\left(\frac{n\pi y}{b}\right) \right] e^{-j\beta_g z}$$

$$(8\text{-}1\text{-}21)$$

will be determined in accordance with the given boundary conditions.

Since $E_x = 0$, $\partial H_z / \partial y = 0$, at $y = 0, b$. Hence $C_n = 0$. Since $E_y = 0$, $\partial H_z / \partial x = 0$, at $x = 0, a$. Hence $A_m = 0$.

It is generally concluded that the normal derivative of H_z must vanish at the conducting surfaces, that is,

$$\frac{\partial H_z}{\partial n} = 0 \qquad \text{at guide walls} \qquad (8\text{-}1\text{-}22)$$

Therefore, the magnetic field in the positive z direction is given by

$$H_z = H_{oz} \cos\left(\frac{m\pi x}{a}\right) \cos\left(\frac{n\pi y}{b}\right) e^{-j\beta_g z} \qquad (8\text{-}1\text{-}23)$$

where H_{0z} is the amplitude constant.

Expansion of the two curl equations of Maxwell's field equations and substitution of $\partial/\partial z = -j\beta_g$ and $E_z = 0$ yield the following six field equations:

$$\beta_g E_y = -\omega\mu H_x \qquad (8\text{-}1\text{-}24)$$

$$\beta_g E_x = \omega\mu H_y \qquad (8\text{-}1\text{-}25)$$

$$\frac{\partial E_y}{\partial x} - \frac{\partial E_x}{\partial y} = -j\omega\mu H_z \qquad (8\text{-}1\text{-}26)$$

$$\frac{\partial H_z}{\partial y} + j\beta_g H_y = j\omega\epsilon E_x \qquad (8\text{-}1\text{-}27)$$

$$-j\beta_g H_x - \frac{\partial H_z}{\partial x} = j\omega\epsilon E_y \qquad (8\text{-}1\text{-}28)$$

$$\frac{\partial H_y}{\partial x} - \frac{\partial H_x}{\partial y} = 0 \qquad (8\text{-}1\text{-}29)$$

Solving the foregoing six equations for E_x, E_y, H_x, and H_y in terms of H_z will give the TE-mode field equations in rectangular waveguides as

$$E_x = \frac{-j\omega\mu}{k_c^2} \frac{\partial H_z}{\partial y} \qquad (8\text{-}1\text{-}30)$$

$$E_y = \frac{j\omega\mu}{k_c^2} \frac{\partial H_z}{\partial x} \qquad (8\text{-}1\text{-}31)$$

$$E_z = 0 \qquad (8\text{-}1\text{-}32)$$

$$H_x = \frac{-j\beta_g}{k_c^2} \frac{\partial H_z}{\partial x} \qquad (8\text{-}1\text{-}33)$$

$$H_y = \frac{-j\beta_g}{k_c^2} \frac{\partial H_z}{\partial y} \qquad (8\text{-}1\text{-}34)$$

$$H_z = \text{Eq. (8-1-23)} \qquad (8\text{-}1\text{-}35)$$

where $k_c^2 = \omega^2 \mu\epsilon - \beta_g^2$ has been replaced.

Differentiating Eq. (8-1-23) with respect to x and y, and then substituting the results in Eqs. (8-1-30) through (8-1-34) will yield a set of field equations for TE$_{mn}$ modes in a rectangular waveguide as

$$E_x = E_{ox} \cos\left(\frac{m\pi x}{a}\right) \sin\left(\frac{n\pi y}{b}\right) e^{-j\beta_g z} \qquad (8\text{-}1\text{-}36)$$

$$E_y = E_{0y} \sin\left(\frac{m\pi x}{a}\right) \cos\left(\frac{n\pi y}{b}\right) e^{-j\beta_g z} \qquad (8\text{-}1\text{-}37)$$

$$E_z = 0 \qquad (8\text{-}1\text{-}38)$$

$$H_x = H_{0x} \sin\left(\frac{m\pi x}{a}\right) \cos\left(\frac{n\pi y}{b}\right) e^{-j\beta_g z} \qquad (8\text{-}1\text{-}39)$$

$$H_y = H_{0y} \cos\left(\frac{m\pi x}{a}\right) \sin\left(\frac{n\pi y}{b}\right) e^{-j\beta_g z} \qquad (8\text{-}1\text{-}40)$$

$$H_z = \text{Eq. (8-1-23)} \qquad (8\text{-}1\text{-}41)$$

where $m = 0, 1, 2, \ldots$
$\quad n = 0, 1, 2, \ldots$
$\quad m = n = 0$ excepted

The characteristic equations for TE modes in a rectangular waveguide are found as follows.

1. The cutoff wave number k_c as defined by Eq. (8-1-15) for the TE_{mn} modes is given by

$$k_c = \sqrt{\left(\frac{m\pi}{a}\right)^2 + \left(\frac{n\pi}{b}\right)^2} = \omega_c\sqrt{\mu\epsilon} \qquad (8\text{-}1\text{-}42)$$

where a and b are in meters, and $k_x = m\pi/a$ and $k_y = n\pi/b$ are substituted.

2. The cutoff frequency as defined in Eq. (8-1-17) for the TE_{mn} modes is

$$f_c = \frac{1}{2\sqrt{\mu\epsilon}}\sqrt{\frac{m^2}{a^2} + \frac{n^2}{b^2}} \qquad (8\text{-}1\text{-}43)$$

3. The propagation constant (or the phase constant here) β_g as defined in Eq. (8-1-16) is expressed by

$$\beta_g = \omega\sqrt{\mu\epsilon}\sqrt{1 - \left(\frac{f_c}{f}\right)^2} \qquad (8\text{-}1\text{-}44)$$

The phase velocity in the positive z direction for the TE_{mn} modes is shown as

$$v_g = \frac{\omega}{\beta_g} = \frac{v_p}{\sqrt{1 - (f_c/f)^2}} \qquad (8\text{-}1\text{-}45)$$

where $v_p = 1/\sqrt{\mu\epsilon}$ is the phase velocity in an unbounded dielectric.

The characteristic wave impedance of TE_{mn} modes in the guide can be derived from Eq. (8-1-24) or (8-1-25) to be

$$Z_g = \frac{E_x}{H_y} = -\frac{E_y}{H_x} = \frac{\omega\mu}{\beta_g} = \frac{\eta}{\sqrt{1 - (f_c/f)^2}} \qquad (8\text{-}1\text{-}46a)$$

where $\eta = \sqrt{\mu/\epsilon}$ is the intrinsic impedance in unbounded dielectric. The wavelength λ_g in the guide for the TE_{mn} modes is given by

$$\lambda_g = \frac{\lambda}{\sqrt{1 - (f_c/f)^2}} \qquad (8\text{-}1\text{-}47)$$

where $\lambda = v_p/f$ is the wavelength in unbounded dielectric.

Equation (8-1-46a) defined the wave impedance or field impedance of TE_{mn} modes in a rectangular waveguide. However, in practical engineering design, the characteristic impedance of the TE_{mn}-mode waveguide is desirable for impedance-matching purposes. The characteristic impedance of a rectangular waveguide is defined as the ratio of the voltage over current in the guide. For example, the characteristic impedance of the TE_{mn}-mode rectangular waveguide can be expressed as

$$Z_{og} = \frac{V}{I} = \frac{-bE_y}{aH_x} = \frac{b}{a}\frac{\eta}{\sqrt{1 - (f_c/f)^2}} \qquad (8\text{-}1\text{-}46b)$$

As the cutoff frequency shown in Eq. (8-1-43) is a function of the modes and guide dimensions, the physical size of the waveguide will determine the propagation of the modes. Table 8-1-1 tabulates the ratio of cutoff frequency of some modes with respect to that of the dominant mode in terms of the physical dimension.

TABLE 8-1-1   MODES OF $(f_c)_{mn}/f_c$ FOR $a \geq b$

Modes f/f_{10} a/b	TE_{10}	TE_{01}	TE_{11} TM_{11}	TE_{20}	TE_{02}	TE_{21} TM_{21}	TE_{12} TM_{12}	TE_{22} TM_{22}	TE_{30}
1	1	1	1.414	2	2	2.236	2.236	2.838	3
1.5	1	1.5	1.803	2	3	2.500	3.162	3.506	3
2	1	2	2.236	2	4	2.828	4.123	4.472	3
3	1	3	3.162	2	6	3.606	6.083	6.325	3
∞	1	∞	∞	2	∞	∞	∞	∞	3

In the rectangular guide, the corresponding TE_{mn} and TM_{mn} modes are always degenerate. In the square guide, the TE_{mn}, TE_{nm}, TM_{mn}, and TM_{nm} modes form a foursome of degeneracy. The rectangular guides are ordinarily constructed in such a size so that only one mode will propagate. The dimensions of the commonly used waveguides is $a = 2b$. The mode with the lowest cutoff frequency in a particular guide is called the *dominant mode*. The dominant mode in a rectangular guide with $a > b$ is the TE_{10} mode. Each mode has a specific mode pattern (or field pattern).

It is normal that all modes do exist simultaneously in a given waveguide. However, it is not so bad as it seems at first. The fact is that only the dominant mode will propagate, and the higher modes near the sources or discontinuities will decay very fast if the guide is excited by a single mode.

Example 8-1-1: TE_{10} Mode in Rectangular Waveguide

An air-filled waveguide with a cross section of 2×1 cm (Fig. 8-1-4) transports energy in the TE_{10} mode at the rate of 0.5 hp. The impressed frequency is 30 GHz. What is the peak value of electric field occurring in the guide?

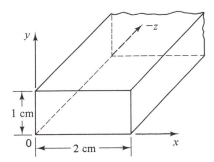

Figure 8-1-4   Rectangular waveguide for Example 8-1-1.

Solution   The field components of the dominant mode TE_{10} may be obtained by substituting $m = 1$ and $n = 0$ in Eqs. (8-1-36) through (8-1-41). Then

$$E_x = 0 \qquad\qquad\qquad H_x = -\frac{E_{0y}}{Z_g} \sin\left(\frac{\pi x}{a}\right) e^{-j\beta_g z}$$

$$E_y = E_{0y} \sin\left(\frac{\pi x}{a}\right) e^{-j\beta_g z} \qquad\qquad H_y = 0$$

$$E_z = 0 \qquad\qquad\qquad H_z = H_{0z} \cos\left(\frac{\pi x}{a}\right) e^{-j\beta_g z}$$

where $Z_g = \omega\mu_0/\beta_g$.

The phase constant β_g may be found from

$$\beta_g = \sqrt{\omega^2\mu_0\epsilon_0 - \frac{\pi^2}{a^2}} = \pi\sqrt{\frac{(2f)^2}{c^2} - \frac{1}{a^2}} = \pi\sqrt{\frac{4 \times 9 \times 10^{20}}{9 \times 10^{16}} - \frac{1}{4 \times 10^{-4}}}$$

$$= 193.5\pi = 608.8093 \text{ rad/m}$$

The power delivered in the z direction by the guide is

$$P = \text{Re}\left[\frac{1}{2}\int_0^b\int_0^a (\mathbf{E} \times \mathbf{H^*})\right] dx\, dy\, \mathbf{u}_z$$

$$= \frac{1}{2}\int_0^b\int_0^a \left[\left(E_{0y}\sin\left(\frac{\pi x}{a}\right)e^{-j\beta_g z}\mathbf{u}_y\right) \times \left(-\frac{\beta_g}{\omega\mu_0}E_{0y}\sin\left(\frac{\pi x}{a}\right)e^{+j\beta_g z}\mathbf{u}_x\right)\right] \cdot dx\, dy\, \mathbf{u}_z$$

$$= \frac{1}{2}E_{0y}^2\frac{\beta_g}{\omega\mu_0}\int_0^b\int_0^a\left[\sin\left(\frac{\pi x}{a}\right)\right]^2 dx\, dy$$

$$= \frac{1}{4}E_{0y}^2\frac{\beta_g}{\omega\mu_0}ab$$

$$373 = \frac{1}{4}E_{0y}^2\frac{193.5\pi(10^{-2})(2 \times 10^{-2})}{2\pi(3 \times 10^{10})(4\pi \times 10^{-7})}$$

$$E_{0y} = 53.8746 \text{ kV/m}$$

The peak value of electric intensity is 53.8746 kV/m.

8-1-3 TM Modes in Rectangular Waveguides

The TM_{mn} modes in a rectangular guide are characterized by $H_z = 0$. In other words, the z component of the electric field E must exist for energy transmission in the guide. Consequently, the Helmholtz equation for E in the rectangular coordinates is given by

$$\nabla^2 E_z = \gamma^2 E_z \qquad\qquad (8\text{-}1\text{-}48)$$

A solution of the Helmholtz equation is in the form of

$$E_z = [A_m \sin(k_x x) + B_m \cos(k_x x)][C_n \sin(k_y y) + D_n \cos(k_y y)]e^{-j\beta_g z} \qquad (8\text{-}1\text{-}49)$$

which is required to be determined in accordance with the given boundary conditions. Figure 8-1-5 shows the coordinates of a rectangular waveguide for TM modes.

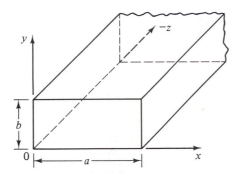

Figure 8-1-5 Coordinates of rectangular guide for TM modes.

The boundary conditions on E_z require that the field vanishes at the waveguide walls since the tangent component of the electric field E_z is zero on the conducting surface. For $E_z = 0$ at $x = 0$ and a,

$$B_m = 0 \quad \text{and} \quad k_x a = m\pi \quad \text{or} \quad k_x = \frac{m\pi}{a}$$

For $E_z = 0$ at $y = 0$ and b,

$$D_n = 0 \quad \text{and} \quad k_y b = n\pi \quad \text{or} \quad k_y = \frac{n\pi}{b}$$

Thus the solution as shown in Eq. (8-1-49) reduces to

$$E_z = E_{0z} \sin\left(\frac{m\pi x}{a}\right) \sin\left(\frac{n\pi y}{b}\right) e^{-j\beta_g z} \tag{8-1-50}$$

where $m = 1, 2, 3, \ldots$
$\quad\quad\quad n = 1, 2, 3, \ldots$

If $m = 0$, or $n = 0$, or $m = n = 0$, the field intensities all vanish. Hence there is no TM_{01}, TM_{10}, or TM_{00} mode in a rectangular waveguide.

For a lossless dielectric, Maxwell's curl equations in the frequency domain are given by

$$\nabla \times \mathbf{E} = -j\omega\mu\mathbf{H} \tag{8-1-51}$$

$$\nabla \times \mathbf{H} = j\omega\epsilon\mathbf{E} \tag{8-1-52}$$

In rectangular coordinates, their components are

$$\frac{\partial E_z}{\partial y} - \frac{\partial E_y}{\partial z} = -j\omega\mu H_x \tag{8-1-53}$$

$$\frac{\partial E_x}{\partial z} - \frac{\partial E_z}{\partial x} = -j\omega\mu H_y \tag{8-1-54}$$

$$\frac{\partial E_y}{\partial x} - \frac{\partial E_x}{\partial y} = -j\omega\mu H_z \tag{8-1-55}$$

$$\frac{\partial H_z}{\partial y} - \frac{\partial H_y}{\partial z} = j\omega\epsilon E_x \tag{8-1-56}$$

$$\frac{\partial H_x}{\partial z} - \frac{\partial H_z}{\partial x} = j\omega\epsilon E_y \tag{8-1-57}$$

$$\frac{\partial H_y}{\partial x} - \frac{\partial H_x}{\partial y} = j\omega\epsilon E_z \tag{8-1-58}$$

with the substitution of $\partial/\partial z = -j\beta_g$ and $H_z = 0$, the equations above are simplified to be

$$\frac{\partial E_z}{\partial y} + j\beta_g E_y = -j\omega\mu H_x \tag{8-1-59}$$

$$j\beta_g E_x + \frac{\partial E_z}{\partial x} = j\omega\mu H_y \tag{8-1-60}$$

$$\frac{\partial E_y}{\partial x} - \frac{\partial E_x}{\partial y} = 0 \tag{8-1-61}$$

$$\beta_g H_y = \omega\epsilon E_x \tag{8-1-62}$$

$$-\beta_g H_x = \omega\epsilon E_y \tag{8-1-63}$$

$$\frac{\partial H_y}{\partial x} - \frac{\partial H_x}{\partial y} = j\omega\epsilon E_z \tag{8-1-64}$$

These equations may be solved simultaneously for E_x, E_y, H_x, and H_y in terms of E_z. The resultant field equations for TM modes are

$$E_x = \frac{-j\beta_g}{k_c^2}\frac{\partial E_z}{\partial x} \tag{8-1-65}$$

$$E_y = \frac{-j\beta_g}{k_c^2}\frac{\partial E_z}{\partial y} \tag{8-1-66}$$

$$E_z = \text{Eq. (8-1-50)} \tag{8-1-67}$$

$$H_x = \frac{j\omega\epsilon}{k_c^2}\frac{\partial E_z}{\partial y} \tag{8-1-68}$$

$$H_y = \frac{-j\omega\epsilon}{k_c^2}\frac{\partial E_z}{\partial x} \tag{8-1-69}$$

$$H_z = 0 \tag{8-1-70}$$

where $\beta_g^2 - \omega^2\mu\epsilon = -k_c^2$ is replaced.

Differentiating Eq. (8-1-50) with respect to x or y and substituting the results in Eqs. (8-1-65) through (8-1-70) yields a new set of TM$_{mn}$-mode field equations in a rectangular waveguide:

$$E_x = E_{0x}\cos\left(\frac{m\pi x}{a}\right)\sin\left(\frac{n\pi y}{b}\right)e^{-j\beta_g z} \tag{8-1-71}$$

$$E_y = E_{0y}\sin\left(\frac{m\pi x}{a}\right)\cos\left(\frac{n\pi y}{b}\right)e^{-j\beta_g z} \tag{8-1-72}$$

$$E_z = \text{Eq. (8-1-50)} \tag{8-1-73}$$

$$H_x = H_{0x}\sin\left(\frac{m\pi x}{a}\right)\cos\left(\frac{n\pi y}{b}\right)e^{-j\beta_g z} \tag{8-1-74}$$

$$H_y = H_{0y}\cos\left(\frac{m\pi x}{a}\right)\sin\left(\frac{n\pi y}{b}\right)e^{-j\beta_g z} \tag{8-1-75}$$

$$H_z = 0 \tag{8-1-76}$$

Some of the TM-mode characteristic equations are identical to those for the TE modes, but some are different. For convenience, all of those are shown here.

Cutoff frequency
$$f_c = \frac{1}{2\sqrt{\mu\epsilon}}\sqrt{\frac{m^2}{a^2}+\frac{n^2}{b^2}} \tag{8-1-77}$$

Phase constant
$$\beta_g = \omega\sqrt{\mu\epsilon}\sqrt{1-\left(\frac{f_c}{f}\right)^2} \tag{8-1-78}$$

Wavelength
$$\lambda_g = \frac{\lambda}{\sqrt{1-(f_c/f)^2}} \tag{8-1-79}$$

Phase velocity
$$v_g = \frac{v_p}{\sqrt{1-(f_c/f)^2}} \tag{8-1-80}$$

Wave impedance
$$Z_g = \frac{\beta_g}{\omega\epsilon} = \eta\sqrt{1-\left(\frac{f_c}{f}\right)^2} \tag{8-1-81a}$$

Characteristic impedance
$$Z_{0g} = \frac{b\eta}{a}\sqrt{1-\left(\frac{f_c}{f}\right)^2} \tag{8-1-81b}$$

8-1-4 Characteristics of Standard Rectangular Waveguides

Rectangular waveguides are commonly used for power transmission at microwave frequencies. Its physical dimensions are regulated by the frequency of the signal being transmitted. For example: at X-band frequencies from 8 to 12 GHz, the outside dimensions of a rectangular waveguide are 2.54 cm (1.0 in.) wide and 1.27 cm (0.5 in.) high, designated as EIA WR (90) by the Electronic Industry Association, but its inside dimensions are 2.286 cm (0.90 in.) wide and 1.016 cm (0.40 in.) high. Table 8-1-2 tabulates the characteristics of the standard rectangular waveguides.

TABLE 8-1-2 CHARACTERISTICS OF STANDARD RECTANGULAR WAVEGUIDES

EIA designation WR( )ᵃ	Physical dimensions [cm (in.)]				Cutoff frequency for air-filled waveguide in (GHz)	Recommended frequency range for TE_{10} mode in (GHz)
	Inside		Outside			
	width a	height b	width a	height b		
2300	58.420 (23.000)	29.210 (11.500)	59.055 (23.250)	29.845 (11.750)	0.257	0.32–0.49
2100	53.340 (21.000)	26.670 (10.500)	53.973 (21.250)	27.305 (10.750)	0.281	0.35–0.53
1800	45.720 (18.000)	22.860 (9.000)	46.350 (18.250)	23.495 (9.250)	0.328	0.41–0.62
1500	38.100 (15.000)	19.050 (7.500)	38.735 (15.250)	19.685 (7.750)	0.394	0.49–0.75
1150	29.210 (11.500)	14.605 (5.750)	29.845 (11.750)	15.240 (6.000)	0.514	0.64–0.98
975	24.765 (9.750)	12.383 (4.875)	25.400 (10.000)	13.018 (5.125)	0.606	0.76–1.15
770	19.550 (7.700)	9.779 (3.850)	20.244 (7.970)	10.414 (4.100)	0.767	0.96–1.46
650	16.510 (6.500)	8.255 (3.250)	16.916 (6.660)	8.661 (3.410)	0.909	1.14–1.73
510	12.954 (5.100)	6.477 (2.500)	13.360 (5.260)	6.883 (2.710)	1.158	1.45–2.20
430	10.922 (4.300)	5.461 (2.150)	11.328 (4.460)	5.867 (2.310)	1.373	1.72–2.61
340	8.636 (3.400)	4.318 (1.700)	9.042 (3.560)	4.724 (1.860)	1.737	2.17–3.30
284	7.214 (2.840)	3.404 (1.340)	7.620 (3.000)	3.810 (1.500)	2.079	2.60–3.95
229	5.817 (2.290)	2.908 (1.145)	6.142 (2.418)	3.233 (1.273)	2.579	3.22–4.90
187	4.755 (1.872)	2.215 (0.872)	5.080 (2.000)	2.540 (1.000)	3.155	3.94–5.99
159	4.039 (1.590)	2.019 (0.795)	4.364 (1.718)	2.344 (0.923)	3.714	4.64–7.05
137	3.485 (1.372)	1.580 (0.622)	3.810 (1.500)	1.905 (0.750)	4.304	5.38–8.17

TABLE 8-1-2 CHARACTERISTICS OF STANDARD RECTANGULAR WAVEGUIDES (*Continued*)

| EIA designation WR( )[a] | Physical dimensions [cm (in.)] | | | | Cutoff frequency for air-filled waveguide in (GHz) | Recommended frequency range for TE$_{10}$ mode in (GHz) |
| | Inside | | Outside | | | |
	width a	height b	width a	height b		
112	2.850 (1.122)	1.262 (0.497)	3.175 (1.250)	1.588 (0.625)	5.263	6.57–9.99
90	2.286 (0.900)	1.016 (0.400)	2.540 (1.000)	1.270 (0.500)	6.562	8.20–12.50
75	1.905 (0.750)	0.953 (0.375)	2.159 (0.850)	1.207 (0.475)	7.874	9.84–15.00
62	1.580 (0.622)	0.790 (0.311)	1.783 (0.702)	0.993 (0.391)	9.494	11.90–18.00
51	1.295 (0.510)	0.648 (0.255)	1.499 (0.590)	0.851 (0.335)	11.583	14.50–22.00
42	1.067 (0.420)	0.432 (0.170)	1.270 (0.500)	0.635 (0.250)	14.058	17.60–26.70
34	0.864 (0.340)	0.432 (0.170)	1.067 (0.420)	0.635 (0.250)	17.361	21.70–33.00
28	0.711 (0.280)	0.356 (0.140)	0.914 (0.360)	0.559 (0.220)	21.097	26.40–40.00
22	0.569 (0.224)	0.284 (0.112)	0.772 (0.304)	0.488 (0.192)	26.362	32.90–50.10
19	0.478 (0.188)	0.239 (0.094)	0.681 (0.268)	0.442 (0.174)	31.381	39.20–59.60
15	0.376 (0.148)	0.188 (0.074)	0.579 (0.228)	0.391 (0.154)	39.894	49.80–75.80
12	0.310 (0.122)	0.155 (0.061)	0.513 (0.202)	0.358 (0.141)	48.387	60.50–91.90
10	0.254 (0.100)	0.127 (0.050)	0.457 (0.180)	0.330 (0.130)	59.055	73.80–112.00
8	0.203 (0.080)	0.102 (0.040)	0.406 (0.160)	0.305 (0.120)	73.892	92.20–140.00
7	0.165 (0.065)	0.084 (0.033)	0.343 (0.135)	0.262 (0.103)	90.909	114.00–173.00
5	0.130 (0.051)	0.066 (0.026)	0.257 (0.101)	0.193 (0.076)	115.385	145.00–220.00
4	0.109 (0.043)	0.056 (0.022)	0.211 (0.083)	0.157 (0.062)	137.615	172.00–261.00
3	0.086 (0.034)	0.043 (0.017)	0.163 (0.064)	0.119 (0.047)	174.419	217.00–333.00

[a] EIA stands for Electronic Industry Association, WR for rectangular waveguide.

Example 8-1-2: *TE$_{10}$ in Rectangular Waveguide*

An air-filled rectangular waveguide of inside dimensions, 3.5×7 cm (Fig. 8-1-6) operates in the dominant TE$_{10}$ mode. Determine (a) the cutoff frequency, (b) the phase velocity of the guided wave in the guide at a frequency of 3.5 GHz, and (c) the guide wavelength at the same frequency.

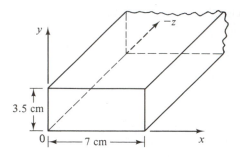

Figure 8-1-6 Rectangular waveguide for Example 8-1-2.

Solution

$$\text{(a) } f_c = \frac{c}{2b} = \frac{3 \times 10^8}{2 \times 7 \times 10^{-2}} = 2.14 \text{ GHz}$$

$$\text{(b) } v_g = \frac{c}{\sqrt{1 - (f_c/f)^2}} = \frac{3 \times 10^8}{\sqrt{1 - (2.14/3.5)^2}} = 3.78 \times 10^8 \text{ m/s}$$

$$\text{(c) } \lambda_g = \frac{\lambda_0}{\sqrt{1 - (f_c/f)^2}} = \frac{(3 \times 10^8)/(3.5 \times 10^9)}{\sqrt{1 - (2.14/3.5)^2}} = 10.8 \text{ cm}$$

8-2 CIRCULAR WAVEGUIDES

A circular waveguide is a tubular, circular conductor. A plane wave propagating through a circular waveguide results in a transverse electric (TE) or transverse magnetic (TM) mode. Several other types of waveguides, such as elliptical and reentrant guides, will also propagate electromagnetic waves.

8-2-1 Solutions of Wave Equations

As described previously for rectangular waveguides, only a sinusoidal steady-state or frequency-domain solution will be attempted for cylindrical waveguides. A cylindrical coordinate system is shown in Fig. 8-2-1.

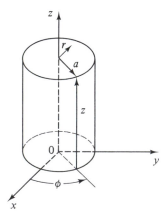

Figure 8-2-1 Cylindrical coordinates.

The scalar Helmholtz equation in cylindrical coordinates is given by

$$\frac{1}{r}\frac{\partial}{\partial r}\left(r\frac{\partial\psi}{\partial r}\right) + \frac{1}{r^2}\frac{\partial^2\psi}{\partial\phi^2} + \frac{\partial^2\psi}{\partial z^2} = \gamma^2\psi \tag{8-2-1}$$

By the method of separation of variables, the solution is assumed in the form of

$$\psi = R(r)\Phi(\phi)Z(z) \tag{8-2-2}$$

where $R(r)$ = function of the r coordinate only
$\Phi(\phi)$ = function of the ϕ coordinate only
$Z(z)$ = function of the z coordinate only

Substitution of Eq. (8-2-2) in Eq. (8-2-1) and division of the resultant by Eq. (8-2-2) yields

$$\frac{1}{rR}\frac{\partial}{\partial r}\left(r\frac{\partial R}{\partial r}\right) + \frac{1}{r^2\Phi}\frac{\partial^2\Phi}{\partial\phi^2} + \frac{1}{Z}\frac{\partial^2 Z}{\partial z^2} = \gamma^2 \tag{8-2-3}$$

The sum of the three independent terms is a constant, so each of the three terms must be a constant. The third term may be set equal to a constant γ_g^2. Hence

$$\frac{d^2 Z}{dz^2} = \gamma_g^2 Z \tag{8-2-4}$$

The solutions of the equation above are given by

$$Z = Ae^{-\gamma_g z} + Be^{\gamma_g z} \tag{8-2-5}$$

where γ_g is the propagation constant of the wave in the guide.

Inserting γ_g^2 for the third term of the left-hand side in Eq. (8-2-3) and multiplying the resultant by r^2 yield

$$\frac{r}{R}\frac{d}{dr}\left(r\frac{dR}{dr}\right) + \frac{1}{\Phi}\frac{d^2\Phi}{d\phi^2} - (\gamma^2 - \gamma_g^2)r^2 = 0 \tag{8-2-6}$$

The second term is a function of ϕ only; hence, equating the second term to a constant $(-n^2)$ yields

$$\frac{d^2\Phi}{d\phi^2} = -n^2\Phi \tag{8-2-7}$$

The solution of Eq. (8-2-7) is also the harmonic function:

$$\Phi = A_n \sin(n\phi) + B_n \cos(n\phi) \tag{8-2-8}$$

Replacing the Φ term by $(-n^2)$ in Eq. (8-2-6) and multiplying through by R yields the result

$$r\frac{d}{dr}\left(r\frac{dR}{dr}\right) + [(k_c r)^2 - n^2]R = 0 \tag{8-2-9}$$

This is Bessel's equation of the order n, in which

$$k_c^2 + \gamma^2 = \gamma_g^2 \qquad (8\text{-}2\text{-}10)$$

Equation (8-2-10) is called the characteristic equation of Bessel's equation for a simple time-harmonic case. For a lossless guide, the characteristic equation reduces to

$$\beta_g = \pm\sqrt{\omega^2\mu\epsilon - k_c^2} \qquad (8\text{-}2\text{-}11)$$

The solutions of Bessel's equation are in the form

$$R = C_n J_n(k_c r) + D_n N_n(k_c r) \qquad (8\text{-}2\text{-}12)$$

where $J_n(k_c r)$ is the nth-order Bessel function of the first kind representing a standing wave of $\cos(k_c r)$ for $r < a$ as shown in Fig. 8-2-2. $N_n(k_c r)$ is the nth-order Bessel function of the second kind representing a standing wave of $\sin(k_c r)$ for $r > a$ as shown in Fig. 8-2-3.

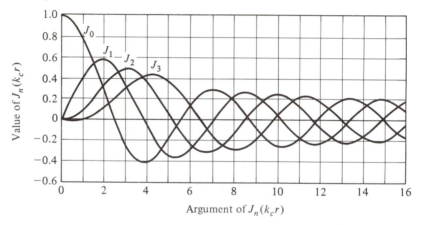

Figure 8-2-2 Bessel functions of the first kind.

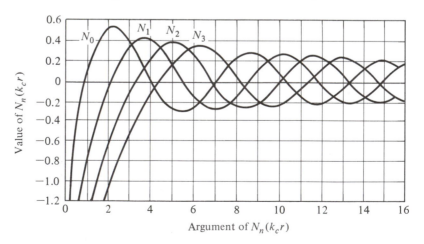

Figure 8-2-3 Bessel functions of the second kind.

Therefore, the total solution of the Helmholtz equation in cylindrical coordinates is given by

$$\psi = [C_n J_n(k_c r) + D_n N_n(k_c r)][A_n \sin (n\phi) + B_n \cos (n\phi)]e^{\pm j\beta_g z} \quad (8\text{-}2\text{-}13)$$

However, at $r = 0$, $k_c r = 0$; then the function N_n approaches to infinity, so $D_n = 0$. This means that at $r = 0$ on the z axis, the field must be finite. Also, by use of trigonometrical manipulations, the two sinusoidal terms will become

$$[A_n \sin (n\phi) + B_n \cos (n\phi)] = \sqrt{A_n^2 + B_n^2} \cos \left[n\phi + \tan^{-1}\left(\frac{A_n}{B_n}\right) \right]$$

$$= F_n \cos (n\phi) \quad (8\text{-}2\text{-}14)$$

Finally, the traveling-wave solution of the Helmholtz equation is reduced to

$$\psi = \psi_0 J_n(k_c r) \cos (n\phi)e^{-j\beta_g z} \quad (8\text{-}2\text{-}15)$$

8-2-2 TE Modes in Circular Waveguides

It is commonly assumed that the waves in a circular waveguide are propagating in the positive z direction. Figure 8-2-4 shows the coordinates of a circular guide. The TE_{np} modes in the circular guide are characterized by $E_z = 0$. This means that the z component of the magnetic field H_z must exist in the guide for electromagnetic energy transmission. A Helmholtz equation for H_z in a circular guide is given by

$$\nabla^2 H_z = \gamma^2 H_z \quad (8\text{-}2\text{-}16)$$

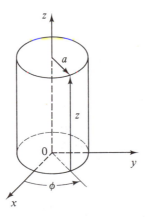

Figure 8-2-4 Coordinates of a circular guide.

Its solution is given in Eq. (8-2-15) by

$$H_z = H_{0z} J_n(k_c r) \cos (n\phi)e^{-j\beta_g z} \quad (8\text{-}2\text{-}17)$$

which is subject to the given boundary conditions.

For a lossless dielectric, Maxwell's curl equations in frequency domain are given by

$$\nabla \times \mathbf{E} = -j\omega\mu\mathbf{H} \tag{8-2-18}$$

$$\nabla \times \mathbf{H} = j\omega\epsilon\mathbf{E} \tag{8-2-19}$$

In cylindrical coordinates, their components are expressed as

$$\frac{1}{r}\frac{\partial E_z}{\partial \phi} - \frac{\partial E_\phi}{\partial z} = -j\omega\mu H_r \tag{8-2-20}$$

$$\frac{\partial E_r}{\partial z} - \frac{\partial E_z}{\partial r} = -j\omega\mu H_\phi \tag{8-2-21}$$

$$\frac{1}{r}\frac{\partial}{\partial r}(rE_\phi) - \frac{1}{r}\frac{\partial E_r}{\partial \phi} = -j\omega\mu H_z \tag{8-2-22}$$

$$\frac{1}{r}\frac{\partial H_z}{\partial \phi} - \frac{\partial H_\phi}{\partial z} = j\omega\epsilon E_r \tag{8-2-23}$$

$$-j\beta_g H_r - \frac{\partial H_z}{\partial r} = j\omega\epsilon E_\phi \tag{8-2-24}$$

$$\frac{1}{r}\frac{\partial}{\partial r}(rH_\phi) - \frac{1}{r}\frac{\partial H_r}{\partial \phi} = j\omega\epsilon E_z \tag{8-2-25}$$

with the differentiation $\partial/\partial z$ being replaced by $(-j\beta_g)$ and the z component of electric field E_z by zero, the TE-mode equations in a circular waveguide are given in terms of H_z by

$$E_r = -\frac{j\omega\mu}{k_c^2}\frac{1}{r}\frac{\partial H_z}{\partial \phi} \tag{8-2-26}$$

$$E_\phi = \frac{j\omega\mu}{k_c^2}\frac{\partial H_z}{\partial r} \tag{8-2-27}$$

$$E_z = 0 \tag{8-2-28}$$

$$H_r = \frac{-j\beta_g}{k_c^2}\frac{\partial H_z}{\partial r} \tag{8-2-29}$$

$$H_\phi = \frac{-j\beta_g}{k_c^2}\frac{1}{r}\frac{\partial H_z}{\partial \phi} \tag{8-2-30}$$

$$H_z = H_{0z}J_n(k_c r)\cos(n\phi)e^{-j\beta_g z} \tag{8-2-31}$$

where $k_c^2 = \omega^2\mu\epsilon - \beta_g^2$ has been replaced.

The boundary conditions require that the ϕ component of the electric field E_ϕ, which is tangential to the inner surface of the circular waveguide at $r = a$, must vanish, or that the r component of the magnetic field H_r, which is normal to the inner surface of $r = a$, must vanish. Consequently,

$$E_\phi = 0 \text{ at } r = a, \qquad \therefore \left.\frac{\partial H_z}{\partial r}\right|_{r=a} = 0$$

or

$$H_r = 0 \text{ at } r = a, \qquad \therefore \left.\frac{\partial H_z}{\partial r}\right|_{r=a} = 0$$

This requirement is equivalent to that expressed in Eq. (8-2-17):

$$\left.\frac{\partial H_z}{\partial r}\right|_{r=a} = H_{0z} J'_n(k_c a) \cos(n\phi) e^{-j\beta_g z} = 0 \tag{8-2-32}$$

Hence

$$J'_n(k_c a) = 0 \tag{8-2-33}$$

where J'_n indicates the derivative of J_n.

Since the J'_n are oscillatory functions, the $J'_n(k_c a)$ are also oscillatory functions. There is an infinite sequence of values of $(k_c a)$ that satisfy Eq. (8-2-33). These points, the roots of Eq. (8-2-33), correspond to the maxima and minima of the curves $J'_n(k_c a)$ as shown in Fig. 8-2-2. Table 8-2-1 tabulates a few roots of $J'_n(k_c a)$ for some lower-order n. The permissible values of k_c may be written as

$$k_c = \frac{X'_{np}}{a} \tag{8-2-34}$$

TABLE 8-2-1 pth ZEROS OF $J'_n(k_c a)$

	n					
p	0	1	2	3	4	5
1	3.832	1.841	3.054	4.201	5.317	6.416
2	7.016	5.331	6.706	8.015	9.282	10.520
3	10.173	8.536	9.969	11.346	12.682	13.987
4	13.324	11.706	13.170			

Substitution of Eq. (8-2-17) in Eqs. (8-2-26) through (8-2-31) yields the complete field equations of the TE_{np} modes in circular waveguides as

$$E_r = E_{0r} J_n\left(\frac{X'_{np} r}{a}\right) \sin(n\phi) e^{-j\beta_g z} \tag{8-2-35}$$

$$E_\phi = E_{0\phi} J'_n\left(\frac{X'_{np} r}{a}\right) \cos(n\phi) e^{-j\beta_g z} \tag{8-2-36}$$

$$E_z = 0 \tag{8-2-37}$$

$$H_r = -\frac{E_{0\phi}}{Z_g} J_n'\left(\frac{X_{np}'\, r}{a}\right) \cos\,(n\phi) e^{-j\beta_{gz}} \tag{8-2-38}$$

$$H_\phi = \frac{E_{0r}}{Z_g} J_n\left(\frac{X_{np}'\, r}{a}\right) \sin\,(n\phi) e^{-j\beta_{gz}} \tag{8-2-39}$$

$$H_z = H_{0z} J_n\left(\frac{X_{np}'\, r}{a}\right) \cos\,(n\phi) e^{-j\beta_{gz}} \tag{8-2-40}$$

where

$Z_g = E_r/H_\phi = -E_\phi/H_r$, has been replaced for the wave impedance in the guide

$n = 0, 1, 2, 3, \ldots$

$p = 1, 2, 3, 4, \ldots$

The first subscript n represents the number of full cycles of field variation in one revolution through 2π radians of ϕ. The second subscript p indicates the number of zeros of E_ϕ; that is, $J_n\,(X_{np}'\, r/a)$ along the radial of a guide, but the zero on the axis is excluded if it exists.

The mode propagation constant is determined by Eqs. (8-2-26) through (8-2-31) and (8-2-34) to be

$$\beta_g = \sqrt{\omega^2 \mu\epsilon - \left(\frac{X_{np}'}{a}\right)^2} \tag{8-2-41}$$

The cutoff wave number of a mode is that for which the mode propagation constant vanishes. Hence

$$k_c = \frac{X_{np}'}{a} = \omega_c \sqrt{\mu\epsilon} \tag{8-2-42}$$

The cutoff frequency for TE modes in a circular guide is then given by

$$f_c = \frac{X_{np}'}{2\pi a \sqrt{\mu\epsilon}} \tag{8-2-43}$$

The phase velocity for TE modes in a circular guide is given by

$$v_g = \frac{\omega}{\beta_g} = \frac{v_p}{\sqrt{1 - (f_c/f)^2}} \tag{8-2-44}$$

where $v_p = 1/\sqrt{\mu\epsilon} = c/\sqrt{\mu_r \epsilon_r}$ is the phase velocity in an unbounded dielectric.

The wavelength and wave impedance for TE modes in a circular guide are given, respectively, by

$$\lambda_g = \frac{\lambda}{\sqrt{1 - (f_c/f)^2}} \tag{8-2-45}$$

and

$$Z_g = \frac{\omega\mu}{\beta_b} = \frac{\eta}{\sqrt{1 - (f_c/f)^2}} \qquad (8\text{-}2\text{-}46a)$$

where $\lambda = v_p/f$ is the wavelength in an unbounded dielectric

$\eta = \sqrt{\mu/\epsilon}$ is the intrinsic impedance in an unbounded dielectric

The characteristic impedance of the TE_{np}-mode circular waveguide is given by

$$Z_{0g} = \frac{aE_r}{2\pi a H_\phi} = \frac{\eta}{2\pi\sqrt{1 - (f_c/f)^2}} \qquad (8\text{-}2\text{-}46b)$$

Example 8-2-1:   TE Mode Cylindrical Waveguide

A TE_{11} mode is propagating through a circular waveguide (Fig. 8-2-5). The radius of the guide is 5 cm, and the guide contains an air dielectric. Determine (a) the cutoff frequency, (b) the wavelength λ_g in the guide for an operating frequency of 3 GHz, and (c) the wave impedance Z_g in the guide.

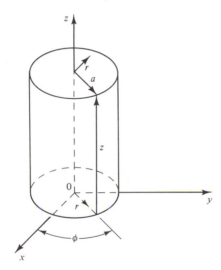

Figure 8-2-5   Diagram for Example 8-2-1.

Solution   (a) From Table 8-2-1 for the TE_{11} mode, $n = 1$, $p = 1$, and

$$X'_{11} = 1.841 = k_c a$$

The cutoff wave number is

$$k_c = \frac{1.841}{a} = \frac{1.841}{5 \times 10^{-2}} = 36.82$$

The cutoff frequency is

$$f_c = \frac{k_c}{2\pi\sqrt{\mu_0\epsilon_0}} = \frac{(36.82)(3 \times 10^8)}{2\pi} = 1.758 \times 10^9 \text{ Hz}$$

(b) The phase constant in the guide is

$$\beta_g = \sqrt{\omega^2 \mu_0 \epsilon_0 - k_c^2}$$

$$= \sqrt{(2\pi \times 3 \times 10^9)^2 (4\pi \times 10^{-7} \times 8.85 \times 10^{-12}) - (36.82)^2}$$

$$= 50.9 \text{ rad/m}$$

The wavelength in the guide is

$$\lambda_g = \frac{2\pi}{\beta_g} = \frac{6.28}{50.9} = 12.3 \text{ cm}$$

(c) The wave impedance in the guide is

$$Z_g = \frac{\omega \mu_0}{\beta_g} = \frac{(2\pi \times 10^{10})(4\pi \times 10^{-7})}{50.9} = 465 \ \Omega$$

8-2-3 TM Modes in Circular Waveguides

The TM_{np} modes in circular guide are characterized by $H_z = 0$. However, the z component of the electric field E_z must exist for wave propagation in the guide. Consequently, the Helmholtz equation for E_z in a cylindrical waveguide is given by

$$\nabla^2 E_z = \gamma^2 E_z \tag{8-2-47}$$

Its solution is given in Eq. (8-2-15) by

$$E_z = E_{0z} J_n(k_c r) \cos(n\phi) e^{-j\beta_g z} \tag{8-2-48}$$

which is subject to the given boundary conditions.

The boundary condition requires that the tangential component of electric field E_z at $r = a$ vanishes. Consequently,

$$J_n(k_c a) = 0 \tag{8-2-49}$$

Since $J_n(k_c r)$ are oscillatory functions as shown in Fig. 8-2-2, there are infinite numbers of roots of $J_n(k_c r)$. Table 8-2-2 tabulates a few roots of $J_n(k_c r)$ for some lower-order n.

TABLE 8-2-2 pth ZEROS OF $j_n(k_c r)$

			n			
p	0	1	2	3	4	5
1	2.405	3.832	5.136	6.380	7.588	8.771
2	5.520	7.106	8.417	9.761	11.065	12.339
3	8.645	10.173	11.620	13.015	14.372	
4	11.792	13.324	14.796			

For $H_z = 0$ and $\partial/\partial z = -j\beta_g$, the field equations in the circular guide after expanding $\nabla \times \mathbf{E} = -j\omega\mu\mathbf{H}$ and $\nabla \times \mathbf{H} = j\omega\epsilon\mathbf{E}$ are given by

$$E_r = \frac{-j\beta_g}{k_c^2} \frac{\partial E_z}{\partial r} \tag{8-2-50}$$

$$E_\phi = \frac{-j\beta_g}{k_c^2} \frac{1}{r} \frac{\partial E_z}{\partial \phi} \tag{8-2-51}$$

$$E_z = \text{Eq. (8-2-48)} \tag{8-2-52}$$

$$H_r = \frac{j\omega\epsilon}{k_c^2} \frac{1}{r} \frac{\partial E_z}{\partial \phi} \tag{8-2-53}$$

$$H_\phi = -\frac{j\omega\epsilon}{k_c^2} \frac{\partial E_z}{\partial r} \tag{8-2-54}$$

$$H_z = 0 \tag{8-2-55}$$

where $k_c^2 = \omega^2\mu\epsilon - \beta_g^2$ has been replaced.

Differentiation of Eq. (8-2-48) with respect to z and substitution of the result in Eqs. (8-2-50) through (8-2-55) yield the field equations of TM$_{np}$ modes in a circular waveguide as

$$E_r = E_{0r}J'_n\left(\frac{X_{np}r}{a}\right)\cos(n\phi)e^{-j\beta_g z} \tag{8-2-56}$$

$$E_\phi = E_{0\phi}J_n\left(\frac{X_{np}r}{a}\right)\sin(n\phi)e^{-j\beta_g z} \tag{8-2-57}$$

$$E_z = E_{0z}J_n\left(\frac{X_{np}r}{a}\right)\cos(n\phi)e^{-j\beta_g z} \tag{8-2-58}$$

$$H_r = \frac{E_{0\phi}}{Z_g}J_n\left(\frac{X_{np}r}{a}\right)\sin(n\phi)e^{-j\beta_g z} \tag{8-2-59}$$

$$H_\phi = \frac{E0_r}{Z_g}J'_n\left(\frac{X_{np}r}{a}\right)\cos(n\phi)e^{-j\beta_g z} \tag{8-2-60}$$

$$H_z = 0 \tag{8-2-61}$$

where $Z_g = E_r/H_\phi = -E_\phi/H_r = \beta_g/(\omega\epsilon)$ and $k_c = X_{np}/a$ have been replaced
  $n = 0, 1, 2, 3, \ldots$
  $p = 1, 2, 3, 4, \ldots$

Some of the TM-mode characteristic equations in the circular guide are identical to that of the TE mode, but some are different. For convenience, all of those are shown here.

Phase constant $\qquad \beta_g = \sqrt{\omega^2 \mu\epsilon - \left(\dfrac{X_{np}}{a}\right)^2}$ $\hfill$ (8-2-62)

Cutoff wave number $\qquad k_c = \dfrac{X_{np}}{a} = \omega_c \sqrt{\mu\epsilon}$ $\hfill$ (8-2-63)

Cutoff frequency $\qquad f_c = \dfrac{X_{np}}{2\pi a \sqrt{\mu\epsilon}}$ $\hfill$ (8-2-64)

Phase velocity $\qquad v_g = \dfrac{\omega}{\beta_g} = \dfrac{v_p}{\sqrt{1 - (f_c/f)^2}}$ $\hfill$ (8-2-65)

Wavelength $\qquad \lambda_g = \dfrac{\lambda}{\sqrt{1 - (f_c/f)^2}}$ $\hfill$ (8-2-66)

Wave impedance $\qquad Z_g = \dfrac{\beta_g}{\omega\epsilon} = \eta \sqrt{1 - \left(\dfrac{f_c}{f}\right)^2}$ $\hfill$ (8-2-67a)

Characteristic impedance $\qquad Z_{0g} = \dfrac{\eta}{2\pi} \sqrt{1 - \left(\dfrac{f_c}{f}\right)^2}$ $\hfill$ (8-2-67b)

It should be noted that the dominant mode, or the mode of lowest cutoff frequency in a circular waveguide, is the mode of TE_{11} that has the smallest value of the product, $k_c a = 1.841$, as shown in Tables 8-2-1 and 8-2-2.

Example 8-2-2:   Wave Propagation in Circular Waveguide

An air-filled circular waveguide has a radius of 2 cm and is to carry energy at a frequency of 10 GHz. Find all the TE_{np} and TM_{np} modes for which energy transmission is possible.

Solution   Since the physical dimension of the guide and the frequency of the wave remain constant, the product of $(k_c a)$ is also constant. Hence

$$k_c a = (\omega_0 \sqrt{\mu_0 \epsilon_0})a = \frac{2\pi \times 10^{10}}{3 \times 10^8} (2 \times 10^{-2}) = 4.18$$

Any mode that has a product of $(k_c a)$ less than or equal to 4.18 will propagate the wave with a frequency of 10 GHz; that is,

$$k_c a \leq 4.18$$

The possible modes are

$$\begin{array}{ll} TE_{11}(1.841) & TM_{01}(2.405) \\ TE_{21}(3.054) & TM_{11}(3.832) \\ TE_{01}(3.832) & \end{array}$$

8-2-4 TEM Modes in Circular Waveguides

As stated previously, the TEM mode or transmission-line mode is characterized by

$$E_z = H_z = 0$$

This means that the electric and magnetic fields are completely transverse to the direction of propagation of wave. This mode cannot exist in hollow waveguides since it requires two conductors, such as the coaxial transmission line and two-open wires.

Analysis of the TEM mode will illustrate an excellent analogous relationship between the method of circuit theory and the analysis based on the field theory. Figure 8-2-6 shows a coaxial line.

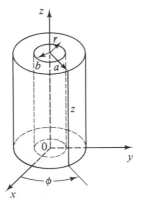

Figure 8-2-6   Coordinates of a coaxial line.

Maxwell's curl equations in cylindrical coordinates

$$\nabla \times \mathbf{E} = -j\omega\mu\mathbf{H} \tag{8-2-68}$$

$$\nabla \times \mathbf{H} = j\omega\epsilon\mathbf{E} \tag{8-2-69}$$

will become

$$B_g E_r = \omega\mu H_\phi \tag{8-2-70}$$

$$B_g E_\phi = \omega\mu H_r \tag{8-2-71}$$

$$\frac{\partial}{\partial r}(rE_\phi) - \frac{\partial E_r}{\partial \phi} = 0 \tag{8-2-72}$$

$$\beta_g H_r = -\omega\epsilon E_\phi \tag{8-2-73}$$

$$\beta_g H_\phi = \omega\epsilon E_r \tag{8-2-74}$$

$$\frac{\partial}{\partial r}(rH_\phi) - \frac{\partial H_r}{\partial \phi} = 0 \tag{8-2-75}$$

where $\partial/\partial r = -j\beta_g$ and $E_z = H_z = 0$ are replaced.

Substitution of Eq. (8-2-71) in Eq. (8-2-73) yields the propagation constant of the TEM mode in a coaxial line.

$$\beta_g = \omega\sqrt{\mu\epsilon} \tag{8-2-76}$$

which is the phase constant of the wave in a lossless transmission line with dielectric.

In comparing the equation above with the characteristic equation of the Helmholtz equation in cylindrical coordinates as given in Eq. (8-2-11) by

$$\beta_g^2 = \omega^2\mu\epsilon - k_c^2 \tag{8-2-77}$$

it is evident that

$$k_c = 0 \qquad (8\text{-}2\text{-}78)$$

This means that the cutoff frequency of the TEM mode in a coaxial line is zero, which is the same as in the ordinary two-wire line.

The phase velocity of the TEM mode can be expressed from Eq. (8-2-76) as

$$v_p = \frac{\omega}{\beta_g} = \frac{1}{\sqrt{\mu\epsilon}} \qquad (8\text{-}2\text{-}79)$$

which is the velocity of light in an unbounded dielectric.

The wave impedance of the TEM mode is found from either Eqs. (8-2-70) and (8-2-73) or Eqs. (8-2-71) and (8-2-74) as

$$\eta(\text{TEM}) = \sqrt{\frac{\mu}{\epsilon}} \qquad (8\text{-}2\text{-}80)$$

which is the wave impedance of the lossless transmission lines in dielectric.

Ampère's law states that the line integral of H about any closed path is exactly equal to the current enclosed by that path; that is

$$\int \mathbf{H} \cdot d\ell = I = I_0 e^{-j\beta_g z} = 2\pi r H_\phi \qquad (8\text{-}2\text{-}81)$$

where I is the complex current that must be supported by the center conductor of a coaxial line. This clearly demonstrates that the TEM mode can only exist in the two-conductor system, but not in the hollow waveguide because the center conductor does not exist.

In summary, the properties of the TEM modes in a lossless medium are:

1. Its cutoff frequency is zero.
2. Its transmission line is a two-conductor system.
3. Its wave impedance is the impedance in an unbounded dielectric.
4. Its propagation constant is the constant in an unbounded dielectric.
5. Its phase velocity is the velocity of light in an unbounded dielectric.

8-2-5 Characteristics of Standard Circular Waveguides

The inner diameter of a circular waveguide is regulated by the frequency of the signal being transmitted. For example, at X-band frequencies from 8 to 12 GHz, the inner diameter of a circular waveguide is 2.383 cm (0.938 in.), designated as EIA WC(94) by the Electronic Industry Association. Table 8-2-3 tabulates the characteristics of the standard circular waveguides.

TABLE 8-2-3 CHARACTERISTICS OF STANDARD CIRCULAR WAVEGUIDES

EIA designation WC( )[a]	Inside diameter, $2a$ [cm (in.)]	Cutoff frequency for air-filled waveguide (GHz)	Recommended frequency range for TE_{11} mode (GHz)
992	25.184 (9.915)	0.698	0.80– 1.10
847	21.514 (8.470)	0.817	0.94– 1.29
724	18.377 (7.235)	0.957	1.10– 1.51
618	15.700 (6.181)	1.120	1.29– 1.76
528	13.411 (5.280)	1.311	1.51– 2.07
451	11.458 (4.511)	1.534	1.76– 2.42
385	9.787 (3.853)	1.796	2.07– 2.83
329	8.362 (3.292)	2.102	2.42– 3.31
281	7.142 (2.812)	2.461	2.83– 3.88
240	6.104 (2.403)	2.880	3.31– 4.54
205	5.199 (2.047)	3.381	3.89– 5.33
175	4.445 (1.750)	3.955	4.54– 6.23
150	3.810 (1.500)	4.614	5.30– 7.27
128	3.254 (1.281)	5.402	6.21– 8.51
109	2.779 (1.094)	6.326	7.27– 9.97
94	2.383 (0.938)	7.377	8.49– 11.60
80	2.024 (0.797)	8.685	9.97– 13.70
69	1.748 (0.688)	10.057	11.60– 15.90
59	1.509 (0.594)	11.649	13.40– 18.40
50	1.270 (0.500)	13.842	15.90– 21.80
44	1.113 (0.438)	15.794	18.20– 24.90
38	0.953 (0.375)	18.446	21.20– 29.10
33	0.833 (0.328)	21.103	24.30– 33.20
28	0.714 (0.281)	24.620	28.30– 38.80
25	0.635 (0.250)	27.683	31.80– 43.60
22	0.556 (0.219)	31.617	36.40– 49.80
19	0.478 (0.188)	36.776	42.40– 58.10
17	0.437 (0.172)	40.227	46.30– 63.50
14	0.358 (0.141)	49.103	56.60– 77.50
13	0.318 (0.125)	55.280	63.50– 87.20
11	0.277 (0.109)	63.462	72.70– 99.70
9	0.239 (0.094)	73.552	84.80–116.00

[a] EIA stands for Electronic Industry Association, WC for circular waveguide.

8-3 RESONANT CAVITIES

In general, a cavity resonator is a metallic enclosure that confines the electromagnetic field. The stored electric and magnetic energies inside the cavity determine its equivalent inductance and capacitance. The energy dissipated by the finite conductivity of the cavity walls determines its equivalent resistance. In practice, rectangular resonator, cylindrical resonator, and reentrant resonators are commonly used as shown in Figure 8-3-1.

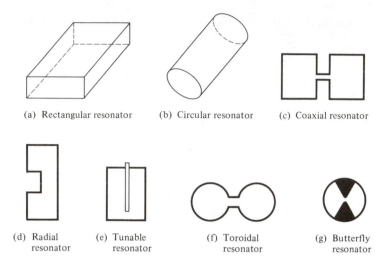

(a) Rectangular resonator   (b) Circular resonator   (c) Coaxial resonator

(d) Radial   (e) Tunable   (f) Toroidal   (g) Butterfly
    resonator   resonator   resonator   resonator

Figure 8-3-1   Resonators.

A given resonator theoretically has an infinite number of resonant modes, and each mode corresponds to a definite resonant frequency. When the frequency of the impressed signal is equal to a resonant frequency, a maximum amplitude of the standing wave occurs; and the peak energy stored in the electric and magnetic fields are equal. The mode having the lowest resonant frequency is known as the *dominant mode*.

8-3-1 Rectangular Cavity

The electromagnetic field inside a cavity should satisfy Maxwell's equations subject to the boundary conditions. The boundary conditions of the field equations are that the electric field tangential to and the magnetic field normal to the metal walls must vanish. The geometry of a rectangular cavity is illustrated in Fig. 8-3-2.

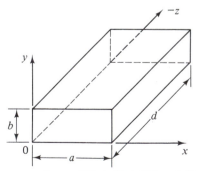

Figure 8-3-2   Coordinates of a rectangular cavity.

The wave equations of Eqs. (8-1-23) and (8-1-50) satisfy the boundary conditions of zero tangential **E** at the walls. It is merely necessary to choose the harmonic functions in z to satisfy this condition on the remaining two end walls. These can be accomplished if

$$H_z = H_{0z} \cos\left(\frac{m\pi x}{a}\right) \cos\left(\frac{n\pi y}{b}\right) \sin\left(\frac{p\pi z}{d}\right) \qquad \text{TE}_{mnp} \qquad (8\text{-}3\text{-}1)$$

where $m = 0, 1, 2, 3, \ldots$
$\quad\quad\ \ n = 0, 1, 2, 3, \ldots$
$\quad\quad\ \ p = 1, 2, 3, 4, \ldots$
and

$$E_z = E_{0z} \sin\left(\frac{m\pi x}{a}\right) \sin\left(\frac{n\pi y}{b}\right) \cos\left(\frac{p\pi z}{d}\right) \qquad \text{TH}_{mnp} \qquad (8\text{-}3\text{-}2)$$

where $m = 1, 2, 3, 4, \ldots$
$\quad\quad\ \ n = 1, 2, 3, 4, \ldots$
$\quad\quad\ \ p = 0, 1, 2, 3, \ldots$

The separation equation for both the TE and TM modes is given by

$$k^2 = \left(\frac{m\pi}{a}\right)^2 + \left(\frac{n\pi}{b}\right)^2 + \left(\frac{p\pi}{d}\right)^2 \qquad (8\text{-}3\text{-}3)$$

For a lossless dielectric, $k^2 = \omega^2 \mu\epsilon$; therefore, the resonant frequency for the TE and TM modes is expressed by

$$f_r = \frac{1}{2\sqrt{\mu\epsilon}} \sqrt{\left(\frac{m}{a}\right)^2 + \left(\frac{n}{b}\right)^2 + \left(\frac{p}{d}\right)^2} \qquad (8\text{-}3\text{-}4)$$

for $b < a < d$, the dominant mode is the TE$_{101}$ mode.

In general, a straight-wire probe inserted at the position of maximum electric intensity is used to excite a desired mode; and a loop coupling placed at the position of maximum magnetic intensity is utilized to launch a specific mode. Figure 8-3-3 shows the methods of excitation for a rectangular resonator.

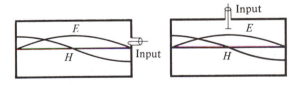

Figure 8-3-3   Methods of exciting a resonator.

8-3-2 Circular Cavity

A circular-cavity resonator is a circular waveguide with two ends closed by metal walls. This is shown in Fig. 8-3-4.

The wave function in circular resonator should satisfy Maxwell's equations subject to the same boundary conditions as described for the rectangular-cavity re-

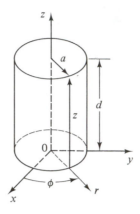

Figure 8-3-4 A circular resonator.

sonator. It is merely necessary to choose the harmonic functions in z for Eqs. (8-2-17) and (8-2-48) to satisfy the boundary conditions at the remaining two end walls. These can be achieved if

$$H_z = H_{0z} J_n\left(\frac{X'_{npr}}{a}\right) \cos{(n\phi)} \sin{\left(\frac{q\pi z}{d}\right)} \qquad \text{TE}_{npq} \qquad (8\text{-}3\text{-}5)$$

where $n = 0, 1, 2, 3, \ldots$
$\quad\quad\;\; p = 1, 2, 3, 4, \ldots$
$\quad\quad\;\; q = 1, 2, 3, 4, \ldots$

and

$$E_z = E_{0z} J_n\left(\frac{X_{npr}}{a}\right) \cos{(n\phi)} \cos{\left(\frac{q\pi z}{d}\right)} \qquad \text{TM}_{npq} \qquad (8\text{-}3\text{-}6)$$

where $n = 0, 1, 2, 3, \ldots$
$\quad\quad\;\; p = 1, 2, 3, 4, \ldots$
$\quad\quad\;\; q = 0, 1, 2, 3, \ldots$

The separation equation for the TE and TM modes are given, respectively, by

$$k^2 = \left(\frac{X'_{np}}{a}\right)^2 + \left(\frac{q\pi}{d}\right)^2 \qquad \text{TE modes} \qquad (8\text{-}3\text{-}7)$$

$$k^2 = \left(\frac{X_{np}}{a}\right)^2 + \left(\frac{q\pi}{d}\right)^2 \qquad \text{TM modes} \qquad (8\text{-}3\text{-}8)$$

Substitution of $k^2 = \omega^2 \mu \epsilon$ in Eqs. (8-3-7) and (8-3-8) yields the resonant frequencies for the TE and TM modes as

$$f_r = \frac{1}{2\pi\sqrt{\mu\epsilon}} \sqrt{\left(\frac{X'_{np}}{a}\right)^2 + \left(\frac{q\pi}{d}\right)^2} \qquad \text{TE modes} \qquad (8\text{-}3\text{-}9)$$

$$f_r = \frac{k}{2\pi\sqrt{\mu\epsilon}} \sqrt{\left(\frac{X_{np}}{a}\right)^2 + \left(\frac{q\pi}{d}\right)^2} \qquad \text{TM modes} \qquad (8\text{-}3\text{-}10)$$

It is interesting to note that for $2a > d$, the TM_{010} mode is dominant; and for $d \geq 2a$, the TE_{111} mode is dominant.

8-3-3 Quality Factor of a Cavity

The quality factor Q is a measure of the frequency selectivity of a resonant or antiresonant circuit, and it is defined as

$$Q \equiv 2\pi \frac{\text{maximum energy stored}}{\text{energy dissipated per cycle}} = \omega \frac{W}{P} \tag{8-3-11}$$

where W = maximum energy stored

P = average power loss

At resonant frequency, the electric and magnetic energies are equal and in time quadrature. When the electric energy is maximum, the magnetic energy is zero; and vice versa. The total energy stored in the resonator is obtained by integrating, respectively, the energy density over the volume of the resonator. That is,

$$W_e = \int_v \frac{\epsilon}{2} |E|^2 \, dv = W_m = \int_v \frac{\mu}{2} |H|^2 \, dv = W \tag{8-3-12}$$

where $|E|$ and $|H|$ are the peak values of the field intensities. The average power loss in the resonator may be evaluated by integrating the power density over the inner surface of the resonator. That is,

$$P = \frac{R_s}{2} \int_s |H_t|^2 \, da \tag{8-3-13}$$

where H_t is the peak value of the tangential magnetic intensity.

Substitution of Eqs. (8-3-12) and (8-3-13) in Eq. (8-3-11) yields

$$Q = \frac{\omega\mu \int_v |H|^2 \, dv}{R_s \int_s |H_t|^2 \, da} \tag{8-3-14}$$

Since the peak value of the magnetic intensity is related to the tangential components by

$$|H|^2 = |H_t|^2 + |H_n|^2 \tag{8-3-15}$$

where $|H_n|$ is the peak value of the normal magnetic intensity, the value of $|H_t|^2$ at the resonator walls is approximately twice the value of $|H|^2$ averaged over the volume. Hence the Q of a cavity resonator as shown in Eq. (8-3-14) can be expressed approximately by

$$Q = \frac{\omega\mu(\text{volumes})}{2R_s(\text{surface areas})} \tag{8-3-16}$$

8-4 REFLECTION AMPLIFIER CONCEPT AND DESIGN

A reflection amplifier is one whose incident signal is amplified by a negative resistance. Because a negative resistance provides a negative reflection coefficient of magnitude greater than unity, it reflects the incident signal and does not dissipate power as the positive resistance does. The amplified power is usually derived from a dc source. Therefore, all reflection amplifiers must have devices with negative resistance available for power reflection. Microwave solid-state devices having negative resistance are tunnel diodes, Gunn diodes, IMPATT diodes, and TRAPATT diodes. Most practical reflection amplifiers provide some means of isolating the input and output signals in order to improve stability and increase gain–bandwidth performance. Four types of reflection amplifier circuits are commonly used as follows:

1. Coaxial-cavity reflection amplifier circuit
2. Reduced-height waveguide reflection amplifier circuit
3. Circulator reflection amplifier circuit
4. Hybrid-coupler reflection amplifier circuit

The power gain of a reflection amplifier can be expressed as the square of the magnitude of the reflection coefficient. That is,

$$g = \left(\frac{R_\ell - R_n}{R_\ell + R_n} \right)^2 \tag{8-4-1}$$

where R_n = negative resistance (the value is negative)
  R_ℓ = load resistance or line resistance

8-4-1 Coaxial-Cavity Reflection Amplifier Design

An IMPATT diode can be designed as an amplifier or oscillator depending on its resonant circuit. In general, the added power in amplifier applications is the same as the output power in oscillator applications. The resonant circuit can be either a waveguide type or a coaxial type. While the proper oscillator load impedance is the complex conjugate of the diode impedance, the amplifier load impedance is related to the power gain g of the amplifier by the equation

$$g = \left(\frac{Z_\ell - Z_d^*}{Z_\ell + Z_d} \right)^2 \tag{*8-4-2}$$

where $Z_\ell = R_\ell + jX_\ell$ is the load impedance
  $Z_d = R_d + jX_d$ is the diode impedance
  $Z_d^* = R_d - jX_d$ is the complex conjugate of the diode impedance

If $X_\ell = -X_d$ is chosen for maximum power transfer, Eq. (8-4-2) becomes

$$g = \left(\frac{R_\ell - R_d}{R_\ell + R_d} \right)^2 \tag{8-4-3}$$

In addition, amplifier design is very similar to oscillator design. In both cases, the load resistance resonates the diode resistance. In amplifier design, the load resistance is larger than the oscillator load resistance.

Design Example 8-4-1:   Coaxial-Cavity Reflection Amplifier Design

A certain IMPATT diode has a diode impedance $Z_d = -0.70 + j11$ at 10 GHz. It is required to design a waveguide-type amplifier for a power gain of 10 dB in TE_{10} mode. The width W of the waveguide is 2.287 cm and its height H is 1.016 cm for X-band operation.

   Design data:

Operating frequency	$f = 10$ GHz
Cutoff frequency	$f_c = 6.56$ GHz for TE_{10} mode
Cutoff wavelength	$\lambda_c = 4.57$ cm
Wavelength in waveguide	$\lambda_g = 4$ cm
Characteristic impedance of TE_{10}-mode rectangular waveguide	$Z_{0g} = \dfrac{H}{W} \dfrac{377}{\sqrt{1 - (f_c/f)^2}} = 221 \ \Omega$

The basic amplifier circuit and its equivalent circuit are shown in Fig. 8-4-1. The coaxial section starts at the center of the waveguide. The waveguide is shorted at one

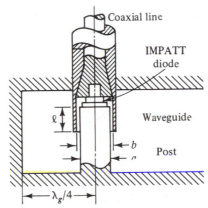

(a) Basic circuit of an IMPATT amplifier (after HP Application Note 968, IMPATT amplifier. Reprinted by permission of Hewlett-Packard, Inc.)

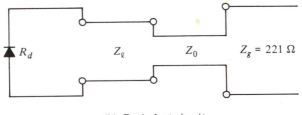

(b) Equivalent circuit

Figure 8-4-1   Circuit diagram and equivalent circuit of an IMPATT amplifier. (From Ref. [1]. Reprinted by permission of Hewlett-Packard Company.)

end. The post that supports the diode is located in the center line at $\lambda_g/4$ from the short end. It is required to design a coaxial transformer matching the load impedance Z_ℓ to the waveguide impedance Z_{og} for maximum power transfer.

Coaxial characteristic impedance

$$Z_0 = \frac{\eta_0}{2\pi\sqrt{\epsilon_r}}\ell n\left(\frac{b}{a}\right) \qquad (8\text{-}4\text{-}4a)$$

$$= 92 \log\frac{b}{a} \ \Omega \ \text{for} \ \epsilon_r = 2.25$$

Free-space intrinsic impedance $\qquad \eta_0 = 377 \ \Omega$

Post diameter $\qquad\qquad\qquad\quad a = 0.363 \ \text{cm}$

Load impedance

$$Z_\ell = Z_0 \frac{Z_g + jZ_0 \tan \beta\ell}{Z_0 + jZ_g \tan \beta\ell} \qquad (8\text{-}4\text{-}4b)$$

where

$$\beta_\ell = \frac{2\pi}{\lambda}\ell = 2\pi\frac{\ell}{\lambda}$$

Figure 8-4-2 shows the graphs for the load resistance and the coaxial characteristic impedance in terms of the coaxial-line length.

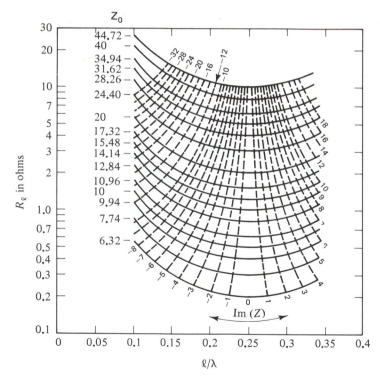

Figure 8-4-2 Data sheet of an IMPATT diode (From Ref. [1]. Reprinted by permission of Hewlett-Packard Company.)

Design procedure:

1. Calculate the load resistance R_ℓ and the load impedance Z_ℓ for maximum power transfer of 10-dB gain.
2. Find the electrical length ℓ of the coaxial line which will match the waveguide impedance to the load resistance from the IMPATT diode data sheet.
3. Determine the characteristic impedance Z_0 of the coaxial line which will transfer the load impedance Z_ℓ to the waveguide impedance Z_g.
4. Compute the values of a and b for the designed coaxial line.

Design Steps

Step 1. From Eq. (8-4-3), the load resistance is found to be

$$R_\ell = 1.35 \ \Omega$$

The load impedance is then

$$Z_\ell = 1.35 - j11 \ \Omega$$

Step 2. From the diode data sheet in Fig. 8-4-2, it is found that

$$\frac{\ell}{\lambda} = 0.132$$

Then the phase angle $\beta\ell$ is found to be $47.52°$. The electrical length ℓ is found as 0.52 cm.

Step 3. The coaxial-line characteristic impedance Z_0 can be found from the diode data graphs in Fig. 8-4-2.

$$Z_0 = 12.20 \ \Omega$$

Step 4. The diameter of the coaxial-line section is given as

$$a = 0.363 \text{ cm}$$

From the coaxial-line equation (8-4-4b), the ratio of the coaxial outer diameter b to its inner diameter a is found to be

$$\frac{b}{a} = 1.40$$

Then $b = 0.49$ cm.

8-4-2 Reduced-Height Waveguide Reflection Amplifier Design

A reduced-height waveguide is the one with a taper transition section to full-height circuit as shown in Fig. 8-4-3.

Design Example 8-4-2:   Reduced-Height Waveguide Amplifier Design

The following design example is based on Fong's work [2]. The reflection amplifier circuit consists of a reduced-height waveguide section of 0.0635-cm (0.025-in.) height with a taper transition to full-height WR-15 waveguide of 0.1879 cm (0.074 in.) at the output. A coaxial line with a center conductor of 0.0762 cm (0.030 in.) and an outer conductor of 0.1854 cm (0.073 in.) is cross-coupled on top of the waveguide. The coaxial section is terminated by a short. The packaged IMPATT diode is mounted under the bias post on the floor of the waveguide as shown in Fig. 8-4-3. The coaxial section above the diode is required to suppress oscillation and provide the desired frequency response in the amplifier. The length of coaxial section is important for tuning the amplifier. Figure 8-4-4 shows an equivalent circuit.

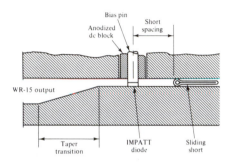

Figure 8-4-3   Reduced-height waveguide circuit. (From T. T. Fong et al. [2]. Copyright © 1976 IEEE. Reprinted by permission of the IEEE, Inc.)

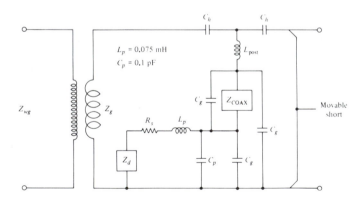

Figure 8-4-4   Equivalent circuit of Fig. 8-4-3. (After T. T. Fong et al. [2]. Copyright © 1976 IEEE. Reprinted by permission of the IEEE, Inc.)

The subscript h in the figure refers to the lumped reactance of the post, g refers to the post gap, and p indicates the device package. In practice, the coaxial length corresponds approximately to one-quarter wavelength of the operating frequency. A 10-dB gain is achieved at 60 GHz by an n-type flat-profile single-drift IMPATT diode with a doping density of 1×10^{17} cm^{-3}. The IMPATT diode has a junction area of 2×10^{-5} cm^2 and was biased at a current density of 17.5 kA/cm^2.

8-4-3 Circulator Reflection Amplifier Design

A circulator is a nonreciprocal phase shifter consisting of a thin slab of ferrite placed in a circular waveguide at a point where the dc magnetic field of the incident wave mode is circularly polarized. Ferrite is a family of $MeO \cdot Fe_2O_3$, where Me is a divalent metal iron. When a dc magnetic field is applied to a ferrite, the unpaired electrons in the ferrite tend to line up with the dc field due to their magnetic dipole moment. However, the nonreciprocal precession of unpaired electrons in the ferrite causes their relative permeabilities (μ_r^+ and μ_r^-) to be unequal and the wave in the ferrite is then circularly polarized.

A microwave circulator is a multiport waveguide junction in which the wave can flow only from the nth port to the $(n + 1)$ th port in one direction, as shown in Fig. 8-4-5. Although there is no restriction on the number of ports, the four-port microwave circulator is the most common. In microwave circuits, the circulator is commonly used for phase shift or Faraday rotation.

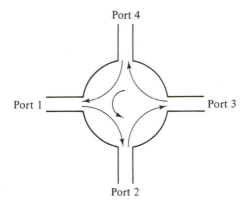

Port 4

Port 1

Port 3

Port 2

Figure 8-4-5 Diagram of a circulator.

Design Example 8-4-3:   Circulator Reflection Amplifier

The following design example is based on Kuno's work [3]. A circulator is incorporated with an IMPATT diode as shown in Fig. 8-4-6. The power gain can be expressed

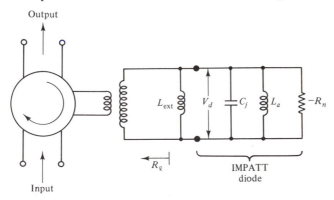

Output

L_{ext} V_d C_j L_a $-R_n$

R_ℓ

IMPATT
diode

Input

Figure 8-4-6 Equivalent circuit of a circulator reflection amplifier. (From H. J. Kuno and D. L. English [3]. Copyright © 1976 IEEE. Reprinted by permission of the IEEE, Inc.)

as the square of the magnitude of the reflection coefficient and it is

$$g = \left(\frac{R_\ell - R_n}{R_\ell + R_n}\right)^2 \tag{8-4-5}$$

where R_n = device negative resistance (the value is negative)
     R_ℓ = load resistance
The power gain at 60 GHz is shown in Fig. 8-4-7.

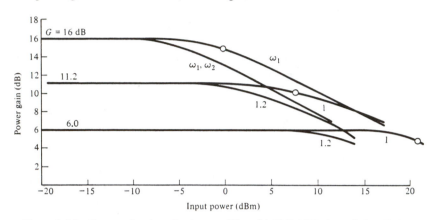

Figure 8-4-7   Power gain of a reflection amplifier with IMPATT diode. (After H. J. Kuno and D. L. English [3]. Copyright © 1976 IEEE. Reprinted by permission of the IEEE, Inc.)

8-4-4 Hybrid-Coupler Reflection Amplifier Design

The hybrid directional coupler used in microwave integrated circuits is the well-known Lange coupler or Wilkinson power divider, described in Chapter 5. The Lange 3-dB and 90° coupler consists of three or more parallel striplines with alternate line tied together. A signal wave incident in port 1 couples equal power into ports 2 and 4 but none into port 3, as shown in Fig. 8-4-8.

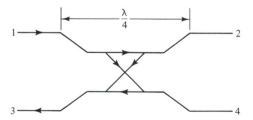

Figure 8-4-8   Hybrid 3-dB and 90° coupler.

On the contrary, a signal wave incident in port 3 couples equal power into ports 2 and 4 but none into port 1. Therefore, port 1 and 3 are isolated. If ports 2 and 4 are terminated in negative-resistance devices; the incident power from port 1 will be amplified and reflected by the negative-resistance devices at ports 2 and 4 and returned to port 3, as shown in Fig. 8-4-9.

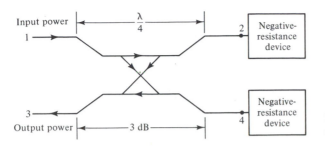

Figure   8-4-9   Hybrid-coupler reflection amplifier.

From the basic transmission-line theory, the power gain for a hybrid coupler is the square of the magnitude of the reflection coefficient and it is given by

$$g = \left(\frac{R_n - R_0}{R_n + R_0}\right)^2 \tag{8-4-6}$$

where R_n = negative resistance (the value is negative)
R_0 = hybrid-coupler characteristic impedance

Example 8-4-4:   Design of a Hybrid-Coupler Reflection Amplifier

A certain tunnel diode has a negative resistance R_n of 25 Ω and is to be used in a hybrid coupler for reflection amplifier. The hybrid coupler is a Lange coupler and has a characteristic impedance R_0 of 50 Ω. An input signal power of 200 mW is fed into port 1. Determine (a) the power gain in dB, and (b) the output power in watts.

Solution   (a) The power gain is

$$g = \left(\frac{-25 - 50}{-25 + 50}\right)^2 = 9 = 9.54 \text{ dB}$$

(b) The output power is

$$P_{\text{out}} = 0.2 \times 9 = 1.8 \text{ W}$$

REFERENCES

1. Hewlett-Packard Company, Inc, Application note 968, IMPATT amplifiers, pp. 1,2.

2. Fong, T. T., et al., Circuit characterization of V-band IMPATT oscillators and amplifiers. *IEEE Trans. Microwave Theory and Techniques,* Vol. MTT-24, No. 11, November 1976, pp. 752–758.

3. Kuno, H. J., and D. L. English, Nonlinear and intermodulation characteristics of millimeter-wave IMPATT amplifiers. *IEEE Trans. Microwave Theory and Techniques,* Vol. MTT-24, No. 11, November 1976, pp. 744–751.

PROBLEMS

8-1 Rectangular waveguides

8-1-1. An air-filled rectangular waveguide has dimensions of $a = 2.286$ cm and $b = 1.016$ cm. The signal frequency is 3 GHz. Compute the following for the TE_{10}, TE_{01}, TE_{11}, and TM_{11} modes.

(a) Cutoff frequency.

(b) Wavelength in the waveguide.

(c) Phase constant and phase velocity in the waveguide.

(d) Group velocity and wave impedance in the waveguide.

8-1-2. Show that the TM_{01} and TM_{10} modes in a rectangular waveguide do not exist.

8-1-3. The dominant mode TE_{10} is propagated in a rectangular waveguide of dimensions $a = 6$ cm and $b = 3$ cm. The distance between a maximum and a minimum is 4.55 cm. Determine the signal frequency of the dominant mode.

8-1-4. A TE mode of 10 GHz is propagated in an air-filled rectangular waveguide. The magnetic field in the z direction is given by

$$H_z = H_0 \cos\left(\frac{\pi x}{a}\right) \cos\left(\frac{\pi y}{a}\right) \qquad \text{A/m}$$

The phase constant is $\beta = 1.0367$ rad/cm, and the quantities x and y are expressed in centimeters and $a = b = \sqrt{6}$ are also in centimeters. Determine the cutoff frequency f_c, phase velocity v_g, guided wavelength λ_g, and the magnetic field intensity in the y direction.

8-1-5. A rectangular waveguide is designed to propagate the dominant mode TE_{10} at a frequency of 5 GHz. The cutoff frequency is 0.8 of the signal frequency. The ratio of the guide height to width is 0.5. The time-average power flowing through the guide is 1 kW. Determine the magnitudes of electric and magnetic intensities in the guide and indicate where these occur in the guide.

8-1-6. An air-filled rectangular waveguide has dimensions of $a = 1.295$ cm and $b = 0.648$ cm. The guide transports energy in the dominant mode TE_{10} at a rate of 1 hp (746 J). If the frequency is 20 GHz, what is the peak value of electric field occurring in the guide?

8-1-7. A rectangular waveguide is filled by dielectric material of $\epsilon_r = 9$ and has inside dimensions of 7×3.5 cm. It operates in the dominant TE_{10} mode.

(a) Determine the cutoff frequency.

(b) Find the phase velocity in the guide at a frequency of 2 GHz.

(c) Find the guided wavelength λ_g at the same frequency.

8-1-8. The electric field intensity of the dominant TE_{10} mode in a lossless rectangular waveguide is

$$E_y = E_{0y} \sin\left(\frac{\pi y}{a}\right) e^{-j\beta z} \qquad \text{for } f > f_c$$

(a) Find the magnetic field intensity $\mathbf{H}$.

(b) Compute the cutoff frequency.

(c) Determine the time-average power transmitted.

8-1-9. An air-filled rectangular waveguide with dimensions of 3 cm $\times$ 1 cm operates in the TE_{10} mode at 10 GHz. The waveguide is perfectly matched and the maximum E field existing everywhere in the guide is 10^3 V/m. Determine the expressions for the waveguide voltage, current, and characteristic impedance.

8-1-10. The dominant mode TE_{10} is propagated in a rectangular waveguide of dimensions $a = 2.28$ cm and $b = 1$ cm. Assume an air dielectric with a breakdown gradient of 30 kV/

cm and a frequency of 10 GHz. There are no standing waves in the guide. Determine the maximum average power that can be carried by the guide.

8-1-11. A rectangular waveguide is terminated in an unknown impedance at $z = 25$ cm. A dominant mode TE_{10} is propagated in the guide and its VSWR is measured as 2.8 at a frequency of 8 GHz. The adjacent minima are located at $z = 9.46$ cm and $z = 12.73$ cm.

    (a) Determine the value of the load impedance in terms of Z_0.

    (b) Find the position closest to the load where an inductive window is placed in order to obtain a VSWR of unity.

    (c) Determine the value of the window admittance.

8-2 Circular Waveguides

8-2-1. An air-filled circular waveguide is to be operated at a frequency of 6 GHz and is to have dimensions such that $f_c = 0.8f$ for the dominant mode. Determine:

    (a) The diameter of the guide.

    (b) The wavelength λ_g and the phase velocity v_g in the guide.

8-2-2. An air-filled circular waveguide of 2 cm inside radius is operated in the TE_{01} mode.

    (a) Compute the cutoff frequency.

    (b) If the guide is to be filled with a dielectric material of $\epsilon_r = 2.25$, to what value must its radius be changed in order to maintain the cutoff frequency at its original value?

8-2-3. An air-filled circular waveguide has a radius of 1.5 cm and is to carry energy at a frequency of 10 GHz. Find all TE and TM modes for which transmission is possible.

8-2-4. A TE_{11} wave is propagating through a circular waveguide. The diameter of the guide is 10 cm, and the guide is air-filled.

    (a) Find the cutoff frequency.

    (b) Find the wavelength λ_g in the guide for a frequency of 3 GHz.

    (c) Determine the wave impedance in the guide.

8-2-5. An air-filled circular waveguide has a diameter of 4 cm and is to carry energy at a frequency of 10 GHz. Determine all TE_{nm} modes for which transmission is possible.

8-2-6. A circular waveguide has a cutoff frequency of 9 GHz in dominant mode.

    (a) Find the inside diameter of the guide if it is air filled.

    (b) Determine the inside diameter of the guide if the guide is dielectric filled. The dielectric constant is $\epsilon_r = 4$.

8-3 Resonant Cavities

8-3-1. A circular waveguide has a radius of 3 cm. It is desired to make the waveguide resonant for a TE_{01} mode at 10 GHz by placing two perfectly conducting plates at its two ends. Determine the minimum distance in centimeters between the two plates.

8-3-2. A coaxial resonator is constructed of a section of coaxial line 5 cm long. The resonator is filled with dielectric material ($\epsilon_r = 2.56$) and open-circuited at both ends. The radius of inner conductor is 1 cm and the radius of the outer conductor is 2.5 cm.

    (a) Determine the resonant frequency of the resonator.

(b) Find the resonant frequency of the same cavity with one end open and one end shorted.

8-3-3. A rectangular-cavity resonator has dimensions of $a = 5$ cm, $b = 2$ cm, and $d = 15$ cm. Compute:

(a) The resonant frequency of the dominant mode for an air-filled cavity.

(b) The resonant frequency of the dominant mode for a dielectric-filled cavity of $\epsilon_r = 2.56$.

8-4 Reflection Amplifiers

8-4-1. A certain tunnel diode has a negative resistance R_n of 20 Ω and is connected to port 2 of a perfect three-port circulator to construct as a reflection amplifier. The circulator has a positive real characteristic impedance R_0 of 15 Ω. An input signal power of 2 mW is fed into port 1 and the output is taken from port 3. Compute:

(a) The power gain of the reflection amplifier.

(b) The output power in mW and dBm.

8-4-2. A certain IMPATT diode has a diode impedance Z_d of $-0.80 + j12$ at 9 GHz. It is required to design a waveguide-type amplifier for a power gain of 9 dB in the TE_{10} mode. The width of the waveguide is 2.287 cm and its height is 1.016 cm for X-band operation.

(a) Determine the cutoff frequency in gigahertz.

(b) Compute the cutoff wavelength in centimeters.

(c) Find the wavelength in waveguide in centimeters.

(d) Determine the waveguide characteristic impedance in ohms.

(e) Calculate the load resistance R_ℓ and the load impedance Z_ℓ for maximum power transfer of a 9-dB gain.

(f) Find the electrical length ℓ of the coaxial line which will match the waveguide impedance to the load resistance from the IMPATT diode data sheet.

(g) Determine the coaxial-line characteristic impedance Z_0 which will transfer the load impedance Z_ℓ to the waveguide impedance Z_g.

(h) Compute the values of a and b for the designed coaxial line.

8-4-3. An IMPATT diode has a negative resistance of 15 Ω and is connected to a circulator to make a reflection amplifier. The circulator has a positive real characteristic impedance of 20 Ω. The reactances of the diode and the circulator circuit are just conjugately canceled. Compute:

(a) The power gain of the circulator reflection amplifier.

(b) The output power in dBm and mW if the input power is 20 mW.

8-4-4. A Gunn diode has a negative resistance of 12 Ω and is connected to a circulator to form a reflection amplifier. The circulator has a positive real characteristic impedance of 15 Ω. The reactances of the diode and the circulator are just conjugately canceled. Determine:

(a) The power gain of the circulator reflection amplifier.

(b) The output power in mW and dBm if the input power is 10 mW.

8-4-5. Two IMPATT diodes are connected to ports 2 and 4 of a 3-dB and 90° hybrid coupler to make a reflection amplifier as shown in Fig. 8-4-9. The coupler has a positive characteristic impedance of 30 Ω and the diode has a negative resistance of 25 Ω. Compute:

    (a) The power gain of the hybrid-coupler reflection amplifier.

    (b) The output power of the amplifier if the input power is 100 mW.

8-4-6. Two TRAPATT diodes are connected to a 3-dB and 90° Lange coupler to form as a reflection amplifier. The coupler has a positive characteristic impedance of 25 Ω and the diode has a negative resistance of 30 Ω. Determine:

    (a) The power gain of the hybrid-coupler reflection amplifier.

    (b) The output power of the amplifier if the input power is 200 mW.

8-4-7. Design a hybrid-coupler reflection amplifier by using two Gunn diodes. The diode has a negative resistance of 25 Ω. Determine the characteristic impedance of the 3-dB and 90° Lange coupler in order to have a power gain of 20 dB.

Chapter 9

Microwave Oscillator Circuits and Oscillator Design

9-0 INTRODUCTION

Microwave oscillator design is similar to microwave amplifier design. The oscillator designed may have the same dc-biasing circuits and the same active device with the same set of S parameters used for the designed amplifier. For example, negative resistance is commonly used for amplifier design and it can also be used for oscillator design. However, the difference between the two is that the amplifier has an ac signal as its input source and the oscillator must have a tuning mechanism for its oscillation. As a result, the normal Smith chart is used for the amplifier design because the input reflection coefficient Γ_{in} and the output reflection coefficient Γ_{out} are less than unity, and a compressed Smith chart must be used for the oscillator design because both Γ_{in} and Γ_{out} are greater than unity. In this chapter, oscillation conditions, oscillator circuits, and microwave oscillator design are discussed.

9-1 OSCILLATION CONDITIONS

As described previously in amplifier design, the input circuit at the source port is called the *input matching network*, and the output circuit at the load port is called the *output matching network*. However, in a negative-resistance oscillator circuit, the matching networks at the two ports are often called the *generator-tuning network* and the *load-matching network*, respectively, as shown in Fig. 9-1-1. The generator-tuning network determines the oscillation frequency and the load-matching network provides the matching function.

320

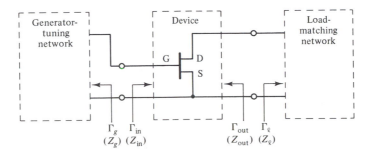

Figure 9-1-1   Two-port oscillator model.

There are three oscillation conditions for two-port oscillators, expressed as follows:

Condition 1: $K < 1$ (9-1-1)
Condition 2: $\Gamma_{in}\Gamma_g = 1$ (9-1-2)
Condition 3: $\Gamma_{out}\Gamma_\ell = 1$ (9-1-3)

where $K = \dfrac{1 + |\Delta|^2 - |S_{11}|^2 - |S_{22}|^2}{2|S_{12}S_{21}|}$ is the stability factor of the active device

$\Delta = S_{11}S_{22} - S_{12}S_{21}$ is the delta factor
Γ_{in} = input reflection coefficient of the active device
Γ_g = generator reflection coefficient of the resonant circuit
Γ_{out} = output reflection coefficient of the active device
Γ_ℓ = load reflection coefficient of the load-matching network

Condition 1 indicates that the stability factor must be less than unity for any possibility of oscillation. If $K > 1$, either the common terminal should be changed or positive feedback should be added. Conditions 2 and 3 say that the passive terminations Z_g and Z_ℓ must be added to resonate the input and output ports of the active device at the oscillation frequency. These two conditions are complementary to each other; that is, if condition 2 is satisfied, condition 3 is also satisfied, and vice versa. The simultaneous oscillation occurs when condition 2 is satisfied if and only if condition 3 is satisfied. These can be proved as follows:

From Eq. (3-4-8), we have

$$\Gamma_{in} = S_{11} + \frac{S_{12}S_{21}\Gamma_\ell}{1 - S_{22}\Gamma_\ell} = \frac{S_{11} - \Delta\Gamma_\ell}{1 - S_{22}\Gamma_\ell} \qquad (9\text{-}1\text{-}4)$$

Then

$$\frac{1}{\Gamma_{in}} = \frac{1 - S_{22}\Gamma_\ell}{S_{11} - \Delta\Gamma_\ell} \qquad (9\text{-}1\text{-}5)$$

Let condition 2 be satisfied, $\Gamma_{in}\Gamma_g = 1$. Then

$$S_{11}\Gamma_g - \Delta\Gamma_\ell\Gamma_g = 1 - S_{22}\Gamma_\ell$$

$$\Gamma_\ell = \frac{1 - S_{11}\Gamma_g}{S_{22} - \Delta\Gamma_g} \tag{9-1-6}$$

From Eq. (3-4-9) we have

$$\Gamma_{out} = S_{22} + \frac{S_{12}S_{21}\Gamma_g}{1 - S_{11}\Gamma_g} = \frac{S_{22} - \Delta\Gamma_g}{1 - S_{11}\Gamma_g} \tag{9-1-7}$$

Finally,

$$\Gamma_{out}\Gamma_\ell = 1 \tag{9-1-8}$$

This verifies that condition 3 is satisfied if and only if condition 2 is satisfied. Similarly, condition 2 is satisfied if and only if condition 3 is satisfied. Since $|\Gamma_g|$ and $|\Gamma_\ell|$ are less than unity, $|\Gamma_{in}|$ and $|\Gamma_{out}|$ must be greater than unity. Therefore, a compressed Smith chart must be used for oscillator design.

When a two-port is potentially unstable, an appropriate load termination Z_ℓ changes the two-port to a one-port negative-resistance device with input impedance Z_g as shown in Fig. 9-1-2. Therefore, from Fig. 9-1-2, there are two other equivalent oscillation conditions for a one-port oscillator:

$$\left. \begin{aligned} R_g + R_\ell = 0 \\ X_g + X_\ell = 0 \end{aligned} \right\} \text{ for series resonance} \qquad \begin{aligned} (9\text{-}1\text{-}9) \\ (9\text{-}1\text{-}10) \end{aligned}$$

or

$$\left. \begin{aligned} G_g + G_\ell = 0 \\ B_g + B_\ell = 0 \end{aligned} \right\} \text{ for shunt resonance} \qquad \begin{aligned} (9\text{-}1\text{-}11) \\ (9\text{-}1\text{-}12) \end{aligned}$$

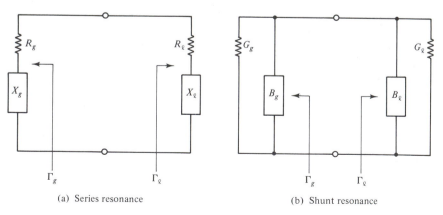

(a) Series resonance   (b) Shunt resonance

Figure 9-1-2 One-port oscillator model.

This means that the active device must provide a negative resistance for the occurrence of oscillation. From transmission-line theory, we have

$$\Gamma_{in} = \Gamma_\ell = \frac{Z_\ell - Z_0}{Z_\ell + Z_0} = \frac{R_\ell + jX_\ell - Z_0}{R_\ell + jX_\ell + Z_0} \qquad (9\text{-}1\text{-}13)$$

$$\Gamma_g = \frac{Z_g - Z_0}{Z_g + Z_0} = \frac{R_g + jX_g - Z_0}{R_g + jX_g + Z_0} \qquad (9\text{-}1\text{-}14)$$

Substitution of Eqs. (9-1-9) and (9-1-10) in Eq. (9-1-14) yields

$$\Gamma_{in}\Gamma_g = 1$$

This verifies Eq. (9-1-2); Eq. (9-1-3) can be verified similarly.

Example 9-1-1:   Calculation of Stability Factor K for Oscillation

A certain GaAs MESFET has the following S parameters measured with a 50-Ω reference at 8 GHz:

$$S_{11} = 0.44 \; \underline{/-56°}$$

$$S_{12} = 0.09 \; \underline{/75°}$$

$$S_{21} = 2.45 \; \underline{/65°}$$

$$S_{22} = 0.89 \; \underline{/-44°}$$

Determine the oscillation condition 1.

Solution   The delta factor is

$$\Delta = S_{11}S_{22} - S_{12}S_{21}$$

$$= 0.44 \; \underline{/-56°} \times 0.89 \; \underline{/-44°} - 0.09 \; \underline{/75°} \times 2.45 \; \underline{/65°}$$

$$= 0.53 \; \underline{/-79.11°}$$

The stability factor K is

$$K = \frac{1 + |\Delta|^2 - |S_{11}|^2 - |S_{22}|^2}{2|S_{12}S_{21}|}$$

$$= \frac{1 + |0.53|^2 - |0.44|^2 - |0.89|^2}{2|0.09 \times 2.45|} = 0.68 < 1$$

The device meets the oscillation condition 1.

9-2 OSCILLATOR-CIRCUIT CONFIGURATIONS

In general, there are two types of oscillator-circuit configurations: one for high microwave frequencies and one for low microwave frequencies.

9-2-1 Oscillator Circuits for High Microwave Frequencies

There are six oscillator-circuit configurations in common use for high microwave frequencies [1] and they are shown in Fig. 9-2-1. It can be seen that the first three series oscillator circuits are merely the resonant circuits with series feedback. The second three shunt oscillator circuits are viewed as the resonant circuits with shunt feedback. The ground node is usually set at the load point. The practical realization and dc-biasing circuit design can be analyzed by means of two-port network theory. The elements of the six circuit configurations are calculated from either the z or the y parameter [2]. If the S parameters of an active device are available, the z or y parameters can be obtained by conversion process from S parameters to either z or y parameters.

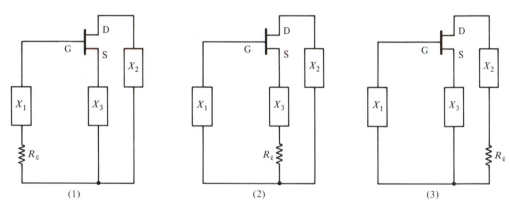

(a) Three series oscillator-circuit configurations

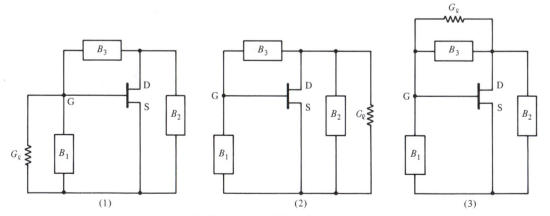

(b) Three shunt oscillator-circuit configurations

Figure 9-2-1 Oscillator-circuit configurations.

The basic equations for the values of the elements in the six oscillator circuits can be expressed in terms of z or y parameters as follows [1]:

Series oscillator circuit 1:

$$R_\ell = D_1 + F_r D_3 + F_i D_4 \tag{9-2-1}$$

$$X_1 = D_2 - (1 + F_r)\left(D_4 + \frac{D_3 F_r}{F_i}\right) \tag{9-2-2}$$

$$X_2 = -\frac{D_3(1 + F_r)}{F_i} - D_4 \tag{9-2-3}$$

$$X_3 = \frac{D_3 F_r}{F_i} + D_4 \tag{9-2-4}$$

where

$$D_1 = -\text{Re}\,(z_{11} + Fz_{12}) \tag{9-2-5}$$

$$D_2 = -\text{Im}\,(z_{11} + Fz_{12}) \tag{9-2-6}$$

$$D_3 = -\text{Re}\,(z_{21} + Fz_{22}) \tag{9-2-7}$$

$$D_4 = -\text{Im}\,(z_{21} + Fz_{22}) \tag{9-2-8}$$

$$F = F_r + jF_i = \frac{z_{21} - Az_{11}}{Az_{12} - z_{22}} \tag{9-2-9}$$

Series oscillator circuit 2:

$$R_\ell = \frac{D_1 + F_r D_3 + F_i D_4}{|F|^2} \tag{9-2-10}$$

$$X_1 = \frac{D_1(1 + F_r)}{F_i} + D_2 \tag{9-2-11}$$

$$X_2 = \frac{D_1(F_r + |F|^2)/F_i - F_i D_3 + F_r D_4}{|F|^2} \tag{9-2-12}$$

$$X_3 = -\frac{D_1}{F_i} \tag{9-2-13}$$

Series oscillator circuit 3:

$$R_\ell = \frac{D_1 + F_r D_3 + F_i D_4}{|1 + F|^2} \tag{9-2-14}$$

$$X_1 = \frac{(D_1 - D_3)F_r}{F_i} + D_2 - D_4 \tag{9-2-15}$$

$$X_2 = \frac{D_1 - D_3}{F_i} \tag{9-2-16}$$

$$X_3 = \frac{(1 + F_r)D_4 - (D_1 - D_3)F_r/F_i - F_iD_1}{|1 + F|^2} \tag{9-2-17}$$

Shunt oscillator circuit 1:

$$G_\ell = C_1 + A_r C_3 + A_i C_4 \tag{9-2-18}$$

$$B_1 = C_2 + (1 - A_r)\left(C_4 + \frac{C_3 A_r}{A_i}\right) \tag{9-2-19}$$

$$B_2 = \frac{C_3(A_r - 1)}{A_i} + C_4 \tag{9-2-20}$$

$$B_3 = -\frac{C_3 A_r}{A_i} - C_4 \tag{9-2-21}$$

where

$$C_1 = -\mathrm{Re}\,(y_{11} + Ay_{12}) \tag{9-2-22}$$

$$C_2 = -\mathrm{Im}\,(y_{11} + Ay_{12}) \tag{9-2-23}$$

$$C_3 = -\mathrm{Re}\,(y_{21} + Ay_{22}) \tag{9-2-24}$$

$$C_4 = -\mathrm{Im}\,(y_{21} + Ay_{22}) \tag{9-2-25}$$

$$A = A_r + jA_i = -\frac{y_{21} + y_{12}^*}{2\,\mathrm{Re}\,y_{22}} \tag{9-2-26}$$

Shunt oscillator circuit 2:

$$G_\ell = \frac{C_1 + A_r C_3 + A_i C_4}{|A|^2} \tag{9-2-27}$$

$$B_1 = \frac{C_1(A_r - 1)}{A_i} + C_2 \tag{9-2-28}$$

$$B_2 = \frac{C_1(A_r - |A|^2)/A_i - A_i C_3 + A_r C_4}{|A|^2} \tag{9-2-29}$$

$$B_3 = \frac{C_1}{A_i} \tag{9-2-30}$$

Shunt oscillator circuit 3:

$$G_\ell = \frac{C_1 + A_r C_3 + A_i C_4}{|1 + A|^2} \tag{9-2-31}$$

$$B_1 = \frac{(C_1 + C_3)A_r}{A_i} + C_2 + C_4 \tag{9-2-32}$$

$$B_2 = -\frac{C_2 + C_3}{A_i} \tag{9-2-33}$$

$$B_3 = \frac{(A_r - 1)[C_4 + (C_1 + C_3)A_r/A_i] + A_iC_1}{|1 + A|^2} \tag{9-2-34}$$

Sometimes, even though the embedding element values for the six oscillator-circuit configurations are available, not all of these circuit configurations can readily be realized in practice. In particular, all of the three shunt oscillator configurations are difficult to realize with a coax packaged device. This is due to the fact that there is no easy way physically to locate a capacitor between the gate and the drain since the source ring gets in the way. The series circuit configurations are more amenable to fabrication.

9-2-2 Oscillator Circuits for Low Microwave Frequencies

At low microwave frequencies, four practical oscillator-circuit configurations are in use: the Armstrong, Clapp, Colpitts, and Hartley oscillator circuits (Fig. 9-2-2).

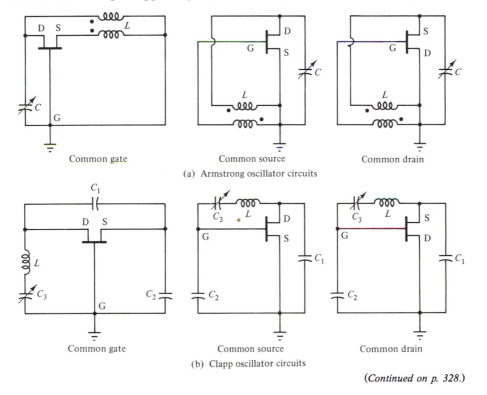

(a) Armstrong oscillator circuits

(b) Clapp oscillator circuits

(*Continued on p. 328.*)

Figure 9-2-2   Four conventional types of oscillator circuits.

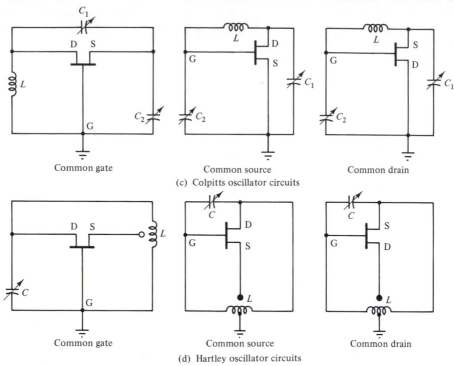

Common gate   Common source   Common drain

(c) Colpitts oscillator circuits

Common gate   Common source   Common drain

(d) Hartley oscillator circuits

Figure 9-2-2 (*Continued*)

The common-gate circuits provide series feedback results; the common-source and common-drain circuits yield shunt feedback functions. When the dc-biasing and load circuits are added to the oscillator circuit, the complete oscillator circuit is usually complicated. Figure 9-2-3 shows two practical Colpitts oscillator circuits in common

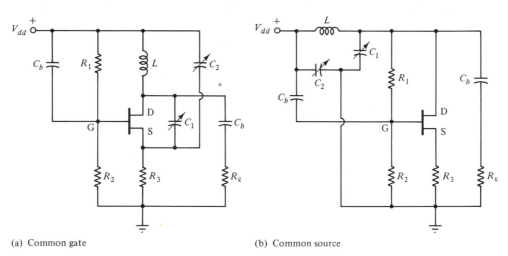

(a) Common gate   (b) Common source

Figure 9-2-3 Practical Colpitts oscillator circuits.

use. The large block capacitor C_b across R_1 effectively grounds the base for the common-gate oscillator, and it also prevents the dc shorting of the drain to gate.

9-3 OSCILLATION-TUNING CIRCUITS

There are four types of oscillation-tuning circuits: (1) fixed-tuning circuits, (2) YIG-tuned circuits, (3) varactor-tuned circuits, and (4) cavity-tuned circuits.

9-3-1 Fixed-Tuning Circuits

As described in Section 9-2, the Armstrong circuit uses a mutual inductor for tuning function, the Clapp circuit utilizes an inductor in series with a capacitor as the tuning device, the Colpitts circuit employs a capacitor voltage divider for tuning purpose, and the Hartley circuit adopts a tapped inductor for tuning mechanism. All of these tuning circuits are fixed.

9-3-2 YIG-Tuned Circuits

YIG stands for yttrium–iron–garnet ($Y_3Fe_5O_{12}$) and is a magnetic crystal. The YIG can provide a high-Q resonance in a magnetic field. Figure 9-3-1 shows the circuits for a YIG-tuned oscillator.

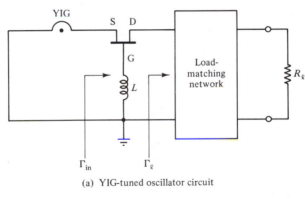

(a) YIG-tuned oscillator circuit

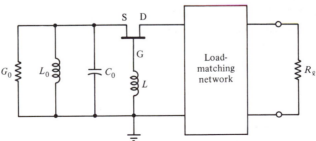

(b) YIG-tuned equivalent oscillator circuit

Figure 9-3-1   YIG-tuned oscillator circuits.

The resonant angular frequency of electron in the dc magnetic field is given by

$$\omega_0 = \gamma H_0 \qquad (9\text{-}3\text{-}1)$$

where $\gamma = 2.21 \times 10^5$ (rad/s)/(A/m) is the gyromagnetic ratio
$H_0 =$ dc magnetic field, amperes per meter

The resonant angular frequency in the magnetization field is expressed by

$$\omega_m = \gamma M_0 \qquad (9\text{-}3\text{-}2)$$

where M_0 is the dc magnetization in webers per square meter.

The YIG crystal has a diameter much less than a quarter-wave long and is fully magnetized. Its equivalent circuit [2] can be expressed by three parallel elements G_0, L_0, and C_0 as

$$G_0 = \frac{d^2}{\mu_0 V \omega_m Q_u} + G_L \qquad (9\text{-}3\text{-}3)$$

$$L_0 = \frac{\mu_0 V \omega_m}{\omega_0 d^2} \qquad (9\text{-}3\text{-}4)$$

$$C_0 = \frac{1}{\omega_0^2 L_0} \qquad (9\text{-}3\text{-}5)$$

where $\mu_0 = 4\pi \times 10^{-7}$ H/m
$V =$ crystal volume, cubic meters
$d =$ coupling loop diameter, meters
$G_L =$ inductor conductance
$G_u = (H_0 - M_s/3)/\Delta H$ is the unloaded quality factor
$M_s =$ saturation magnetization, webers per square meter
$\Delta H =$ resonance line width, webers

9-3-3 Varactor-Tuned Circuits

The third type of oscillator-tuning circuit is the varactor-tuned or voltage-tuned circuit shown in Fig. 9-3-2. The varactor may be inserted in series with the drain terminal or in series with the gate terminal. The output power of both oscillators is taken from the drain terminal and the power capacity of 50 to 200 mW can be achieved from 8 to 12 GHz. This type of oscillator circuit, often called a voltage-controlled oscillator (VCO), provides a low-Q resonance.

9-3-4 Cavity-Tuned Circuits

The fourth type of oscillator-tuning circuit is the cavity-tuned circuit, of which there are three types: (1) full-height waveguide cavity, (2) reduced-height waveguide cavity, and (3) coaxial cavity.

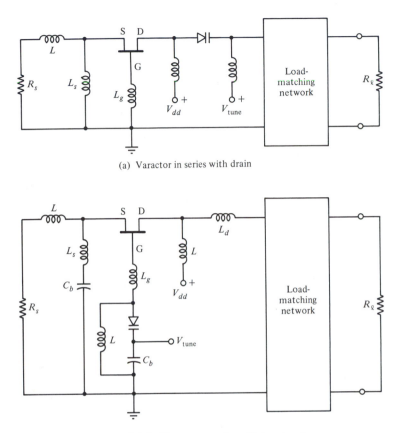

(a) Varactor in series with drain

Figure 9-3-2 Varactor-tuned oscillator circuits.

Full-height waveguide cavity.   A section of full-height waveguide with a sliding plunger can be used as a cavity tuning circuit for an oscillator, as shown in Fig. 9-3-3. The full-height waveguide cavity consists of a section of a TE_{10}-mode rectangular waveguide shorted at two ends to form a TE_{101}-mode resonant cavity. Its equivalent circuit is shown in Fig. 9-3-4,

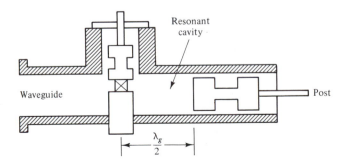

Figure 9-3-3 Full-height waveguide cavity.

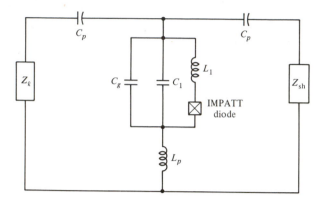

Figure 9-3-4 Equivalent circuit of full-height waveguide cavity.

where

L_1 = diode package inductance
C_1 = diode package capacitance
L_p = resonant post inductance
C_p = resonant post capacitance
C_g = gap capacitance
Z_ℓ = load impedance
Z_{sh} = $jZ_0 \tan (\beta_g d)$ for short circuit
$Z_0 = \dfrac{H}{W} \dfrac{377}{\sqrt{1 - (f_c/f)^2}}$ is the characteristic impedance of waveguide with one end
       shorted
H = waveguide height
β_g = phase constant in waveguide
W = waveguide width
d = distance between the post and the cavity side wall

Reduced-height waveguide cavity.   The reduced-height waveguide cavity is one whose height is gradually reduced with a taper transition to a lower height at one end for tuning purposes, with its other end serving as the power output port, as shown in Fig. 9-3-5. This type of cavity-tuned oscillator circuit is frequently used

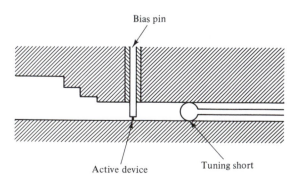

Bias pin

Active device   Tuning short

Figure 9-3-5 Reduced-height waveguide cavity.

for high-power oscillators and its equivalent circuit is shown in Fig. 9-3-6. The waveguide used for the reduced-height cavity is usually the WR-15 type and its dimensions are 0.3759 cm by 0.1879 cm. The subscript h in Fig. 9-3-6 refers to the lumped reactance of the post, g refers to the post gap, and p indicates the device package.

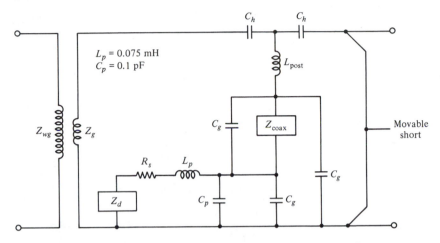

Figure 9-3-6 Equivalent circuit of reduced-height waveguide cavity. (After T. T. Fong et al. [3]. Copyright © 1976 IEEE. Reprinted by permission of the IEEE, Inc.)

Coaxial cavity.   A section of coaxial line is also used as a waveguide cavity for tuning purposes as shown in Fig. 9-3-7. The device impedance Z_d is always known. The first step is to determine the load impedance Z_ℓ for oscillation. After the load impedance is found it is necessary to design the required dimensions of the coaxial cavity for tuning purpose. The equivalent circuit of a coaxial cavity is shown in Fig. 9-3-8.

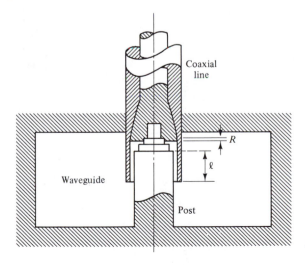

Figure 9-3-7 Coaxial cavity.

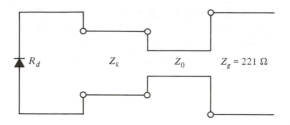

Figure 9-3-8 Equivalent circuit of a coax-ial cavity.

9-4 ONE-PORT OSCILLATOR DESIGN

Two commonly used one-port oscillator circuits are shown in Fig. 9-4-1. One is the series resonant circuit and the other is the shunt resonant circuit. Two-terminal negative-resistance diodes are the commonly used active devices.

From the oscillation conditions described earlier, the active device must produce a negative resistance for oscillation and the load circuit must provide an impedance for matching purpose. This means, from Eqs. (9-1-9) and (9-1-10), that

$$R_g + R_\ell = 0 \tag{9-4-1}$$

and $\left. \begin{array}{c} \\ \end{array} \right\}$ for series resonance

$$X_g + X_\ell = 0 \tag{9-4-2}$$

and

$$G_g + G_\ell = 0 \tag{9-4-3}$$

$\left. \begin{array}{c} \\ \end{array} \right\}$ for shunt resonance

$$B_g + B_\ell = 0 \tag{9-4-4}$$

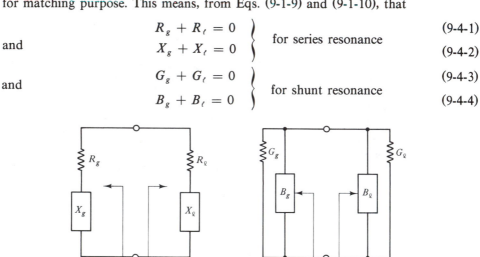

Figure 9-4-1 One-port oscillator circuits.

However, for startup of oscillation in a series resonant circuit, the negative resistance of the active device must exceed the load resistance by about 20%. That is,

$$R_g > 1.20R_\ell \quad \text{for startup of oscillation in series resonant circuit} \tag{9-4-5}$$

Similarly, for a shunt resonant circuit, the following condition must be maintained:

$$G_g < 1.20G_\ell \quad \text{for startup of oscillation in a shunt resonant circuit} \tag{9-4-6}$$

The compressed Smith chart includes the normal Smith chart for $|\Gamma| \leq 1$ plus a portion of the negative resistance region for $|\Gamma| > 1$. As described in Chapter 2, one alternative method of obtaining negative resistance for $|\Gamma| > 1$ is to plot $1/\Gamma^*$ in the normal Smith chart and take the values of the resistance circles as being negative and the reactance circles as labeled.

Design Example 9-4-1:   One-Port Oscillator Design

A one-port oscillator is to be designed to have a generator reflection coefficient $\Gamma_g = 1.24 \,\underline{/30°}$ with a 50-Ω reference. The oscillation frequency is 10 GHz. Determine

1. the generator network
2. the load-matching network
3. the complete one-port oscillator circuit diagram

Design Steps:

Step 1. Plot $1/\Gamma_g^* = 0.81 \,\underline{/30°}$ in the normal Smith chart as shown in Fig. 9-4-2. Read $z_g = -1.40 + j3.20$ and $Z_g = -70 + j160.00$.

Step 2. $R_\ell = -R_g = 70$ and $L_g = 2.55$ nH; $X_\ell = -X_g = -160$ and $C_\ell = 0.10$ pF.

Step 3. The complete oscillator circuit is shown in Fig. 9-4-3.

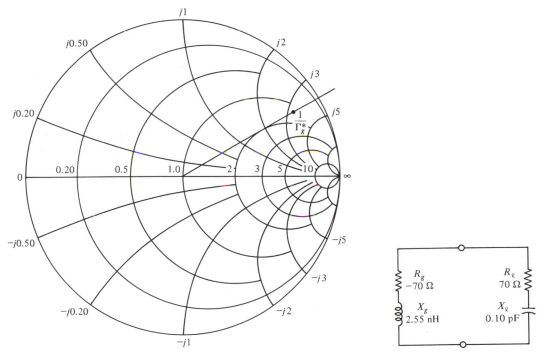

Figure 9-4-3 Designed oscillator circuit.

Figure 9-4-2 Graphical solution for Example 9-4-1.

9-5 TWO-PORT OSCILLATOR DESIGN

Two-port oscillator requires active devices with three terminals. A commonly used two-port oscillator circuit is shown in Fig. 9-5-1.

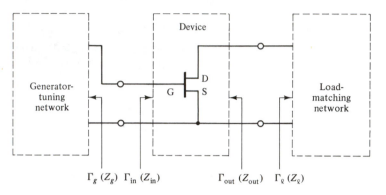

$$\Gamma_g\ (Z_g)\quad \Gamma_{in}\ (Z_{in})\qquad\qquad \Gamma_{out}\ (Z_{out})\quad \Gamma_\ell\ (Z_\ell)$$

Figure 9-5-1   Two-port oscillator circuit.

9-5-1 Maximum Efficient Power Gain

The maximum output power of a two-port oscillator can be approximated by the equation [4]

$$P_{osc(max)} = P_{sat}\left(1 - \frac{1}{G} - \frac{\ell n G}{G}\right) \tag{9-5-1}$$

where P_{sat} = saturated amplifier output power, watts
      $G = |S_{21}|^2$ is the small-signal common-source transducer power gain
Equation (9-5-1) can be derived from a common-source amplifier. The output power of an amplifier is given by

$$P_{out} = P_{sat}\left[1 - \exp\left(\frac{-GP_{in}}{P_{sat}}\right)\right] \tag{9-5-2}$$

where P_{in} is the amplifier input power in watts. For maximum output power,

$$d(P_{out} - P_{in}) = 0 \tag{9-5-3}$$

then

$$\frac{\partial P_{out}}{\partial P_{in}} = 1 \tag{9-5-4}$$

Differentiation of Eq. (9-5-2) with respect to P_{sat} results in

$$\exp\left(\frac{-GP_{in}}{P_{sat}}\right) = \frac{1}{G} \tag{9-5-5}$$

Then

$$P_{in} = P_{sat}\frac{\ell n\ G}{G} \tag{9-5-6}$$

Substitution of Eq. (9-5-5) in Eq. (9-5-2) yields the maximum amplifier output power as

$$P_{out} = P_{sat}\left(1 - \frac{1}{G}\right) \tag{9-5-7}$$

The maximum oscillator output power can be expressed as

$$\begin{aligned} P_{osc(max)} &= P_{out} - P_{in} \\ &= P_{sat}\left(1 - \frac{1}{G} - \frac{\ell n\ G}{G}\right) \end{aligned} \tag{9-5-8}$$

Then the maximum efficient power gain of an oscillator can be calculated from Eqs. (9-5-6) and (9-5-7) as

$$G_{osc(max)} = \frac{P_{out}}{P_{in}} = \frac{G-1}{\ell n\ G} \tag{9-5-9}$$

It is seen from Eq. (9-5-9) that the saturated output power of an active device is not needed for obtaining the oscillator maximum power gain. The maximum efficient power gain is often used to determine at what gain level the large-signal S parameters are to be used for oscillator application. In practical design, all that is needed is the S-parameter S_{21} at different frequencies.

Example 9-5-1:   Maximum Efficient Power Gain of a Two-Port Oscillator

A GaAs MESFET is to be used in a common-source oscillator and it has the following parameters:

Small-signal transducer gain at 9 GHz	$G = |S_{21}|^2 = 6$ dB
Saturated amplifier output power	$P_{sat} = 1.20$ W

Compute (a) the maximum efficient oscillator power gain in decibels, and (b) the maximum oscillator power output in megawatts.

Solution   (a) $G = 6$ dB $= 3.981$. From Eq. (9-5-9), the maximum efficient oscillator gain is

$$G_{osc(max)} = \frac{G-1}{\ell n\ G} = \frac{3.981 - 1}{\ell n\ (3.981)} = 2.16 = 3.34\text{ dB}$$

(b) From Eq. (9-5-8), the maximum oscillator output power is

$$P_{osc(max)} = P_{sat}\left(1 - \frac{1}{g} - \frac{\ell n\, G}{G}\right)$$

$$= 1.20\left[1 - \frac{1}{3.981} - \frac{\ell n\,(3.981)}{3.981}\right]$$

$$= 1.20(1 - 0.251 - 0.347)$$

$$= 0.48\text{ W}$$

9-5-2 Design Example

A design example of two-port oscillator is described as follows.

Design Procedures

1. Check the stability factor $K < 1$ for oscillation.
2. Convert the S parameters to the z parameters for series oscillator-circuit realization.
3. Convert the y parameters from the S parameters.
4. Compute the embedding element values for the series circuit no. 1.
5. Draw the equivalent series-oscillator circuit.
6. Check the maximum efficient output power gain.

Design Example 9-5-2:   Two-Port Oscillator Design

A certain common-source packaged GaAs MESFET has the following S parameters measured at $V_{ds} = 3$ V and $I_{ds} = 30$ mA at 4 GHz:

$$S_{11} = 0.64\ \underline{/-148°}$$

$$S_{12} = 0.06\ \underline{/11°}$$

$$S_{21} = 3.46\ \underline{/29°}$$

$$S_{22} = 0.61\ \underline{/-104°}$$

Design a series resonant circuit for the oscillator.

Design Steps

Step 1. Check the stability $K < 1$ for oscillation.

$$\Delta = S_{11}S_{22} - S_{12}S_{21}$$

$$= 0.64\ \underline{/-148°} \times 0.61\ \underline{/-104°} - 0.06\ \underline{/11°} \times 3.46\ \underline{/29°}$$

$$= 0.37\ \underline{/139.4°}$$

$$K = \frac{1 + |\Delta|^2 - |S_{11}|^2 - |S_{22}|^2}{2|S_{12}S_{21}|}$$

$$= \frac{1 + |0.37|^2 - |0.64|^2 - |0.61|^2}{2|0.06 \times 3.46|}$$

$$= 0.86 < 1$$

The device is potentially unstable.

Step 2. Convert the S parameters to z parameters.

$$z_{11} = \frac{(1 + S_{11})(1 - S_{22}) + S_{12}S_{21}}{(1 - S_{11})(1 - S_{22}) - S_{12}S_{21}}$$

$$= \frac{(1 + 0.64 \angle{-148°})(1 - 0.61 \angle{-104°}) + 0.06 \angle{11°} \times 3.46 \angle{29°}}{(1 - 0.64 \angle{-148°})(1 - 0.61 \angle{-104°}) - 0.06 \angle{11°} \times 3.46 \angle{29°}}$$

$$= 0.49 \angle{-39°}$$

$$= 0.38 - j0.31$$

$$z_{12} = \frac{2S_{12}}{(1 - S_{11})(1 - S_{22}) - S_{12}S_{21}}$$

$$= \frac{2 \times 0.06 \angle{11°}}{(1 - 0.64 \angle{-148°})(1 - 0.61 \angle{-104°}) - 0.06 \angle{11°} \times 3.46 \angle{29°}}$$

$$= 0.07 \angle{-28.7°}$$

$$= 0.06 - j0.03$$

$$z_{21} = \frac{2S_{21}}{(1 - S_{11})(1 - S_{22}) - S_{12}S_{21}}$$

$$= \frac{2 \times 3.46 \angle{29°}}{(1 - 0.64 \angle{-148°})(1 - 0.61 \angle{-104°}) - 0.06 \angle{11°} \times 3.46 \angle{29°}}$$

$$= 3.78 \angle{-10.7°}$$

$$= 3.71 - j0.70$$

$$z_{22} = \frac{(1 - s_{11})(1 + S_{22}) + S_{12}S_{21}}{(1 - S_{11})(1 - S_{22}) - S_{12}S_{21}}$$

$$= \frac{(1 - 0.64 \angle{-148°})(1 + 0.61 \angle{-104°}) + 0.06 \angle{11°} \times 3.46 \angle{29°}}{(1 - 0.64 \angle{-148°})(1 - 0.61 \angle{-104°}) - 0.06 \angle{11°} \times 3.46 \angle{29°}}$$

$$= 0.95 \angle{-56.1°}$$

$$= 0.53 - j0.79$$

Step 3. Calculate the y parameters from the S parameters:

$$y_{11} = \frac{(1 - S_{11})(1 + S_{22}) + S_{12}S_{21}}{(1 + S_{11})(1 + S_{22}) - S_{12}S_{21}}$$

$$= \frac{(1 - 0.64\ \underline{/-148°})(1 + 0.61\ \underline{/-104°}) + 0.06\ \underline{/11°} \times 3.46\ \underline{/29°}}{(1 + 0.64\ \underline{/-148°})(1 + 0.61\ \underline{/-104°}) - 0.06\ \underline{/11°} \times 3.46\ \underline{/29°}}$$

$$= 2.52\ \underline{/70.75°}$$

$$= 0.83 + j2.38$$

$$y_{12} = \frac{-2S_{12}}{(1 + S_{11})(1 + S_{22}) - S_{12}S_{21}}$$

$$= \frac{-2 \times 0.06\ \underline{/11°}}{(1 + 0.64\ \underline{/-148°})(1 + 0.61\ \underline{/-104°}) - 0.06\ \underline{/11°} \times 3.46\ \underline{/29°}}$$

$$= -0.17\ \underline{/98.1°}$$

$$= 0.02 - j0.17$$

$$y_{21} = \frac{-2S_{21}}{(1 + S_{11})(1 + S_{22}) - S_{12}S_{21}}$$

$$= \frac{-2 \times 3.46\ \underline{/29°}}{(1 + 0.64\ \underline{/-148°})(1 + 0.61\ \underline{/-104°}) - 0.06\ \underline{/11°} \times 3.46\ \underline{/29°}}$$

$$= -10.03\ \underline{/116.1°}$$

$$= 4.41 - j9.01$$

$$y_{22} = \frac{(1 + S_{11})(1 - S_{22}) + S_{12}S_{21}}{(1 + S_{11})(1 + S_{22}) - S_{12}S_{21}}$$

$$= \frac{(1 + 0.64\ \underline{/-148°})(1 - 0.61\ \underline{/-104°}) + 0.06\ \underline{/11°} \times 3.46\ \underline{/29°}}{(1 + 0.64\ \underline{/-148°})(1 + 0.61\ \underline{/-104°}) - 0.06\ \underline{/11°} \times 3.46\ \underline{/29°}}$$

$$= 1.29\ \underline{/87.73°}$$

$$= 0.05 + j1.23$$

Step 4. Compute the embedding element values for series oscillator circuit no. 1 as shown in Fig. 9-2-1(a). The A factor is

$$A = A_r + jA_i = -\frac{y_{21} + y_{12}^*}{2\ Re\ y_{22}}$$

$$= -\frac{4.41 - j9.01 + 0.02 + j0.17}{2 \times 0.05}$$

$$= 98.90 \ \underline{/116.62°}$$

$$= -44.31 + j88.42$$

The F factor is

$$F = F_r + jF_i = \frac{z_{21} - Az_{11}}{Az_{12} - z_{22}}$$

$$= \frac{3.71 - j0.70 - 98.90 \ \underline{/116.62°} \times 0.49 \ \underline{/-39°}}{98.90 \ \underline{/116.62°} \times 0.07 \ \underline{/-28.7°} - (0.53 - j0.79)}$$

$$= 6.28 \ \underline{/170°} = 6.28 \ \underline{/-190°}$$

$$= -6.18 + j1.09$$

$$D_1 = -\mathrm{Re}(z_{11} + Fz_{12})$$

$$= -\mathrm{Re}(0.49 \ \underline{/-39°} + 6.28 \ \underline{/-190°} \times 0.07 \ \underline{/-28.7°}$$

$$= -\mathrm{Re}(0.04 - j0.03)$$

$$= -0.04$$

$$D_2 = -\mathrm{Im}(z_{11} + Fz_{12})$$

$$= -\mathrm{Im}(0.04 - j0.03)$$

$$= 0.03$$

$$D_3 = -\mathrm{Re}(z_{21} + Fz_{22})$$

$$= -\mathrm{Re}(3.78 \ \underline{/-10.7°} + 6.28 \ \underline{/-190°} \times 0.95 \ \underline{/-56.1°})$$

$$= -\mathrm{Re}(1.29 + j4.76)$$

$$= -1.29$$

$$D_4 = -\mathrm{IM}(z_{21} + Fz_{22}$$

$$= -\mathrm{Im}(1.29 + j4.76)$$

$$= -4.76$$

The embedding element values are

$$R_\ell = D_1 + F_r D_3 + F_i D_4$$

$$= -0.04 + (-6.18)(-1.29) + (1.09)(-4.76)$$

$$= 2.74 \; \Omega$$

$$X_1 = D_2 - (1 + F_r)\left(D_4 + \frac{D_3 F_r}{F_i}\right)$$

$$= 0.03 - [1 + (-6.18)]\left[-4.76 + \frac{(-1.29)(-6.18)}{(1.09)}\right]$$

$$= 13.24 \; \Omega$$

$$L_1 = \frac{13.24}{2\pi \times 4 \times 10^9}$$

$$= 0.53 \; \text{nH}$$

$$X_2 = -\frac{D_3(1 + F_r)}{F_i} - D_4$$

$$= -\frac{(-1.29)\,|1 + (-6.18)|}{1.09} - (-4.76)$$

$$= -1.37 \; \Omega$$

$$C_2 = \frac{1}{2\pi \times 4 \times 10^9 \times 1.37}$$

$$= 29 \; \text{pF}$$

$$X_3 = \frac{D_3 F_r}{F_i} + D_4$$

$$= \frac{(-1.29)(-6.18)}{1.09} + (-4.76)$$

$$= 2.55 \; \Omega$$

$$L_3 = \frac{2.55}{2\pi \times 4 \times 10^9}$$

$$= 0.10 \; \text{nH}$$

Step 5. Draw the oscillator circuit (Fig. 9-5-2).

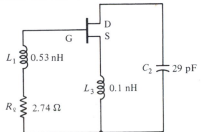

Figure 9-5-2 Oscillator circuit for Example 9-5-1.

Step 6. Check the maximum efficient power gain.

$$\text{Gain} = |S_{21}|^2 = |3.46|^2 = 11.97 = 10.78 \text{ dB}$$

Then from Eq. (9-5-9), we have

$$G_{\text{osc(max)}} = \frac{G - 1}{\ell \text{n } G}$$

$$= \frac{11.97 - 1}{\ell \text{n } (11.97)} = 4.42$$

$$= 6.46 \text{ dB}$$

9-6 HIGH-POWER OSCILLATOR DESIGN

A high-power oscillator generally requires large signal measurements. Kotzebue [1] and Johnson [4] did some basic work for the high-power oscillator design. After the large-signal S parameters have been measured the embedding element values for the six oscillator circuit configurations as shown in Fig. 9-2-1 may be computed and realized.

Design Procedure

1. Measure the large-signal S parameters.
2. Check the stability factor $K < 1$ for oscillation.
3. Convert the S parameters to z and y parameters.
4. Compute the embedding element values R_ℓ, X_1, X_2, and X_3 for a desired oscillator circuit.
5. Realize the oscillator circuit.

Design Example 9-6-1:   High-Power Oscillator Design

A MSX801G FET has its large-signal S parameters measured, z and y parameters converted, and embedding element values plotted as a function of power gain or frequency as shown in Fig. 9-6-1. Design a 5-dB high-power oscillator for 10 GHz by using series oscillator circuit 1 in Fig. 9-2-1.

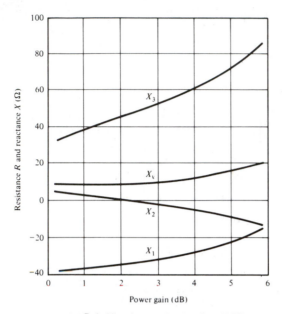

(a) Embedding element values vs. gain at 10 GHz

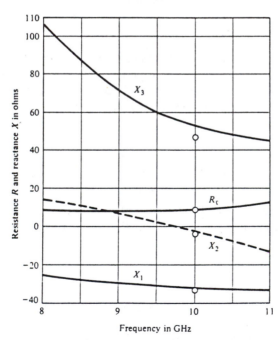

Frequency in GHz

(b) Embedding element values vs. frequency
for maximum power output

Figure 9-6-1 Optimum embedding element values. (After Kenneth M. Johnson [4]. Copyright © 1979 IEEE. Reprinted by permission of the IEEE, Inc.)

Design Steps

Step 1. Large-signal S parameters were measured.

Step 2. The stability factor $K < 1$ was checked.

Step 3. The z and y parameters were converted and plotted.

Step 4. From Fig. 9-6-1(a), we obtain the embedding element values for 5 dB at 10 GHz as

$$R_\ell = 15.43 \ \Omega$$

$$X_1 = -22.57 \ \Omega$$

$$C_1 = \frac{1}{2\pi \times 10^{10} \times 22.57} = 0.71 \ \text{pF}$$

$$X_2 = -9.63 \ \Omega$$

$$C_2 = \frac{1}{2\pi \times 10^{10} \times 9.63} = 1.65 \ \text{pF}$$

$$X_3 = +71.87 \ \Omega$$

$$L_3 = \frac{71.87}{2\pi \times 10^{10}} = 1.14 \ \text{nH}$$

Step 5. Draw equivalent series oscillator circuit no. 1 as shown in Fig. 9-2-1(a) and realize the circuit (Fig. 9-6-2). A coaxial cavity was used for realization as shown in Fig. 9-6-3. The inductance L from the source lead was provided by a short length of line to the outer conductor wall of the coaxial cavity. The series capacitance and load inductance were obtained by transforming down a shortened half-wave 50-Ω line to a point of zero reactance, and using a quarter-wave transformer to provide the correct series resistance of 15.43 Ω. The drain capacitance was supplied by a short line on each side between the outer wall and the base. The quarter-wave transformer was made movable to test the tunability of the cavity. Since the source was grounded to the cavity wall, a dc block was used from the output post, which was not shown. The measured output power at 10 GHz was 100 mW [4].

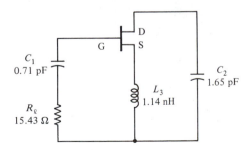

Figure 9-6-2 Equivalent series oscillator circuit.

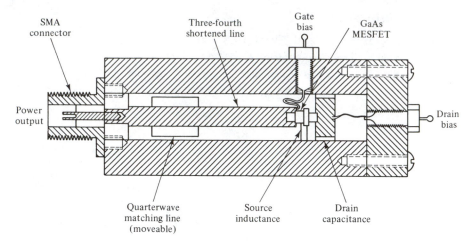

Figure 9-6-3   Coaxial cavity-tuned oscillator. (After Kenneth M. Johnson [4]. Copyright © 1979 IEEE. Reprinted by permission of the IEEE, Inc.)

9-7 BROADBAND OSCILLATOR DESIGN

A broadband oscillator must have a stability factor K less than unity for the entire frequency range and its resonant circuit must be electrically or mechanically tunable. The load reflection coefficient Γ_ℓ in the output load circuit must fall in the unstable region of the stability circle of the required broadband. The YIG resonator is commonly used for the resonant circuit. Its tuning time is in the millisecond range but its quality factor Q is very high. Figure 9-7-1 shows an equivalent circuit for a YIG-tuned common-gate GaAs MESFET broadband oscillator. L_c is the coupling loop inductance; G_0, L_0, and C_0 are for the parallel elements of the YIG sphere; and L_g is the gate inductance.

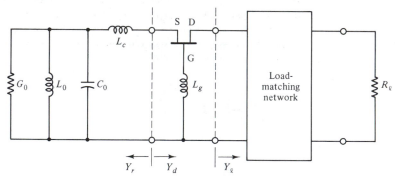

Figure 9-7-1   YIG-tuned oscillator circuit for a common-gate GaAs MESFET. (From Robert J. Trew [5]. Copyright © 1979 IEEE. Reprinted by permission of the IEEE, Inc.)

9-7-1 Oscillation Conditions

According to oscillation conditions as shown in Eqs. (9-1-1), (9-1-11), and (9-1-12), the stability factor K must be less than unity and the sum of the admittances of the resonator plus the active circuit must be equal to zero. They are

$$K < 1 \tag{9-7-1}$$

$$Y_r + Y_d = 0 \tag{9-7-2}$$

or

$$G_r(\omega) + G_d(V_{rf}, \omega) = 0 \tag{9-7-3}$$

and

$$B_r(\omega) + B_d(V_{rf}, \omega) = 0 \tag{9-7-4}$$

where the active circuit consists of the GaAs MESFET and the negative feedback elements. In addition to the oscillation conditions indicated by Eqs. (9-7-3) and (9-7-4), the stable oscillations are possible if and only if the stability criterion is satisfied. That is

$$+ V_{rf} \frac{\partial G_d}{\partial V_{rf}} \frac{dG_r}{d\omega} (\tan\theta + \tan\phi) = 0 \tag{9-7-5}$$

where

$$\tan\theta = \frac{dB_r}{d\omega} \bigg/ \frac{dG_r}{d\omega} \tag{9-7-6}$$

$$\tan\phi = \frac{\partial B_d}{\partial V_{rf}} \bigg/ \frac{\partial G_r}{\partial V_{rf}} \tag{9-7-7}$$

Equations (9-7-3) through (9-7-5) comprise the necessary and sufficient conditions for steady-state oscillations.

9-7-2 YIG Resonator

As shown in Fig. 9-7-1, the equivalent circuit of a YIG resonator consists of a parallel resonant circuit in series with the coupling-loop inductance. Then the resonator input admittance is given by

$$Y_r = \frac{1}{Z_r} \tag{9-7-8}$$

where

$$Z_r = j\omega L_c + \frac{j\omega\omega_0/(G_u G_0)}{\omega_0^2 - \omega^2 + j\omega\omega_0/Q_u} \tag{9-7-9}$$

$$G_0 = \frac{d^2}{\mu_0 V \omega_m Q_u} \qquad (9\text{-}7\text{-}10)$$

$$L_0 = \frac{\mu_0 V \omega_m}{\omega_0 d^2} \qquad (9\text{-}7\text{-}11)$$

$$C_0 = \frac{1}{L_0 \omega_0^2} \qquad (9\text{-}7\text{-}12)$$

$$Q_u = \frac{H_0 - M_s/3}{\Delta H} \qquad (9\text{-}7\text{-}13)$$

$$f_0 = \gamma H_0 \qquad (9\text{-}7\text{-}14)$$

All of these elements are described in Section 9-3. The equivalent parallel inductance L_0 and parallel capacitance C_0 of the YIG resonator are functions of frequency and, also, functions of magnetic biasing field and tuning current. They make the YIG resonator a high-Q, linear, broadband tuning device.

In addition, the resonant conductance G_0 is dependent on the material and geometry of the YIG sphere and coupling structure. Also, the conductance G_0 is inversely proportional to the unloaded Q_u, which increases with frequency as shown in Fig. 9-7-2.

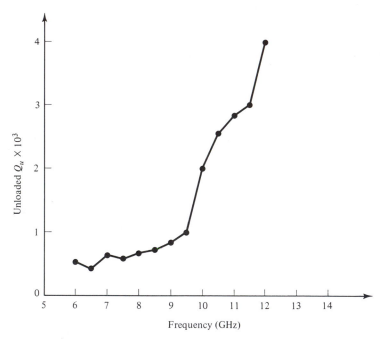

Figure 9-7-2   Unloaded Q_u of a YIG resonator versus frequency. (After Robert J. Trew [5]. Copyright © 1979 IEEE. Reprinted by permission of the IEEE, Inc.)

Therefore, the YIG resonator conductance G_0 is decreased with increasing frequency. On the other hand, the negative conductance of the active device is also decreasing at the high end of their operating frequencies. This is the reason the YIG-tuned oscillator can support a broadband operation because conductances in both the resonator and active circuits are decreased simultaneously with frequency.

9-7-3 Load-Matching Circuit

To increase the bandwidth and the negative conductance magnitude, a load-matching network is required for a broadband oscillator. As described previously, the input reflection coefficient Γ_{in} of a device having a load reflection coefficient connected can be expressed as

$$\Gamma_{in} = S_{11} + \frac{S_{12}S_{21}}{1 - S_{22}} \tag{9-7-15}$$

Then the device admittance as shown in Fig. 9-7-1 is given by

$$Y_d = Y_0 \frac{1 - \Gamma_{in}}{1 + \Gamma_{in}} \tag{9-7-16}$$

or

$$y_d = \frac{1 - \Gamma_{in}}{1 + \Gamma_{in}} \tag{9-7-17}$$

Since the input reflection S_{11} of the device and the gate feedback circuit provide a negative conductance over a certain frequency range, it is necessary to design a load-matching network so that the load reflection coefficient yields a feedback that adds in phase with that of s_{11}. This requires that

$$\Gamma_{in} = S_{11} + \frac{1}{S_{11}^*} \tag{9-7-18}$$

This can be accomplished by setting

$$\frac{1}{S_{11}^*} = \frac{S_{12}S_{21}\Gamma_\ell}{1 - S_{22}\Gamma_\ell} \tag{9-7-19}$$

Then the ideal drain reflection coefficient can be expressed as

$$\Gamma_{ideal} = \frac{1}{S_{11}^* S_{12}S_{21} + S_{22}} \tag{9-7-20}$$

Figure 9-7-3 shows plots of the ideal drain reflection coefficient Γ_{ideal} by a solid line and the actual drain reflection coefficient Γ_{actual} by a dashed line for a circuit designed to operate from 8 to 12 GHz. The Γ_{ideal} curve indicates a filter circuit because it provides a small magnitude of the reflection coefficient in the midband and increasing

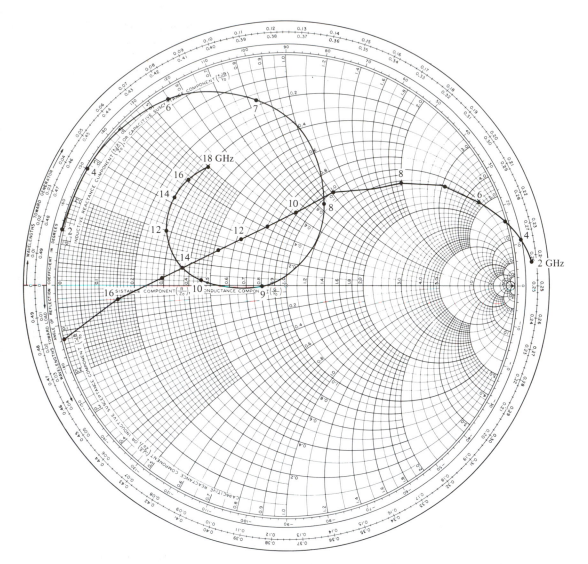

Figure 9-7-3 Ideal and actual drain filter reflection coefficients. (After Robert J. Trew [5]. Copyright © 1979 IEEE. Reprinted by permission of the IEEE, Inc.)

reflection coefficient magnitudes at the two end bands. Low-pass, high-pass, and bandpass filters are found suitable to these variations, but the high-pass filter works best. A constant-*K* filter design is adequate to enhance the negative conductance over the frequency range as shown in Fig. 9-7-4.

The element values of the high-pass filter are given by

$$R = \sqrt{\frac{L}{C}} \qquad (9\text{-}7\text{-}21)$$

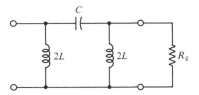

Figure 9-7-4   Device drain filter circuit.

$$L = \frac{R}{4\pi f_c} \qquad (9\text{-}7\text{-}22)$$

$$C = \frac{1}{4\pi f_c R} \qquad (9\text{-}7\text{-}23)$$

$$f_c = \frac{1}{4\pi\sqrt{LC}} \qquad (9\text{-}7\text{-}24)$$

where f_c is the cutoff frequency equal to the lower end frequency.

In practical design, the gate feedback inductor L_g is selected to provide a negative conductance at the high-end frequency and the filter circuit produces a negative feedback at the low-end frequency. The resistance R of the filter can be adjusted to maximize the magnitude of negative conductance. The cutoff frequency f_c has been set equal to the lower-end frequency at 8 GHz, so that the filter feedback adds to S_{11} in phase as shown in Eq. (9-7-18). The admittance Y_d presented to the YIG resonator by the designed load matching filter is shown in Fig. 9-7-5.

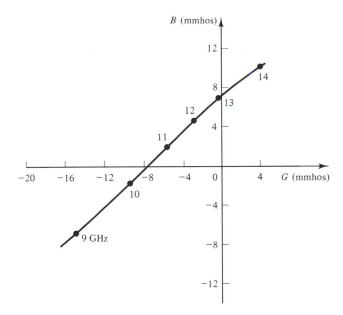

Figure 9-7-5   Admittance Y_d of active device circuit. (After Robert J. Trew [5]. Copyright © 1979 IEEE. Reprinted by permission of the IEEE, Inc.)

9-7-4 Design Example

Design Procedure. The design procedure for a YIG-tuned oscillator can be described as follows:

1. Choose the source inductance for $K < 1$ over the desired frequency range.
2. Predict the YIG parallel-resonant elements for a high-Q resonance.
3. From the oscillation condition 3 ($\Gamma_{\text{out}}\Gamma_\ell = 1$), compute Γ_{out} from the given S parameters and then determine the load reflection coefficient Γ_ℓ.
4. Determine the load impedance Z_ℓ from Γ_ℓ by using a Smith chart or calculation.
5. Realize the load-matching network with Z_ℓ.
6. Check the constant transducer power gain $|S_{21}|^2$ for the desired frequency range.

Design Example 9-7-1: YIG-Tuned Oscillator Design

An X-band oscillator was designed by using the circuit as shown in Fig. 9-7-6, where the element values are

$$L_g = \text{short-circuited 200-}\Omega\text{ line}$$

$$C = 0.50 \text{ pF}$$

$$2L = \text{short-circuited 100-}\Omega\text{ line}$$

$$R_\ell = 25 \ \Omega \text{ of the filter}$$

The circuit and GaAs MESFET chip were assembled on a substrate beryllia (BeO). The oscillator generated an output power of 10-mW minimum from 8 to 14 GHz and a peak power of 25 mW at 8 GHz. The conversion efficiency was about 17% at 8 GHz and 6% from 9 to 14 GHz [7].

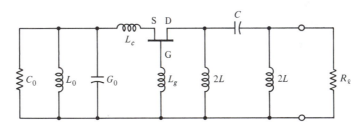

Figure 9-7-6 Designed X-band oscillator.

9-7-5 Spurious Oscillation Considerations

As shown in Fig. 9-7-5, there is observed occurrence of spurious oscillation when the active circuit admittance Y_d changes from an inductive to a capacitive susceptance while maintaining a negative conductance. Under these circumstances, an oscillation can occur even without YIG bias. To eliminate the occurrence of spurious oscillations, the loop inductance L_c should be kept as small as possible so that the spurious oscillation frequency happens out of the negative conductance region; and the gate feedback inductance L_g should be minimized to a small value because it is inversely proportional to the frequency at which the change from inductive to capacitive susceptance occurs. Also, the parasitic reactances, particularly parasitic capacitances, must be maintained at minimum values.

9-8 GUNN-DIODE OSCILLATOR DESIGN

Gunn-diode oscillators are commonly used as radar local oscillators and as transmitters for low-power radars such as police radar and intrusion alarms. The device is a compact, lightweight, mechanically tuned waveguide resonator structure, shown in Fig. 9-8-1.

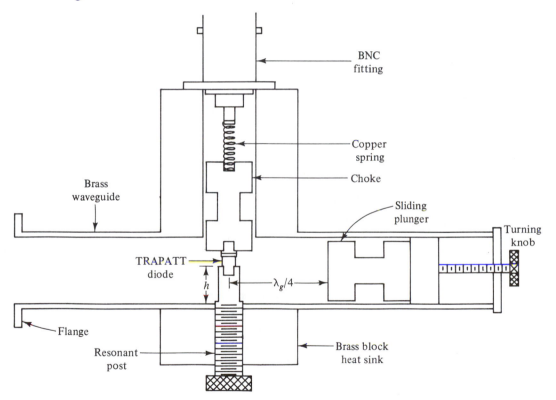

Figure 9-8-1 Schematic diagram of a Gunn-diode oscillator.

The Gunn-diode oscillator can be built by mounting the diode into the cavity of the rectangular waveguide of the TE_{10} mode. The oscillator is designed to be made of brass, which will provide a good heat sink for the diode. The interior surface of the waveguide should be silver-plated to avoid any transmission-line discontinuities and the reduction of power into the cavity. A BNC fitting is mounted on top of the waveguide block for the reversed bias dc feedthrough. Other components, such as the copper spring, the quarter-wavelength choke, the Gunn diode, the resonant post, and the heat sink are placed beneath the BNC fitting. The resonant cavity is located at the end of the waveguide and is terminated by a sliding plunger. The resonant frequecy can be adjusted by moving the plunger along the end of the waveguide. The basic function of the choke is to provide the RF short circuit to the waveguide so as to prevent the loss of any RF power through the bias post. The length of each portion of the choke must be a quarter wavelength of the operating frequency in order to achieve a complete short circuit of the element. The diameter of the resonant post for 10 GHz is about 5 mm.

Basically, the real part of the oscillating diode's negative impedance must be exactly equal to the circuit load resistance at the frequency of oscillation, but the diode reactance must cancel the circuit reactance at the resonant frequency. In a waveguide resonant cavity for TE_{101}-mode resonance, the main function of the resonant post is to introduce some shunt capacitance or shunt inductance across the waveguide in order to cancel the circuit reactance for conjugate matching purpose. As a result, resonance occurs and maximum power is delivered.

The diode impedance is given by

$$Z_d = R_d + jX_d \qquad (9\text{-}8\text{-}1)$$

and the circuit load impedance can be expressed as

$$Z_\ell = R_\ell + jX_\ell \qquad (9\text{-}8\text{-}2)$$

For oscillation, the real parts of the load and diode impedance must be equal and their reactances must be conjugately matched. This means that

$$R_\ell = R_d \qquad (9\text{-}8\text{-}3)$$

$$X_\ell = -X_d \qquad (9\text{-}8\text{-}4)$$

In addition, oscillator design is similar to amplifier design. In both cases, the load resistance resonates the diode resistance. However, in amplifier design the load resistance must not equal the oscillator load resistance.

The position of the diode and the length of the resonant post are adjustable upward or downward inside the main cavity by turning the bottom screw clockwise or counterclockwise direction. The height of the resonant post h is slightly less than $\lambda_0/4$ and is given by

$$h \simeq \frac{\lambda_0}{4} - \frac{d - r}{2 \ell n \, (2d/r)} \qquad (9\text{-}8\text{-}5)$$

where λ_0 = free-space wavelength
$\quad d$ = distance between the post surface and the side wall of the cavity
$\quad r$ = 0.254 cm is the radius of the post for 10 GHz

The short-circuit reactance is expressed as

$$Z_{sh} = jZ_0 \tan (\beta_g \ell) \qquad (9\text{-}8\text{-}6)$$

where

$$Z_0 = \frac{H}{W} \frac{377}{\sqrt{\epsilon_r} \sqrt{1 - (f_c/f)^2}} \qquad (9\text{-}8\text{-}7)$$

ϵ_r = relative dielectric constant of the medium
β_g = phase constant in the guide
$\ell = \lambda_g/2$ is the distance between the plane of the resonant post
   and the effective plane of the short-circuited plunger
λ_g = wavelength in the guide
H = height of the waveguide
W = width of the waveguide
f = resonant frequency
f_c = cutoff frequency

Since $\beta_g \ell$ is approximately equal to π radians,

$$\tan (\beta_g \ell) = -\tan (\pi - \beta_g \ell) \qquad (9\text{-}8\text{-}8)$$
$$\simeq \beta_g \ell - \pi$$

Finally, Eq. (9-8-6) becomes

$$Z_{sh} = jZ_0(\beta_g \ell - \pi) \qquad (9\text{-}8\text{-}9)$$

The equivalent circuit of a full-height waveguide cavity is shown in Fig. 9-3-4.

Design Example 9-8-1:   Gunn-Diode Oscillator Design

A Gunn-diode oscillator is to be designed and its specifications are as follows:

Waveguide	WR(90)
Cavity	Coaxial cavity
Wave mode	TE_{101} mode
Resonant frequency	$f = 10$ GHz
Waveguide width	$W = 2.286$ cm
Waveguide height	$H = 1.016$ cm

Design Data

1. the operating wavelength is

$$\lambda_0 = 3 \text{ cm}$$

2. The cutoff frequency is

$$f_c = 6.56 \text{ GHz}$$

3. The wavelength in the guide is

$$\lambda_g = 4 \text{ cm}$$

4. The phase constant in the guide is

$$\beta_g = 1.57 \text{ rad/cm}$$

5. Choke dimensions:
   Top and bottom parts:

$$\text{diameter} = 0.615 \text{ cm}$$
$$\text{length} = 0.750 \text{ cm}$$

   Center part:

$$\text{diameter} = 0.400 \text{ cm}$$
$$\text{length} = 0.750 \text{ cm}$$

6. The guide characteristic impedance is

$$Z_0 = 221.11 \ \Omega$$

7. The height of the resonant post is computed from Eq. (9-8-5) as

$$h = 0.587 \text{ cm}$$

9-8-1 Tuning Mechanisms

The electronic performance of a Gunn-diode oscillator has recently improved its power output, tuning bandwidth, and noise figure. There are three commonly used tuning mechanisms to achieve the improved performances.

1. *Mechanical-tuning mechanism*. The mechanical-tuning method can tune the frequency by inserting a dielectric rod into the middle of a half-wavelength coaxial resonator to add some capacitive reactance to the resonant circuit. The diameter of the dielectric rod is determined by the equation for the cutoff wavelength (λ_c) of a circular rod with a relative dielectric constant of ϵ_r and it is given by

$$d = \frac{\lambda_c}{1.706 \sqrt{\epsilon_r}} \tag{9-8-10}$$

   This type of tuning method can operate up to 50 GHz with a tuning bandwidth of 10 GHz. The output power is about 100 mW with a dc power supply of 10 V at 2 A.

2. *Biasing-tuning mechanism*. The bias-tuning circuit is used for high power in 50-GHz frequency applications. The circuit design uses a standard disk resonator cavity with a sliding plunger to optimize the power output. There are two outputs of equal power from a single oscillator when the sliding plunger is not used and the disk resonator is presented with an equal load in both directions. A single-diode Gunn oscillator at 35 GHz can produce 400 mW of power and can be bias-tuned over a bandwidth of 100 to 200 MHz.

3. *Varactor-tuning mechanism*. The varactor-tuning method uses varactor to couple the Gunn-diode circuit to the resonator circuit. A flat power can be achieved

by using a stepped eccosorb backshort instead of the standard noncontacting metallic backshort.

9-8-2 Low-Temperature Performance

In recent design development a Gunn-diode oscillator can operate properly at low temperature of $-145°C$ without turn-on problem.

9-9 WAVEGUIDE-CAVITY IMPATT OSCILLATOR DESIGN

Waveguide-cavity IMPATT oscillators can be constructed in several configurations.

9-9-1 Waveguide-Cavity Oscillator

When a waveguide output from an IMPATT oscillator is desired, a waveguide cavity may be incorporated with an IMPATT diode as shown in Fig. 9-9-1. The window opening serves as a load. The oscillation frequency is adjusted by inserting a dielectric rod or a metallic rod into the cavity. A dielectric rod is often preferred because the cutoff waveguide effect of the thin dielectric rod itself can be used to prevent RF leakage. The distance between the window and the diode is approximately a half of the guided wavelength.

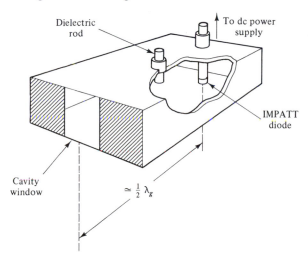

Figure 9-9-1   Waveguide-cavity oscillator circuit.

9-9-2 Single-Tuned Single-Device Oscillator

When a small circuit loss is allowable, oscillators with well-behaved tuning characteristics can be designed as shown in Fig. 9-9-2. The IMPATT diode is mounted at one end of a coaxial line which is coupled to a waveguide cavity. The cavity is then

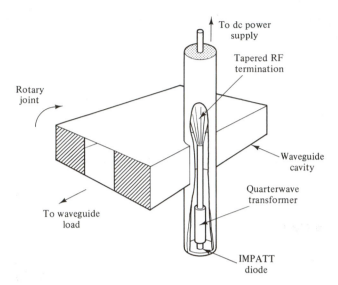

To dc power
supply

Tapered RF
termination

Rotary
joint

Waveguide
cavity

Quarterwave
transformer

To waveguide
load

IMPATT
diode

Figure 9-9-2 Single-tuned single-device oscillator circuit. (After F. M. Magalhaes and K. Kurokawa [6]. Copyright © 1970 IEEE. Reprinted by permission of the IEEE, Inc.)

coupled to the load through a rotary point. A quarter-wave transformer is connected between the diode and the coaxial line. The other end of the coaxial line is RF terminated but dc insulated, and the dc power supply is connected to the center conductor [6]. The equivalent circuit is shown in Fig. 9-9-3.

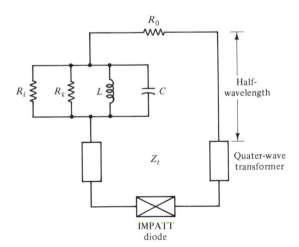

R_0

R_i R_ℓ L C

Half-wavelength

Z_t

Quater-wave
transformer

IMPATT
diode

Figure 9-9-3 Equivalent circuit of the oscillator.

The LC resonant circuit represents the waveguide cavity and the impedance z_t is the characteristic impedance of the coaxial-line transformer section. At the resonant frequency, the circuit resonant impedance is given by

$$R_r = \frac{R_0 + R_i R_t}{R_i + R_t} \qquad (9\text{-}9\text{-}1)$$

where $R_0 = $ coaxial termination
$R_i = $ internal cavity loss
$R_t = $ load resistance transformed by the window and the rotary joint
In practice, the length of the transformer section is chosen first to ensure oscillation. Once the length is properly chosen, the rotary joint and resonant frequency can be adjusted independently.

A waveguide cavity can resonate in many different modes at different frequencies. Several circuit configurations can be designed, such as the radial-choke circuit and high-Q cavity circuit as shown in Fig. 9-9-4. In Fig. 9-9-4(a), the radial choke is used to determine the oscillation frequency and the adjustable short is to change the load. For Fig. 9-9-4(b), at the resonant frequency, the waveguide is essentially open-circuited at the cavity position and the termination is decoupled. At the detuned frequencies, the cavity is primarily short-circuited and termination prevents oscillation.

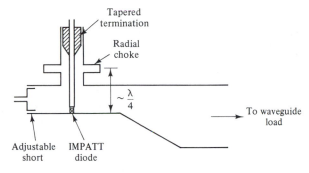

(a) Radial choke circuit

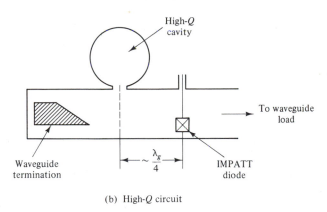

(b) High-Q circuit

Figure 9-9-4 Single-tuned oscillator circuits. (After F. M. Magalhaes and K. Kurokawa [6]. Copyright © 1970 IEEE. Reprinted by permission of the IEEE, Inc.)

9-9-3 Single-Tuned Multiple-Device Oscillator

When a desired power cannot be supplied by a single device a combination of multiple devices must be designed with a combining circuit. Figure 9-9-5 shows a schematic diagram for a single-tuned multiple-device IMPATT oscillator. The coaxial lines are

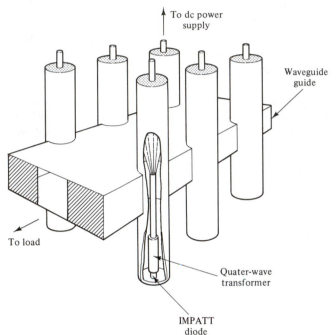

Figure 9-9-5 Single-tuned multiple-IMPATT oscillator circuit. (After K. Kurokawa and F. M. Magalhaes [7]. Copyright © 1971 IEEE. Reprinted by permission of the IEEE, Inc.)

coupled to the waveguide cavity at the locations where the magnetic field intensities are maximum. There are two different causes which bring about mode instabilities. One is involved with undesired resonant modes of the cavity and the other is inherent to multiple devices. For the desired resonant mode of the cavity, there are N possible modes of oscillation, where N is the number of active devices. However, all but the zeroth mode will be suppressed because of the phase cancellations of all ith modes $(i \neq 0)$ [7].

REFERENCES

1. Kotzebue, K. L., and W. J. Parrish, The use of large-signal S-parameters in microwave oscillator design. *Proc. 1975 IEEE Int. Microwave Symposium on Circuits and Systems*.

2. Kotzebue, K. L., A new technique for the direct measurement of y-parameters at microwave frequencies. *IEEE Trans. Instrum. Meas.*, Vol. IM-26, No. 6, June 1977, pp. 119–123.

3. Fong, T. T., et al., Circuit characterization of V-band IMPATT oscillators and amplifiers. *IEEE Trans. Microwave Theory and Techniques*, Vol. MTT-24, No. 11, November 1976, pp. 752–758.

4. Johnson, Kenneth M., Large signal GaAs MESFET oscillator design. *IEEE Trans. Microwave Theory and Techniques*, Vol. MTT-27, No. 3, March 1979, pp. 217–227.

5. Trew, Robert J., Design theory for broadband YIG-tuned FET oscillators. *IEEE Trans. Microwave Theory and Techniques*, Vol. MTT-27, No. 1, January 1979, pp. 8–14.

6. Magalhaes, F. M., and K. Kurokawa, A single-tuned oscillator for IMPATT characteristics. *Proc. IEEE*, Vol. 58, 1970, pp. 831–832.

7. Kurokawa, K., and F. M. Magalhaes, An X-band 10-watt multiple-IMPATT oscillator. *Proc. IEEE*, Vol. 59, 1971, pp. 102–103.

PROBLEMS

9-1 Oscillation Conditions

9-1-1. A certain microwave transistor has the following S parameters measured with a 50-Ω reference at 4 GHz:

$$S_{11} = 0.542 \ \underline{/-155°}$$
$$S_{12} = 0.090 \ \underline{/70°}$$
$$S_{21} = 3.549 \ \underline{/170°}$$
$$S_{22} = 0.584 \ \underline{/-25°}$$

Determine:
(a) The delta factor Δ for the transistor.
(b) Oscillation condition 1 for the transistor.

9-1-2. A GaAs MESFET has the following S parameters measured with a 50-Ω reference at 8 GHz:

$$S_{11} = 0.732 \ \underline{/-30°}$$
$$S_{12} = 0.074 \ \underline{/40°}$$
$$S_{21} = 3.458 \ \underline{/-160°}$$
$$S_{22} = 0.695 \ \underline{/40°}$$

Determine:
(a) The delta factor Δ.
(b) Oscillation condition 1 for the MESFET.

9-4 One-Port Oscillator Design

9-4-1. A one-port oscillator is to be designed to have a generator source reflection coefficient $\Gamma_g = 1.75 \ \underline{/40°}$ with a 50-Ω reference. The oscillation frequency is 8 GHz.
(a) Determine the normalized and total generator impedances.
(b) Find the total load impedance required for oscillation.
(c) Compute the required load circuit element values.
(d) Draw the complete circuit diagram for the designed oscillator.

9-4-2. Design a one-port circuit for an oscillator and the generator source reflection coefficient is $\Gamma_g = 2.00 \ \underline{/50°}$ measured with a 50-Ω reference at 10 GHz.
(a) Determine the normalized and total generator source impedance.
(b) Find the total load impedance required for oscillation.
(c) Compute the required load circuit element values.
(d) Draw the complete circuit diagram for the oscillator designed.

9-5 Two-Port Oscillator Design

9-5-1. A GaAs MESFET is to be used for a common-source oscillator and it has the following parameters:

Small-signal transducer power gain at 10 GHz $G = |S_{21}|^2 = 8$ dB
Saturated amplifier output power $P_{sat} = 1.50$ W

Calculate:
(a) The maximum efficient oscillator power gain in decibels.
(b) The maximum oscillator power output in watts.

9-5-2. A two-port oscillator is to be designed with a GaAs MESFET in the common-source mode. The MESFET has the following S parameters measured at $V_{ds} = 6$ V and $I_{ds} = 50$ mA at 8 GHz.

$$S_{11} = 0.63 \; \underline{/-150°}$$
$$S_{12} = 0.07 \; \underline{/-30°}$$
$$S_{21} = 3.10 \; \underline{/40°}$$
$$S_{22} = 0.70 \; \underline{/-100°}$$

Design series-resonant circuit no. 1 for the oscillator as shown in Fig. 9-2-1.
(a) Check the delta factor Δ and the stability factor K for oscillation.
(b) Convert the S parameters to z parameters for series oscillator-circuit realization.
(c) Compute the y parameters from the S parameters.
(d) Calculate the embedding element values for the series circuit.
(e) Check the maximum efficient output power gain.
(f) Draw the equivalent series-oscillator circuit.

9-5-3. Design a two-port oscillator and the oscillator should be in common-source mode and use the series-resonant circuit no. 3 configuration shown in Fig. 9-2-1. The GaAs MESFET to be used for the oscillator has the following S parameters at 10 GHz.

$$S_{11} = 0.79 \; \underline{/-140°}$$
$$S_{12} = 0.10 \; \underline{/50°}$$
$$S_{21} = 2.75 \; \underline{/130°}$$
$$S_{22} = 0.75 \; \underline{/-40°}$$

(Refer to Problem 9-5-2 for the design steps.)

9-6 High-Power Oscillator Design

9-6-1. The large-signal S parameters of MSX801G FET were measured, the z and y parameters were converted, and the embedding element values were calculated as plotted in Fig. 9-6-1. Design a maximum 4-dB high-power oscillator for 9 GHz by using series oscillator circuit no. 1 shown in Fig. 9-2-1.
(a) Determine the embedding element values from the plots.
(b) Draw equivalent series oscillator circuit no. 1.

9-6-2. Repeat Problem 9-6-1 for a 5-dB high-power oscillator at 10 GHz.

9-8 Gunn-diode Oscillator Design

9-8-1. Design a Gunn-diode oscillator with the following specifications:

Waveguide	WR(90)
Cavity	Coaxial cavity
Resonant frequency	$f = 8$ GHz
Waveguide width	$W = 2.286$ cm
Waveguide height	$H = 1.016$ cm
Wave mode	TE_{101} mode

Compute:

(a) The operating wavelength.

(b) The cutoff frequency.

(c) The wavelength in the guide.

(d) The phase constant in the guide.

(e) The characteristic impedance of the guide.

(f) The choke dimensions.

(g) The height of the resonant post.

Chapter 10

Optical - Fiber Waveguides and Light Modular Design

10-0 INTRODUCTION

The optical fiber is one of the latest additions to the signal links in the submillimeter-wave technology. Its transmission loss is extremely low and its major applications are in computer links, industrial automation, medical instruments, telecommunications, and military communications. In this chapter we discuss the operational mechanisms, modes, structures, and characteristics of the optical-fiber waveguides and light modulator design.

10-1 OPTICAL-FIBER WAVEGUIDES

An optical-fiber cable is an assembly of several optical fibers. The optical-fiber waveguide is simply a circular waveguide system. The center part is a core with a greater refractive index and the outer part is a cladding with a smaller refractive index. If a beam of electromagnetic energy impinges on this system through one end face of the fiber, the light beam will travel through the fiber and emerge from the other end face. The basic difference in signal transmission between a metal transmission line and an optical-fiber cable is in the carriers. In a metal line the carriers are electrons, and in an optical fiber the carriers are photons. Light-wave transmission on an optical fiber has reached a fully commercial stage with carrier wavelengths in the region 0.82 to 0.85 μm. Future systems are expected to operate with carrier wavelengths in the range 1.2 to 1.6 μm.

The advantages for using optical fibers over metal lines in long-distance communication systems are the low transmission loss and attractive bandwidths. Installed optical-fiber cables have shown losses in the neighborhood of 4 dB/km at wavelengths of 0.82 to 0.85 μm. Some experiments have demonstrated that the optical fiber of 30 km or longer lengths had a loss below 0.7 dB/km near 1.3 μm. In a data communication system, a bit rate of 50 Mbits/s for a repeater span of at least 10 km was achieved for multimode fiber.

The key components needed for a light-wave system are the optical fibers, the light-wave sources [lasers or light-emitting diodes (LEDs)], and the light-wave detectors (avalanche photodiodes or *p-i-n*-photodiodes). Optical-fiber cables have been used in underwater intercontinental communication systems, in intercity and metropolitan telephone systems, in data communication systems, in computer and switching link systems, in automation process systems, and in military command and control systems.

10-1-1 Materials and Fabrications

Materials.   Three major materials are used in the production of optical fibers.

1. *Silica fibers.* Silica fibers are basically made of silicon dioxide (SiO_2), with other metal–oxides to establish a difference in refractive index between the core and the cladding. Various dopants have been used to increase or decrease the refractive index of silica, such as TiO_2, Al_2O_3, GeO_2, and P_2O_5. The cladding has a lower refractive index than the core, but its coefficient of thermal expansion is higher than that of the core.
2. *Glass fibers.* Glass fibers are made of compound glasses with low melting temperatures and long-term chemical stability.
3. *Plastic fibers.* Plastic fibers are made of plastics having a higher attenuation loss than silica and glass. They are commonly used for short-distance links in computers.

Fabrications.   There are four fabrication processes for making optical fibers.

1. *Outside vapor-phase oxidation (OVPO) process.* This method was first developed by the Corning Glass Works to deposit high-silica glass from vapor-phase sources. When the mandrel is removed, the "soot" tube is sintered and subsequently drawn to a fiber.
2. *Modified chemical vapor deposition (MCVD) process.* This process was invented by the Bell Laboratories. Glass of the desired composition is deposited inside a silica tube, layer by layer, via a flame-hydrolysis process, over the length of a mandrel. When deposition is completed, the material is sintered and the mandrel removed. The consolidated tube is then fused and collapsed simultaneously into a perform rod.
3. *Vapor-phase axial deposition (VAD) process.* This method, introduced by the Nippon Telegraph and Telephone Public Corporation, is similar to the OVPO process except that deposition occurs on the end of a growing "soot" cylinder.

4. *Plasma chemical vapor deposition (PCVD)* process. The Philips company pioneered the use of microwave plasma to excite reactants and deposit vitreous material directly inside a silica tube.

All four processes fabricate fibers by using compound SiO_2 (silica) with relatively amounts of Ge (germanium), P (phosphorus), and sometimes B (boron) as dopants to alter the refractive index and to lower the working temperature somewhat below that for pure SiO_2.

10-1-2 Physical Structures

An optical fiber is simply a circular dielectric waveguide system. It consists of a core at the center and a cladding outside. The core is a cylinder of transparent dielectric rod with a greater refractive index n_1, and the cladding is second dielectric sheathing or covering, usually glass fused to the core, with lower refractive index n_2. In applications, the optical fibers are commonly classified into three types:

1. *Monomode step-index fiber:* having a core radius of 1 to 16 μm and a cladding radius of 50 to 100 μm
2. *Multimode step-index fiber:* having a core radius of 25 to 60 μm and a cladding radius of 50 to 150 μm
3. *Multimode graded-index fiber:* having a core radius of 10 to 35 μm and a cladding radius of 50 to 80 μm

Figure 10-1-1 shows the diagrams of the three fiber types with their index profiles.

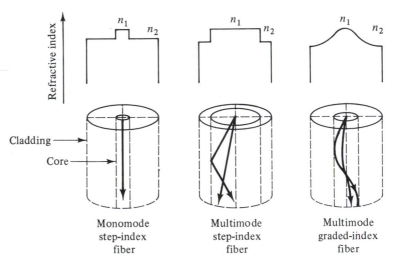

Figure 10-1-1   Physical structures of optical-fiber waveguide. (After T. G. Giallorenzi [1]. Copyright © 1978 IEEE. Reprinted by permission of the IEEE, Inc.)

An optical-fiber cable is an assembly of several optical fibers incorporated in a protective material. There are many different structures of optical-fiber cables. Figure 10-1-2 shows several typical cable configurations. For high-capacity channels, one fiber cable may contain over 100 fibers. Figure 10-1-3 shows a cable made up of 12 stacked ribbons, each containing 12 optical fibers.

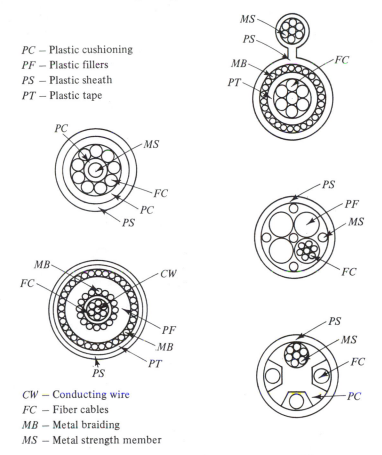

PC − Plastic cushioning
PF − Plastic fillers
PS − Plastic sheath
PT − Plastic tape

CW − Conducting wire
FC − Fiber cables
MB − Metal braiding
MS − Metal strength member

Figure 10-1-2 Fiber-cable configurations. (Reprinted from Ref. [2], by courtesy of Marcel Dekker, Inc.)

10-1-3 Losses

In an ideal case, the fiber core is considered a perfectly transparent material and the cladding a shield confining the electromagnetic energy to the core region by total reflection. Otherwise, there is absorption loss, transmission loss, some scattering loss, and microbending loss.

Absorption losses. If the core material is not perfectly transparent due to im-

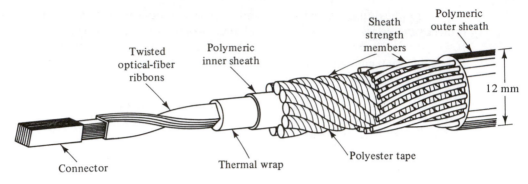

(a) Bell system lightguide cable

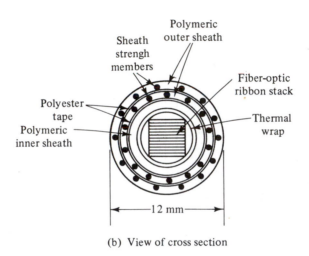

(b) View of cross section

Figure 10-1-3 Optical-fiber cable with 144 fibers. (After Ira Jacobs and S. E. Miller [3]. Copyright © 1977 IEEE. Reprinted by permission of the IEEE, Inc.)

purities, the absorption losses for a length ℓ of the light path through the core glass is given by

$$\text{absorption loss} = e^{-\alpha\ell} \qquad (10\text{-}1\text{-}1)$$

where α = absorption coefficient per unit length
$\quad\quad\;\; \ell = n_1 (n_1^2 - \sin^2\theta_m)^{-1/2}$ is the light-path length
$\quad\quad\;\; \theta_m$ = maximum acceptance angle

Scattering losses. The fiber waveguide scattering losses are caused mainly by geometric irregularities at the core–cladding interface.

Microbending losses. The microbending of an optical fiber causes radiation losses if the fiber is distorted at joints separated on the order of 1 mm longitudinally.

In particular, the plastic optical fibers have higher absorption losses than silica and glass fibers. They are commonly used for short-distance applications such as in links in computers.

Example 10-1-1:   Absorption Loss of an Optical Waveguide

An optical fiber has the following parameters:

Maximum acceptance angle	$\theta_m = 30°$
Core refractive index	$n_1 = 1.40$
Absorption coefficient	$\alpha = 0.47$ N/km

Compute (a) the light-path length, and (b) the absorption loss.

Solution   (a) The light-path length is

$$\ell = \frac{1.40}{\sqrt{(1.40)^2 - \sin^2(30)}} = \frac{1.40}{1.31}$$

$$= 1.07 \text{ km}$$

(b) The absorption loss is

$$\text{loss} = e^{-\alpha\ell} = e^{-0.47 \times 1.07} = e^{-0.50} = 0.61$$

$$= 10 \log (0.61) = -2.20 \text{ dB/km}$$

10-1-4 Characteristics

In the twenty-first century, microwave electronic devices and circuits will be based heavily on electrooptics. The characteristics of optical fibers are described below.

1. *Small size and light weight.* An optical-fiber cable is much smaller than a copper-wire line. Their features of small size and light weight are especially important for underwater cables and in areas of overcrowded transmission lines.

2. *Low cost and low loss.* The cost of optical-fiber cable and its installation is much lower than that of metal cables. The transmission loss of the fibers is very low at 4 dB/km in conjunction with the coaxial lines at 150 dB for 0.5 km at 1 GHz or the microstrip lines at 150 dB for 0.4 km at 10 GHz.

3. *Immunity of interference.* Optical-fiber cable does not generate or receive any electrical or electromagnetic noise or interferences.

4. *High reliability and durability.* Optical-fiber cables are safe to use in any explosive environments and eliminate the hazards of short circuits in wire lines.

5. *High bandwidth.* Optical-fiber cable can carry more channels than do either the coaxial or microstrip lines.

6. *High security capability.* Optical-fiber cables are immune to electrical or electromagnetic noise or grounding such as crosstalk or jamming influence, so they enhance the security capability of data transmission. This feature is very important in military communications.

10-2 OPERATIONAL MECHANISMS OF OPTICAL-FIBER WAVEGUIDES

Light-wave propagation in an optical fiber with a nondissipative core of radius a and refractive index n_1 embedded in a nondissipative cladding medium of radius b and refractive index n_2 can be analyzed by solving Maxwell's field equations for specific boundary conditions. The resultant waves in an optical fiber are neither transverse electric, nor transverse magnetic, but hybrid. This means that there is no nonzero component of both electric E and magnetic H waves in the direction of propagation. A cylindrical coordinate system (r, ϕ, z) for an optical fiber is shown in Fig. 10-2-1.

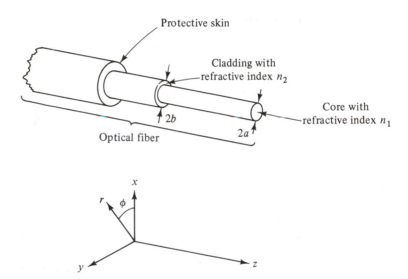

Figure 10-2-1 Schematic diagram of an optical fiber and its cylindrical coordinate system (r, ϕ, z).

The optical properties of a material are usually characterized by its complex refractive index N, which contains two constants, the refractive index n and the extinction index k. That is,

$$N = n - jk \qquad (10\text{-}2\text{-}1)$$

The extinction index k is related to the exponential decay of the wave as it passes through the medium. For nondissipative optical fibers, the extinction index k is equal to zero. The refractive index n is defined as the ratio of the velocity of light in vacuum to the velocity of light in a given medium. For nonmagnetic medium the refractive index is expressed as

$$n = \frac{v_0}{v_\epsilon} = \frac{\sqrt{\mu_0 \epsilon}}{\sqrt{\mu_0 \epsilon_0}} = \sqrt{\epsilon_r} \qquad (10\text{-}2\text{-}2)$$

where ϵ_r is the relative dielectric constant of the medium. Typical values of n are 1.00 for air, 1.50 for polyethylene, 1.60 for polystyrene, and 8.94 for fresh water.

10-2-1 Wave Equations

As shown in Fig. 10-2-1, the optical-fiber waveguide consists of a core of higher refractive index n_1 and radius a surrounded by a cladding of lower refractive index n_2 and radius b. Both regions are assumed to be perfect insulators with free-space magnetic permeability μ_0. Such a structure can support an infinite number of modes, but for a given values of n_1, n_2, a, and b, only a finite number of these are waveguide modes which have their fields localized in the vicinity of the core. The other unbound modes would correspond, for example, to light striking the core from the side, passing on through the core, and emerging from the other side.

The z axis of the cylindrical coordinate system (r, ϕ, z) is chosen to be along the fiber core axis. A waveguide mode is a coherent distribution of light, which is localized in the core by total internal reflection, and which propagates through the guide with a well-defined phase velocity. The z components of the electric field E and magnetic field H parallel to the fiber guide are given by the following expressions in the assumption that exponential factor $e^{j(\omega t - \beta_g z)}$ is implied.

In core region:

$$E_z = AJ_n(kr) \cos(n\phi) \tag{10-2-3}$$

$$H_z = BJ_n(kr) \sin(n\phi) \tag{10-2-4}$$

In the cladding region:

$$E_z = CH_n(\chi r) \cos(n\phi) \tag{10-2-5}$$

$$H_z = DH_n(\chi r) \sin(n\phi) \tag{10-2-6}$$

where
$J_n(kr) = n$th-order Bessel function of the first kind
$H_n(\chi r) = n$th-order Hankel function of the first kind
k = transverse propagation constant in the core region
χ = transverse propagation constant in the cladding region
$k^2 = \beta_1^2 - \beta_g^2$ is the separation equation in the core region
$\chi^2 = \beta_g^2 - \beta_2^2$ is the separation equation in the cladding region
$\beta_g = \omega\sqrt{\mu_0\epsilon_0}$ is the phase constant in free space
$\beta_1 = \omega\sqrt{\mu_1\epsilon_1}$ is the phase constant in the core region
$\beta_2 = \omega\sqrt{\mu_2\epsilon_2}$ is the phase constant in the cladding region

The Bessel function J_n exhibits oscillatory behavior for real k, as do the sinusoidal functions. Hence this function represents a cylindrical standing wave in the core region for $r < a$. The Hankel function H_n represents a traveling wave for real χ in the cladding region for $r > a$, as do the exponential functions. For a nondissipative medium the Hankel function H_n becomes the modified Bessel function. For electric and magnetic fields to be evanescent in the cladding region, χ must be imaginary.

From Eqs. (10-2-3) and (10-2-4) the ϕ components of the field equations in the core region are

$$E_\phi = \left[A \frac{j\beta_g n}{k^2 r} J_n(kr) + B \frac{j\omega\mu_1}{k^2} J_n'(kr) \right] \sin(n\phi) \qquad (10\text{-}2\text{-}7)$$

$$H_\phi = -\left[A \frac{j\omega\epsilon_1}{k^2} J_n'(kr) + B \frac{j\beta_g n}{k^2 r} J_n(kr) \right] \cos(n\phi) \qquad (10\text{-}2\text{-}8)$$

Similarly, from Eqs. (10-2-5) and (10-2-6) for the cladding region, they are

$$E_\phi = -\left[C \frac{j\beta_g n}{\chi^2 r} H_n(\chi r) + D \frac{j\omega\mu_2}{\chi^2} H_n'(\chi r) \right] \sin(n\phi) \qquad (10\text{-}2\text{-}9)$$

$$H_\phi = \left[C \frac{j\omega\epsilon_2}{\chi^2} H_n'(\chi r) + D \frac{j\beta_g n}{\chi^2 r} H_n(\chi r) \right] \cos(n\phi) \qquad (10\text{-}2\text{-}10)$$

where the primes on J_n and H_n refer to differentiation with respect to their arguments (kr) and (χr), respectively.

At the interface $r = a$, the tangential field components should be continuous and therefore

$$AJ_n(ka) = CH_n(\chi a) \qquad (10\text{-}2\text{-}11)$$

$$BJ_n(ka) = DH_n(\chi a) \qquad (10\text{-}2\text{-}12)$$

$$A \frac{j\beta_g n}{k^2 a} J_n(ka) + B \frac{j\omega\mu_1}{k^2} J_n'(ka) = -C \frac{j\beta_g n}{\chi^2 a} H_n(\chi a) - D \frac{j\omega\mu_2}{\chi^2} H_n'(\chi a) \qquad (10\text{-}2\text{-}13)$$

$$A \frac{j\omega\epsilon_1}{k^2} J_n'(ka) + B \frac{j\beta_g n}{k^2 a} J_n(ka) = -C \frac{j\omega\epsilon_2}{\chi^2} H_n'(\chi a) - D \frac{j\beta_g n}{\chi^2 a} H_n(\chi a) \qquad (10\text{-}2\text{-}14)$$

After the constants A, B, C, and D are determined, the electric and magnetic field equations can finally be derived from Maxwell's equations.

Example 10-2-1: Wave Propagation in an Optical Waveguide

A monomode step-index fiber has the following parameters:

Carrier wavelength	$\lambda = 0.82 \ \mu\text{m}$
Carrier frequency	$f = 3.66 \times 10^{14} \ \text{Hz}$
Core radius	$a = 2 \ \mu\text{m}$
Core refractive index	$n_1 = 1.50$
Cladding radius	$b = 80 \ \mu\text{m}$
Cladding refractive index	$n_2 = 1.35$

Compute (a) the phase constant in the core region, (b) the phase constant in the cladding region, (c) the phase constant in free space, (d) the transverse propagation constant in the core region, (e) the transverse propagation constant in the cladding region, (f) the value of the Bessel function $J_1(ka)$, and (g) the value of the Hankel function $H_1(\chi b)$.

Solution   (a) The phase constant in the core region is

$$\beta_1 = \omega\sqrt{\mu_1\epsilon_1} = \omega\frac{n_1}{c} = 2\pi \times 3.66 \times 10^{14} \times \frac{1.5}{3 \times 10^8}$$

$$= 1.15 \times 10^7 \text{ rad/m}$$

(b) The phase constant in the cladding region is

$$\beta_2 = 2\pi \times 3.66 \times 10^{14} \times \frac{1.35}{3 \times 10^8}$$

$$= 1.03 \times 10^7 \text{ rad/m}$$

(c) The phase constant in free space is

$$\beta_g = \frac{2\pi \times 3.66 \times 10^{14}}{3 \times 10^8} = 7.67 \times 10^6 \text{ rad/m}$$

(d) The transverse propagation constant in the core region is

$$k = \sqrt{(1.15 \times 10^7)^2 - (7.67 \times 10^6)^2} = 8.6 \times 10^6 \text{ rad/m}$$

(e) The transverse propagation constant in the cladding region is

$$\chi = \sqrt{(7.67 \times 10^6)^2 - (1.03 \times 10^7)^2} = j6.9 \times 10^6 \text{ rad/m}$$

(f) The Bessel function $J_1(ka)$ is

$$J_1(ka) = J_1(8.6 \times 10^6 \times 2 \times 10^{-6}) = J_1(17.20)$$

$$= -0.12814$$

(g) The Hankel function $H_1(\chi b)$ is

$$H_1(\chi b) = H_1(j6.9 \times 10^6 \times 80 \times 10^{-6}) = H_1(j552)$$

The Hankel function vanishes in the cladding region because its argument is imaginary.

10-2-2 Wave Modes and Cutoff Wavelengths

Wave modes.   For metallic circular waveguides the wave modes are usually designated by TE_{np} and TM_{np} for transverse electric and transverse magnetic waves, respectively. In the dielectric cylindrical waveguides only the cylindrically symmetric ($n = 0$) modes are either transverse electric (TE_{0p}) or transverse magnetic (TM_{0p}). The other modes are all hybrid; that means both electric field E_z and magnetic field H_z are coexistent.

The designation of the hybrid modes are based on the relative contributions of the waves E_z and H_z to a transverse component of the field at some reference point. If the electric wave E_z makes the larger contribution, the mode is considered E-like and designated EH_{np}. On the contrary, if the magnetic wave H_z makes the dominant contribution, the mode is designated HE_{np}. The method of designation is arbitrary, for it does not depend on any particular transverse component of the chosen field, the reference point, and how far the wavelength is from the cutoff. However, the use of two letters, such as EH and HE, is merely to imply the hybrid nature of these modes.

The subscripts on EH_{np} and HE_{np} modes refer to the nth order of the Bessel function and the pth rank, where the rank gives the successive solutions of the boundary condition equation involving the Bessel function of the first kind, J_n.

Cutoff wavelengths. For cutoff condition, $J_n(ka) = 0$, which means that the limit of the argument of the Hankel function (χa) approaches zero [4]. Then the separation equation in cladding region as shown in Eq. (10-2-6) becomes

$$\beta_g^2 = \omega^2 \mu_2 \epsilon_2 \tag{10-2-15}$$

and the separation equation in the core region is

$$k^2 = \omega^2 \mu_1 \epsilon_1 - \omega^2 \mu_2 \epsilon_2 \tag{10-2-16}$$

Let the argument $(k_{np} a)$ of the Bessel function of the first kind be X_{np}. Then

$$k_{np} = \frac{X_{np}}{a} \tag{10-2-17}$$

The cutoff condition is given by

$$X_{np} = \frac{2\pi a}{\lambda_0} (n_1^2 - n_2^2)^{1/2} \tag{10-2-18}$$

and the free-space cutoff wavelength is expressed to be

$$\lambda_0 = \frac{2\pi a}{X_{np}} (n_1^2 - n_2^2)^{1/2} \tag{10-2-19}$$

For a given core refractive index n_1, a cladding refractive index n_2, and a core radius a, the free-space cutoff wavelength for monomode EH_{01} operation is

$$\lambda_{0c} = \frac{2\pi a}{2.405} (n_1^2 - n_2^2)^{1/2} \tag{10-2-20}$$

The cutoff parameter X_{np} (or $K_{np} a$) is usually called the *V number* of the fiber. The number of propagating modes in the step-index fiber is proportional to its V number.

Example 10-2-2: Cutoff Wavelength of Monomode Optical Fibers

A monomode optical fiber has the following parameters:

Core radius	$a = 5 \ \mu m$
Core refractive index	$n_1 = 1.40$
Cladding refractive index	$n_2 = 1.05$

Calculate the cutoff wavelength for the monomode EH_{01} operation.

Solution   The argument of the Bessel function of the first kind for EH_{01} mode is 2.405; the cutoff wavelength is

$$\lambda_{0c} = \frac{2\pi \times 5 \times 10^{-6}}{2.405} [(1.40)^2 - (1.05)^2]^{1/2}$$

$$= 12.11 \ \mu m$$

This means that the signal with a wavelength larger than 12.11 μm will be cutoff.

The wave modes that can propagate in an optical fiber are those for which X_{np} are less than the values as determined by Eq. (10-2-18). Since X_{np} forms an increasing sequence for fixed n and increasing p or for a fixed p and increasing n, the number of allowed modes increases as the square of the radius a. That is, the total number of modes for an optical fiber is given by [5, p. 100]

$$\text{Modes} = \frac{16}{\lambda_0^2} (n_1^2 - n_2^2)a^2 \qquad (10\text{-}2\text{-}21)$$

The electromagnetic field is guided only partially within the core region, whereas outside the core the electromagnetic field is evanescent in a direction normal to propagation. Among the electromagnetic wave modes there is one, the HE_{11} mode, which has no cutoff wavelength. Only this HE_{11} mode can propagate in an optical fiber for wavelength greater than the highest cutoff wavelength of the other modes.

The modes of operation for optical fibers are commonly classified into a single mode, such as monomode step-index fiber, or a multimode, such as multimode step-index and multimode graded-index fiber. Figure 10-2-2 shows a plot of Bessel functions

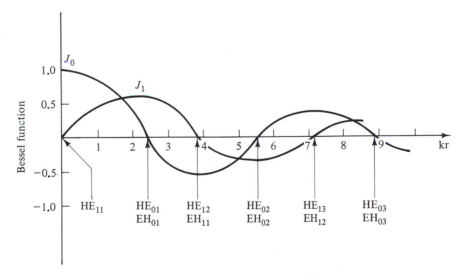

Figure 10-2-2   Cutoff modes of optical fibers.

used for determining cutoff conditions of optical-fiber modes. The cutoff V numbers of the step- and parabolic-index optical fibers for a few lower-order modes are tabulated in Table 10-2-1.

TABLE 10-2-1 CUTOFF V NUMBERS OF THE STEP- AND PARABOLIC-INDEX FIBERS FOR A FEW LOWER-ORDER MODES

Hybrid modes	Cutoff V numbers Step-index	Cutoff V numbers Parabolic-index	Hybrid modes	Cutoff V numbers Step-index	Cutoff V numbers Parabolic-index
HE_{11}	0	0	HE_{13}	7.016	9.158
$HE_{01}, EH_{01}, HE_{21}$	2.405	3.518	EH_{12}, HE_{32}	7.016	9.645
HE_{12}	3.832	5.068	EH_{41}, HE_{61}	7.588	11.938
EH_{11}, HE_{31}	3.832	5.744	EH_{22}, HE_{42}	8.417	11.760
EH_{21}, HE_{41}	5.136	7.848	$HE_{03}, EH_{03}, HE_{23}$	8.654	11.424
$HE_{02}, EH_{02}, HE_{22}$	5.520	7.451	EH_{51}, HE_{71}	8.771	13.959
EH_{31}, HE_{51}	6.380	9.158	EH_{32}, HE_{52}	9.761	13.833

10-2-3 Total Internal Reflection and Numerical Aperture

Total internal reflection. If light wave in medium 1 propagates into medium 2, *Snell's law* states that

$$\frac{\sin \phi_2}{\sin \phi_1} = \frac{\beta_1}{\beta_2} = \frac{v_2}{v_1} = \sqrt{\frac{\epsilon_1}{\epsilon_2}} = \frac{n_1}{n_2} \qquad \text{for } \mu_1 = \mu_2 = \mu_0 \qquad (10\text{-}2\text{-}22)$$

where ϕ_1 = incident angle in medium 1
        ϕ_2 = transmission angle in medium 2
        β_1 = incident phase constant in medium 1
        β_2 = transmission phase constant in medium 2
        v_1 = wave velocity in medium 1
        v_2 = wave velocity in medium 2
        ϵ_1 = dielectric permittivity of medium 1
        ϵ_2 = dielectric permittivity of medium 2
        n_1 = refractive index of medium 1
        n_2 = refractive index of medium 2

The total reflection occurs at $\phi_2 = 90°$ and then the incident angle in medium 1 is given by

$$\phi_1 = \phi_c = \arcsin\left(\frac{n_2}{n_1}\right) \qquad (10\text{-}2\text{-}23)$$

The angle specified by Eq. (10-2-23) is called the *critical incident angle* for total reflection. A wave incident on the interface of the core and cladding in an optical fiber at an angle equal to or greater than the critical angle will be totally reflected.

There is a real critical angle only if $n_1 > n_2$. Hence the total internal reflection occurs only if the wave propagates from the core into the cladding. This is because the value of $\sin \phi_c$ must be equal to or less than unity.

Numerical aperture (NA).   In an optical-fiber waveguide, the light wave is incident on one end face of the fiber. Figure 10-2-3 shows a diagram for the light wave impinging on a fiber. Several terminologies are defined below.

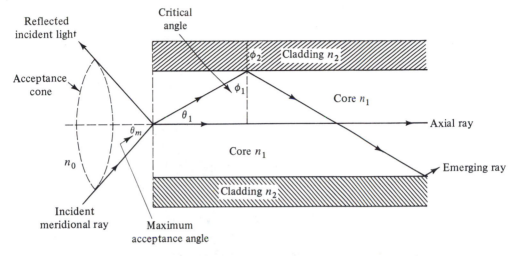

Figure 10-2-3   Wave propagation in a fiber.

1. *Acceptance angle.* The acceptance angle is defined as any angle measured from the longitudinal center line up to the maximum acceptance angle of an incident ray that will be accepted for transmission along a fiber.

2. *Acceptance cone.* The acceptance cone is a cone whose angle is equal to twice the acceptance angle.

3. *Critical angle.* The critical angle is the smallest angle made by a meridional ray in an optical fiber that can be totally reflected from the innermost interface and thus determines the maximum acceptance angle at which a meridional ray can be accepted for transmission along a fiber.

4. *Maximum acceptance angle.* The maximum acceptance angle is dependent on the refractive indices of the two media that determine the critical angle.

5. *Numerical aperture* (*NA*). A numerical aperture is defined as a number that expresses the light-gathering power of a fiber.

The most important parameter for an optical fiber is the numerical aperture, which is expressed as

$$NA = \sin \theta \qquad (10\text{-}2\text{-}24)$$

where θ is the acceptance angle.

From Snell's law,

$$n_0 \sin \theta = n_1 \sin \theta_1 = n_1 \cos \phi_1 \qquad (10\text{-}2\text{-}25)$$

and

$$n_1 \sin \phi_1 = n_2 \sin \phi_2 \qquad (10\text{-}2\text{-}26)$$

For $n_1 > n_2$, the total reflection occurs at

$$\sin \phi_1 > \frac{n_2}{n_1} \qquad (10\text{-}2\text{-}27)$$

Then

$$\cos \phi_1 < \left(1 - \frac{n_2^2}{n_1^2}\right)^{1/2} \qquad (10\text{-}2\text{-}28)$$

Substitution of Eq. (10-2-28) into Eq. (10-2-27) by using the trigonometrical identity yields

$$n_0 \sin \theta < (n_1^2 - n_2^2)^{1/2} \qquad (10\text{-}2\text{-}29)$$

or

$$\sin \theta < \frac{1}{n_0}(n_1^2 - n_2^2)^{1/2} \qquad (10\text{-}2\text{-}30)$$

Equation (10-2-30) determines the incident acceptance angle of a meridional ray for total internal reflection in an optical fiber. Then the maximum acceptance angle is related to the following expression:

$$\sin \theta_m = \frac{1}{n_0}(n_1^2 - n_2^2)^{1/2} \qquad \text{for } n_1^2 < (n_2^2 + n_0^2) \qquad (10\text{-}2\text{-}31)$$

$$= 1.0 \qquad \text{for } n_1^2 > (n_2^2 + n_0^2) \qquad (10\text{-}2\text{-}32)$$

Normally, the optical fiber is immersed in free space or in a vacuum, $n_0 = 1.0$. Then Eqs. (10-2-31) and (10-2-32) become, respectively,

$$\sin \theta_m = (n_1^2 - n_2^2)^{1/2} \qquad \text{for } n_1^2 < (n_2^2 + 1) \qquad (10\text{-}2\text{-}33)$$

$$= 1.0 \qquad \text{for } n_1^2 > (n_2^2 + 1) \qquad (10\text{-}2\text{-}34)$$

Equation (10-2-33) is a measure of the numerical aperture of an optical fiber:

$$\text{NA} = (n_1^2 - n_2^2)^{1/2} \qquad (10\text{-}2\text{-}35)$$

The value of the numerical aperture also indicates the acceptance of impinging light, the degree of openness, the light-gathering ability, and the acceptance cone. The light-gathering power for a meridional ray in an optical fiber is given by

$$P = (\text{NA})^2 = n_1^2 - n_2^2 \qquad (10\text{-}2\text{-}36)$$

Example 10-2-3:   Characteristics of a Multimode Step-Index Fiber

A certain multimode step-index fiber has the following parameters:

Core refractive index	$n_1 = 1.54$
Cladding refractive index	$n_2 = 1.49$
Air refractive index	$n_0 = 1.00$

Determine (a) the critical angle for total reflection in the core region, and (b) the maximum incident angle at the interface between air and the fiber end. (c) Compute the light-gathering power.

Solution   (a) From Eq. (10-2-23), the critical angle is

$$\Phi_c = \arcsin\left(\frac{1.49}{1.54}\right) = 75.36°$$

(b) From Eq. (10-2-33), the maximum incident angle is

$$\phi_m = \arcsin\left[\frac{(1.54)^2 - (1.49)^2}{1}\right]^{1/2}$$

$$= \arcsin(0.39)$$

$$= 22.79°$$

(c) From Eq. (10-2-36), the light-gathering power is

$$P = (1.54)^2 - (1.49)^2 = 0.15$$

10-2-4 Light-Gathering Power

In an optical-fiber system the light source is considered a Lambertian source. That is a source whose intensity is given by

$$I = I_0 \cos \theta \qquad (10\text{-}2\text{-}37)$$

where $I_0 =$ light intensity at the normal direction
   $\theta =$ angle between the normal to the source and the direction of measurement
Figure 10-2-4 shows a diagram for the light output from a Lambertian source through a spherical surface of radius r.

   The amount of light collected by a fiber that is normal to a Lambertian source is defined by the dielectric boundary of the fiber or the maximum acceptance angle θ_m and is given by

$$\text{light} = \int_0^{\theta_m} I_0 \cos \theta \, dA = \int_0^{\theta_m} I_0 \cos \theta \, 2\pi \sin \theta \, d\theta$$

$$= \pi I_0 \sin^2 \theta_m \qquad (10\text{-}2\text{-}38)$$

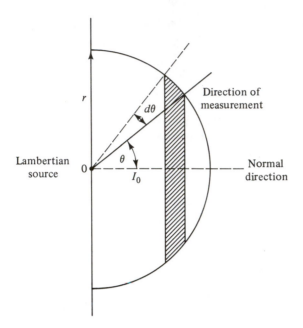

Figure 10-2-4 Light from a Lambertian source.

The total light emitted by the Lambertian source is πI_0 and the fraction of light collected by the fiber is the light-collecting efficiency:

$$\eta_m = \sin^2 \theta_m = (NA)^2 \qquad \text{for } NA < 1.0$$

$$= 1.0 \qquad \text{for } NA = 1.0 \tag{10-2-39}$$

In practice, the light-collecting efficiency is much less than unity due to many factors, such as the angled end-face effect, the fiber-curvature effect, and the varying diameter effect.

The light-gathering power of an optical fiber is given by [6]

$$P = n_0^2 - \frac{2}{\pi} \left\{ (n_1^2 - n_2^2)^{1/2} \, [n_0^2 - (n_1^2 - n_2^2)]^{1/2} \right. \tag{10-2-40}$$

$$\left. + [n_0^2 - 2(n_1^2 - n_0^2)] \cos^{-1} \left(\frac{n_1^2 - n_2^2}{n_0^2} \right)^{1/2} \right\}$$

$$= 1 - \frac{2}{\pi} \{ (NA)[1 - (NA)^2]^{1/2} + [1 - 2(NA)^2] \cos^{-1} (NA) \} \tag{10-2-41}$$

for $n_0 = 1.0$

where $(NA)^2 = n_1^2 - n_2^2$ is the meridional light-gathering power.

10-3 STEP-INDEX FIBERS

A step-index fiber is one that has an abrupt change in refractive indices between the core and cladding materials. There are two modes that can exist in a step-index fiber: monomode fiber and multimode fiber. If the core radius is very small in comparison with the wavelength of the light source, say of the order of 1 to 16 μm, only a single mode is propagated. If the core radius is large enough, say 30 μm, multimodes are coexistent.

10-3-1 Monomode Step-Index Fibers

A monomode step-index fiber is a low-loss optical fiber with a very small core. The fiber requires a laser source for the input signals because of the very small acceptance aperture (or acceptance cone). When the small core radius approaches the wavelength of the light source, only a single mode is propagated.

The core radius is from 1 to 16 μm and the differential in refractive indices between the core and cladding is about 0.6%. The bit rate is from 20 Mbits/s-km to 19 Gbits/s-km. It is ideally suitable for long-haul and high-bandwidth applications such as telecommunication systems.

From Eq. (10-2-18), the value of X_{np} required for a monomode operation in a step-index fiber must be less than 2.405. That is,

$$\mathbf{X}_{np} = \frac{2\pi a}{\lambda_0}\,(n_1^2 - n_2^2)^{1/2} < 2.405 \qquad (10\text{-}3\text{-}1)$$

Figure 10-3-1 shows the diagrams for a monomode step-index fiber, including the light-path and signal-waveform profiles.

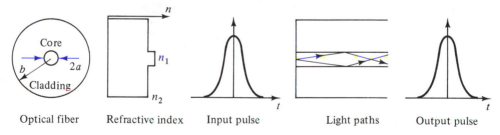

Figure 10-3-1   Monomode step-index fiber. (After T. G. Giallorenzi [1]. Copyright © 1978 IEEE. Reprinted by permission of the IEEE, Inc.)

Example 10-3-1:   Core Radius of a Monomode Optical Waveguide

A monomode step-index optical fiber has the following parameters

Core refractive index	$n_1 = 1.51$
Cladding refractive index	$n_2 = 1.49$
Light source—Nd^{3+}:YAG laser	$\lambda_0 = 1.064\ \mu$m

Determine the core radius that will support only a single-mode operation.

Solution   From Eq. (10-3-1), the maximum core radius is

$$a \leq \frac{2.405 \times 1.064 \times 10^{-6}}{2\pi\sqrt{(1.51)^2 - (1.49)^2}}$$

$$\leq 1.66 \ \mu m$$

10-3-2 Multimode Step-Index Fibers

A multimode step-index fiber has a large core and large numerical aperture (NA), so it can couple efficiently to a light-source LED. The core radius is from 25 to 60 μm and the differential in refractive indices is from 1 to 10%. The bit rate is under 100 Mbits/s-km. It is suitable for low-bandwidth, short-haul, and low-cost applications such as for links in computers. Figure 10-3-2 shows the diagrams for a multimode step-index fiber, including the light-path and signal-waveform profiles.

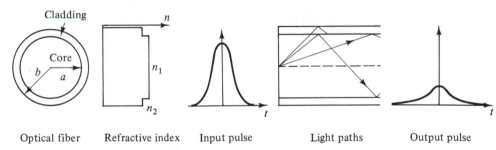

Optical fiber   Refractive index   Input pulse   Light paths   Output pulse

Figure 10-3-2   Multimode step-index fiber. (After T. G. Giallorenzi [1]. Copyright © 1978 IEEE. Reprinted by permission of the IEEE, Inc.)

Example 10-3-2:   Computations of Multimode Step-Index Fiber

A multimode step-index fiber has the following parameters:

Core refractive index	$n_1 = 1.45$
Numerical aperture	$NA = 0.35$
V number	$X_{np} = 100$
Wavelength of LED source	$\lambda_0 = 0.87 \ \mu m$

Calculate (a) the cladding refractive index n_2, (b) the core radius a, and (c) the cladding radius b.

Solution   (a) The cladding refractive index is

$$n_2 = (n_1^2 - NA^2)^{1/2} = \sqrt{(1.45)^2 - (0.35)^2} = 1.41$$

(b) From Eq. (10-3-1), the core radius is

$$a = \frac{\lambda_0 X_{np}}{2\pi(NA)} = \frac{0.87 \times 100}{6.283 \times 0.35} = 39.56 \ \mu m$$

(c) For electric and magnetic fields to be evanescent in the cladding region, the propagation constant χ of Hankel function must be imaginary. That is,

$$\chi = j\sqrt{\beta_2^2 - \beta_g^2} = j\sqrt{\omega^2 \mu_2 \epsilon_2 - \omega^2 \mu_0 \epsilon_0}$$

$$= j\frac{2\pi}{\lambda_0}(n_2^2 - 1)^{1/2} = j\frac{6.2832}{0.87}\sqrt{(1.40)^2 - 1}$$

$$= j7.18$$

From Appendix I, the argument of Hankel function could be $4X_{np} = 400$. Then the cladding radius is

$$b = \frac{400}{7.18} = 55.71 \ \mu m$$

10-4 GRADED-INDEX FIBERS

A graded-index fiber is the one that the refractive index of the core region is decreased monotonically from the center and converged into a flat at the cladding region. Most of the longer path lengths are through the lower index of refractive material. Hence the increased velocity over this part of the path compensates somewhat for the longer path lengths, and the spread in group velocity between modes is not as great as for the step-index fiber. To meet bandwidth requirements for telecommunication applications, multimode fibers are usually fabricated with graded-index profiles to reduce the intermodal dispersion. The graded-index fiber has a relative large core, so it can support multimode operation. This type of fiber is suitable for high-bandwidth and medium-haul applications. Table 10-4-1 lists some typical parameters for the graded-index fibers. Figure 10-4-1 shows the diagrams for a multimode graded-index fiber, including the light-path and signal-waveform profiles.

TABLE 10-4-1   PARAMETERS OF GRADED-INDEX FIBER

Parameter	Values
Bit rate	140–1000 Mbits/sec-km
Core radius	10–35 μm
Cladding radius	50–80μm
Losses	2 –5 dB/km
Numerical aperture (NA)	0.15–0.25
Number of modes at 0.9 μm	140–900
Pulse dispersion/mode	0.1–4.0 ns/km
Refractive index n_1	1.47–1.50
Deviation index Δ	0.7–30%

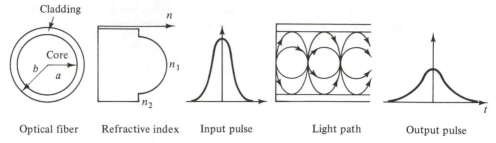

Optical fiber   Refractive index   Input pulse   Light path   Output pulse

Figure 10-4-1 Graded-index fiber. (After T. G. Giallorenzi [1]. Copyright © 1978 IEEE. Reprinted by permission of the IEEE, Inc.)

10-4-1 Refractive-Index Profiles

In a multimode graded-index fiber, the refractive index is expressed by [4, p. 83]

$$n(r) = n_1 \left[1 - 2\Delta \left(\frac{r}{a} \right)^\alpha \right]^{1/2} \qquad \text{for } r \leqslant a \qquad (10\text{-}4\text{-}1)$$

$$n(r) = n_1 (1 - 2\Delta)^{1/2} \qquad \text{for } r > a \qquad (10\text{-}4\text{-}2)$$

where n_1 = refractive index at the center of the core region

$$\Delta = \frac{n_1 - n_2}{n_2} \doteq \frac{n_1 - n_2}{n_1} \doteq \frac{n_1^2 - n_2^2}{2n_1^2} \text{ is the deviation index}$$

α = a parameter which is also called the *power-law coefficient* between 1 and ∞

Figure 10-4-2 shows a cross-sectional diagram of a circular symmetric index profile in a multimode fiber. All profiles reach a constant cladding index n_2 at $r =$

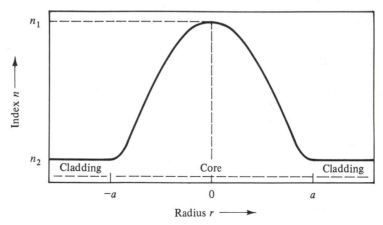

Figure 10-4-2 Cross-sectional diagram for graded refractive index. (After D. Gloge and E.A.J. Marcatili [7]. Reprinted with permission from *The Bell System Technical Journal*. Copyright 1973, AT&T.)

a, while it is n_1 for $r = 0$ at the core center. The core profile has a cone shape or nearly parabolic figure for $\alpha = 1$ and becomes convergent to the case of the step profile for $\alpha = \infty$. Figure 10-4-3 illustrates a few of the index profiles as defined by Eq. (10-4-1) for a small deviation index Δ.

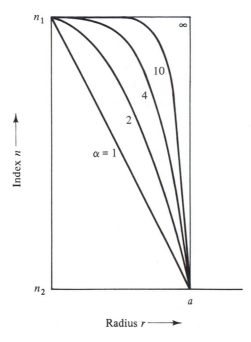

Figure 10-4-3 Graded-index profiles. (After D. Gloge and E.A.J. Marcatili [7]. Reprinted with permission from *The Bell System Technical Journal*. Copyright 1973, AT&T.)

The total number of modes in a graded-index fiber is given by

$$N = \left(\frac{2\pi a}{\lambda_0}n_1\right)^2 \Delta\left(\frac{\alpha}{\alpha + 2}\right) \tag{10-4-3}$$

For a parabolic profile at $\alpha = 2$, the number of modes in a graded-index fiber is equal to half the number that exist in a step-index fiber ($\alpha = \infty$). That is,

$$N(\text{parabolically graded}) = \frac{N(\text{step})}{2} \tag{10-4-4}$$

then the total number of modes in a graded-index fiber can be expressed as

$$N = \frac{V^2}{2}\left(\frac{\alpha}{\alpha + 2}\right) \tag{10-4-5}$$

where $V = X_{np}$ is defined by Eq. (10-2-18).

Example 10-4-1: Number of Modes in a Graded-Index Fiber

A graded-index fiber has the following parameters:

Core radius	$a = 20 \ \mu m$
Power-law coefficient	$\alpha = 2$
Deviation index	$\Delta = 0.03$
Core refractive index	$n_1 = 1.52$
Wavelength of light source LED	$\lambda_0 = 0.87 \ \mu m$

Calculate (a) the total number of modes for the graded-index fiber, (b) the number of modes for a step-index profile, and (c) the graded refractive index.

Solution (a) From Eq. (10-4-3), the number of modes for the graded-index fiber is

$$N = \left(\frac{6.283 \times 20 \times 10^{-6}}{0.87 \times 10^{-6}} \times 1.52 \right)^2 \times 0.03 \left(\frac{2}{2 + 2} \right) = 723 \text{ modes}$$

(b) From Eq. (10-4-4), the number of modes for a step-index profile is

$$N = 2 \times 723 = 1446 \text{ modes}$$

(c) From Eq. (10-4-2), the graded refractive index is

$$n_2 = 1.52(1 - 2 \times 0.03)^{1/2} = 1.47$$

10-4-2 Wave Patterns

Graded-index fibers offer multimode propagation in a relatively large core fiber coupled with low mode dispersion. The typical bit rate is 1400 Mbits/s over a link of 8 to 10 km in conjunction with a GaAs light source, compared to an upper limit of perhaps

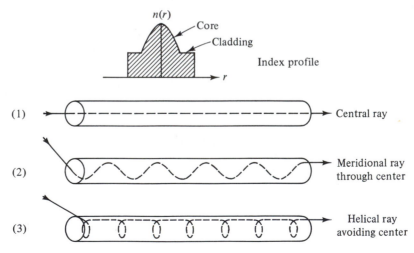

Figure 10-4-4 Wave patterns in graded-index fibers. (After John E. Midwinter [5]. Copyright © 1979 John Wiley & Sons, Inc. Reprinted by permission of John Wiley & Sons, Inc.)

10 to 20 Mbits/s-km using a step-index fiber. There are three wave patterns in a graded-index fiber, [4, p. 570], as shown in Fig. 10-4-4.

1. *Center ray.* Since the index distribution is graded, with a high index in the center surrounded by a steadily decreased index, the medium behaves similarly to a series of lenses. A ray traversing the center of the medium follows the axis, traveling in high-index material all the way but traversing the shortest possible physical path from end to end.

2. *Meridional ray.* A ray leaving the center of the fiber at an angle to the axis is bent by the index profile (lenslike structure) to curve around after some characteristic distance and cross the axis once again. It continues in a sinusoidal path to the end. It thus travels a greater distance than the center ray, but much of its path is in low-index material. This ray, called the meridional ray, passes through the axis of the fiber while being internally reflected.

3. *Skew ray (helical ray).* The third ray is a helical wave, formed when light is launched at a skew angle along the surface of a constant-radius cylinder at some intermediate radius and index. Once again, the physical path length is longer than that taken by the center ray, but it traverses in lower-index material. This ray, called a skew ray or helical ray, never intersects the axis of the fiber while being internally reflected.

10-5 OPTICAL-FIBER COMMUNICATION SYSTEMS

An optical-fiber communication system is essentially a dielectric-filled cylindrical waveguide with some electronic components and circuits, as shown in Fig. 10-5-1. An incoming binary-digital signal switches the light-source laser or LED ON and OFF. When the light is ON, the light impinges on one end of the optical fiber, passes through it by total internal reflection at the core–cladding interface, and emerges at the far end. Then the emerged light impinges on the photodetector, which converts the photon light to electron–hole pairs in the detector and a current pulse will be generated. Consequently, the receiver circuits will reproduce the information signal in a binary-digital form 0 and 1.

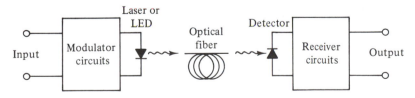

Figure 10-5-1   Optical-fiber communication system.

The design of an optical-fiber communication system is based on four basic requirements: the desired data rate (bandwidth), the signal-to-noise ratio (SNR), the distance between terminals, and the type of source information (digital or analog).

10-5-1 Light Sources

Light sources for optical-fiber communication systems require certain characteristics, such as long lifetime, high efficiency, low cost, high capacity, and sufficient output power. There are two light sources: lasers and LEDs. The power needed for various fiber communications is usually no more than a few milliwatts. At that power level, the fibers are linear. However, when the power reaches the level of 100 mW, single-mode fibers begin to exhibit nonlinearity. For multimode fibers a laser power of about 2.5 W marks the onset of nonlinearity.

LED light sources. Since an LED emits light randomly from its junction in all directions, the light is incoherent. Therefore, transmission of LED-generated signals inherently involves multimodes. For the wavelength range 0.82–0.85 μm, GaAs LED is an adequate choice for data links.

Laser light sources. A laser beam is coherent and it is an ideal light source for monomode fibers. Since the light is very narrow it can be coupled efficiently into the fibers and reduces the effect of the intrinsic chromatic dispersion.

For optical-fiber communications, gas lasers such as CO_2 and helium–neon are in general too large, too expensive, and low in efficiency. At the wavelength 1.06 μm, the Nd^{3+}:YAG (neodymium:yttrium–aluminum–garnet) laser is one of the most useful light sources for single-mode fibers. For the range 0.82 to 0.85 μm, ternary AlGaAs and InGaAs lasers are commonly used. Figure 10-5-2 shows a light-source AlGaAs laser coupled to a glass fiber.

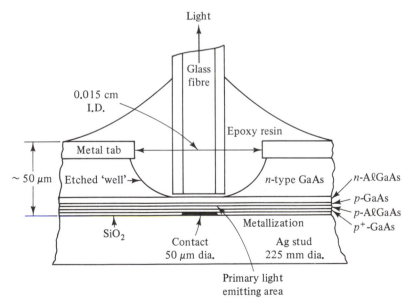

Figure 10-5-2 AlGaAs laser coupled to a fiber. (C. A. Burrus and B. I. Miller [8]. Reprinted by permission of Optics Communications, North-Holland Physics Publishing.)

10-5-2 Light Detectors

The functions of light detectors in an optical-fiber communication system is to detect the optical signal from the fiber and convert it into an electrical signal. At the wavelength region 0.82 to 0.85 μm, the GaAs photodetector is the suitable choice. For the region 1.3 to 1.5 μm, PIN photodetectors, such as InGaAs and InGaAsP, are commonly used. At the receiver side, a low-noise MESFET (GaAs FET) preamplifier usually follows immediately after the photodetector for a better signal-to-noise ratio at the output terminals.

10-5-3 Applications

For communication applications with multimode optical fibers, several assumptions could be made:

1. The index profile of the fibers is circular symmetry about the core axis.
2. The core diameter is large enough, say of the order of 100 wavelengths, so that many modes will be expected to propagate.
3. The index difference is small enough, so that modes can be considered transverse electromagnetic (TEM).
4. The index variation is very small over distances of a wavelength, so that local plane-wave approximation can be made.

The major applications of optical-fiber cables are in computers, industrial automation, medical instruments, the military, and telecommunications.

1. *Computer applications.* Computer terminals and their internal links require very high data rates, up to several Gbits/s-km. The optical fibers can easily meet these requirements. Since the fibers offer the advantage of freedom from electromagnetic interference and grounding problems, the signal quality is much improved.
2. *Industrial automation.* Optical-fiber applications in industrial automation are in process controls, discrete manufacturing automation, and transportation areas. The transportation areas include the airways, shipping, highways, railways, and others.
3. *Medical instruments.* Historically, the medical profession was the first to use optical fibers for diagnosis and treatments. In the late 1930s, a guided light was employed to provide illumination for simple medical inspection instruments. Medical instruments that utilize optical fibers include the cardioscope, colonoscope, endoscope, gastroscope, and opthalmoscope.
4. *Military applications.* The major advantages of optical-fiber cables are the absence of crosstalk, radio-frequency interference, grounding problems, and outside jamming influences. The fibers greatly enhance the security level. So optical fibers

are commonly used for communication links in aircraft, missiles, submarines, ocean surface vessels, and military command and control systems.

5. *Telecommunications.* Optical-fiber applications in telephone and telegraph constitute a major use of fibers. Optical-fiber telecommunications systems can solve many inherent problems that occur in wire systems, such as ringing, echoes, and crosstalk. They are commonly used in the areas of voice telephones, video phones, telegraph services, and various news broadcast systems.

Table 10-5-1 lists applications and characteristics of the three major optical-fiber systems.

TABLE 10-5-1 APPLICATIONS AND CHARACTERISTICS OF THREE MAJOR OPTICAL-FIBER SYSTEMS

	Monomode step-index fiber	Multimode graded-index fiber	Multimode step-index fiber
	Cladding / Core / Protective plastic coating		
Source	Requires laser	Laser or LED	Laser or LED
Bandwidth	Very, very large, > 3 GHz-km	Very large, 200 MHz- to 3 GHz-km	Large, < 200 MHz-km
Splicing	Very difficult due to small core	Difficult but doable	Difficult but doable
Example of application	Submarine cable system	Telephone trunk between central offices	Data links
Cost	Less expensive	Most expensive	Least expensive

Source: After Allen H. Cherin [9]. *An Introduction to Optical Fibers.* Copyright © 1983 McGraw-Hill Book Company. Reprinted by permission.

10-5-4 Design Example

The choice of a light source for a specific optical-fiber system is an essential part of the design process. The output power of a light source is usually expressed in terms of its input current. In practice, the light source is terminated in a fiber connector. For a LED light source and a 50-μm core fiber, the coupling loss may be as high as 20 dB, while for a laser to the same fiber, the coupling loss can be as low as 3 dB.

Example 10-5-1:   Optical-Fiber System

An optical-fiber system is to be designed to meet the following specifications:

LED source to fiber coupling loss	20 dB
Laser source to fiber coupling loss	3 dB
Fiber length	4 km
Fiber loss	4 dB/km
Number of splices	1 splice/0.5 km
Splice loss	0.5 dB/splice
Fiber-to-detector coupling loss	0.1 dB
Minimum power required for receiver	−40 dBm

Solution: For the LED source

1. Choose as the light source the LED 1A83 (ASEA-HAFO shown in Appendix E), which has an output power 10 mW from a drive current of 0.1 A.

$$P_{\text{in}} = 10 \text{ mW} = 10 \text{ dBm}$$

2. The total link loss is

$$\text{loss} = 20 \text{ dB} + 4 \times 4 \text{ dB} + 8 \times 0.5 \text{ dB} + 0.1 \text{ dB} = 40.1 \text{ dB}$$

3. The power available at the receiver is

$$P_{\text{av}} = 10 \text{ dBm} - 40.1 \text{ dB} = -30.1 \text{ dBm}$$

4. The power margin at the receiver is

$$P = -30.1 \text{ dBm} - (-40 \text{ dBm}) = 9.9 \text{ dBm}$$
$$= 9.772 \text{ mW}$$

For the laser source:

1. A CW-laser diode has an output power of 10 mW from a drive current of 0.2 A.

$$P_{\text{in}} = 10 \text{ mW} = 10 \text{ dBm}$$

2. The total link loss is

$$\text{loss} = 3 \text{ dB} + 4 \times 4 \text{ dB} + 8 \times 0.5 \text{ dB} + 0.1 \text{ dB} = 23.1 \text{ dB}$$

3. The power available at the receiver is

$$P_{\text{av}} = 10 \text{ dBm} - 23.1 \text{ dB} = -13.1 \text{ dBm}$$

4. The power margin at the receiver is

$$P = -13.1 \text{ dBm} - (-40 \text{ dBm}) = 26.9 \text{ dBm}$$
$$= 489.8 \text{ mW}$$

10-6 LIGHT MODULATOR DESIGN

Microwave modulation of light wave is becoming increasingly useful in both military and commercial applications. In general, there are three types of light modulators: electrooptic modulator, magnetooptic modulator, and traveling-wave electrooptic modulator. Most crystals are dielectric in nature. This means that they do not carry electric current. However, there are some crystals that have a sizable electrooptic effect and do carry current when a certain level of electric voltage is applied to them. All light modulators use some type of solid or liquid crystals to produce either an electrooptic or a magnetooptic effect. The selection of a particular crystal for any application involves consideration of such requirements as the wavelength of the light source, the beam diameter, the degree of retardation, and the frequency of modulation.

10-6-1 Electrooptic Modulator

The electrooptic modulator uses the electrooptic effect. The application of an electric field or voltage across a crystal or liquid medium causes a small change in its refractive index. The velocity of light in a medium is inversely proportional to the refractive index of that medium. If the index is decreased, the light travels faster through the medium; if it is increased, the light travels more slowly. In other words, under the influence of an electric field, the medium becomes birefringent; that is, the light beams of two different linear polarizations travel through the medium with different velocities. Consequently, the phases of the two waves will not be in time phase as they travel through the medium. There are two types of electrooptic-modulator configurations, often referred to as the longitudinal mode and transverse mode, depending on whether the applied electric field or voltage is parallel or perpendicular to the direction of light propagation.

Longitudinal mode: Pockels effect and Pockels cell.   When a modulating voltage is applied parallel to the optic axis of a crystal, the crystal will exhibit the Pockels effect. This type of device is called the *longitudinal electrooptic modulator* (LEOM).

> *Pockels effect.* The pockels effect, named after F. Pockels, who studied the effect in the late nineteenth century, is a linear electrooptic effect exhibited in certain crystals which have the property of advancing or retarding the phase of the polarized light waves when a voltage is applied parallel to the optic axis of the crystal. The effect is linearly proportional to the first power of the applied voltage.
>
> *Pockels cell.* A Pockels cell is a device that uses the Pockels effect as a fast light modulator or shutter for laser pulsing. A Pockels cell is usually built with a crystal placed between two crossed polarizers, aligned on its optic axis to the direction of light propagation, and a modulating voltage applied parallel to the optic axis of the crystal as shown in Fig. 10-6-1.

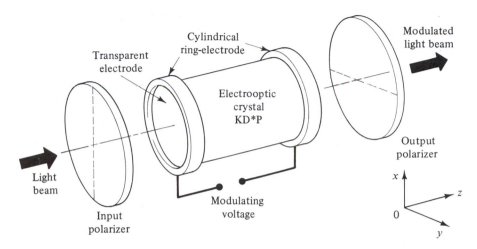

Figure 10-6-1   Pockels cell.

Since the crystals are anisotropic, their properties vary in different directions and they must be described by a group of terms referred to collectively as the *second-rank electrooptic tensor* r_{ij}. Fortunately, for an uniaxial crystal, only the component r_{63} is needed. Table 10-6-1 lists the electrooptic constants and refractive indexes of several commonly used crystals for longitudinal electrooptic modulators.

TABLE 10-6-1   CHARACTERISTICS OF CRYSTALS FOR LONGITUDINAL ELECTROOPTIC MODULATORS

Crystal	Electrooptic constant r_{63} ($\times 10^{-12}$ m/V)	Half-wave voltage (kV)		Refractive index n_o
		$\lambda_o = 0.5461\ \mu m$	$\lambda_o = 1.064\ \mu m$	
ADA (NH$_4$H$_2$AsO$_4$) Ammonium dihydrogen arsenate	5.5	12.59	24.52	1.58
ADP (NH$_4$H$_2$PO$_4$) Ammonium dihydrogen phosphate	8.5	9.15	17.82	1.52
KDP (KH$_2$PO$_4$) Potassium dihydrogen phosphate	10.6	7.48	14.59	1.51
KD*P (KD$_2$PO$_4$) Potassium dideuterium phosphate	26.4	2.95	5.74	1.52
KDA (KH$_2$AsO$_4$) Potassium dihydrogen arsenate	13.0	5.43	10.57	1.57
RDP (RbH$_2$AsO$_4$) Rubidium dihydrogen arsenate	11.0	7.35	14.33	1.50

The crystals used for longitudinal electrooptic modulators are normally uniaxial in the absence of an applied electric voltage. That is, there is only one value of refractive index in the direction of light propagation through the optic axis. However, when an electric voltage is applied parallel to the crystal optic axis and if the light propagation is assumed to be in the z direction as shown in Fig. 10-6-1, the refractive indexes in the x-y plane will be changed under the influence of the crystal birefringence and oriented as a shape of ellipsoid. The light waves propagated through the optic axis will be polarized in the directions of the two new refractive indices and travel with different velocities. At zero voltage, the two indices are orthogonal and define the radius of a circle projected from the index ellipsoid. As the applied voltage is increased, the index circle will be elongated in a direction parallel to one of the induced axes. When a collimated laser light is sent through the input polarizer, the light will be resolved linearly into two orthogonal components in the plane defined by the axis of the polarization of the polarizer, and the light transmittance at the output polarizer is zero with zero applied voltage. As the applied voltage is increased, the two components will travel at different speeds through the crystal and gradually approach being in time phase, and the light transmittance will be increased accordingly. The retardation or the phase change of the two polarized waves at the output polarizer is given by

$$\theta = \frac{2\pi n_0^3 r_{63} V}{\lambda_0} \qquad \text{rad} \qquad (10\text{-}6\text{-}1)$$

where n_0 = refractive index of crystal for the ordinary wave
λ_0 = wavelength in vacuum, meters
V = applied voltage, volts
r_{63} = electrooptic constant of crystal, meters per volt

If maximum light transmittance through the output polarizer is required, the phase change of the two linearly polarized light waves through the crystal must be one π or a half-wavelength. Therefore, the half-wave voltage is expressed by

$$V(\lambda/2) = \frac{\lambda_0}{2n_0^3 r_{63}} \qquad \text{volts} \qquad (10\text{-}6\text{-}2)$$

The output light intensity can be written as

$$I_o = I_i \sin^2\left(\frac{\pi V}{2V_{\lambda/2}}\right) \qquad (10\text{-}6\text{-}3)$$

where I_i is the input light intensity to the input polarizer.

Early Pockels cells were made of ammonium dihydrogen phosphate $(NH_4H_2PO_4)$ or ADP, or of potassium dihydrogen phosphate (KH_2PO_4) or KDP; both crystals are still in use. However, since potassium dideuterium phosphate (KD_2PO_4), known as KD*P, has the highest electrooptic coefficient of any known isomorph and yields the same retardation with a voltage of less than half of that needed by KDP and ADP, at present the crystal KD*P has been widely used in Pockels cells for laser modulation. In addition, these three major crystals have quite different relative dielectric constants, such as 50 for KD*P, 42 for KDP, and 12 for ADP[10,11].

Example 10-6-1:   Half-Wave Voltage of KD*P Crystal

The dielectric crystal KD*P is widely used in Pockels cells for laser modulation because its half-wave voltage is much less than that of other crystals. Determine its half-wave voltage for the Nd^{3+}:YAG laser source.

Solution   From Eq. (10-6-2) the half-wave voltage of the KD*P crystal is

$$V(\lambda/2) = \frac{1.064 \times 10^{-6}}{2 \times (1.52)^3 \times 26.4 \times 10^{-12}} = 5.74 \text{ kV}$$

That means that the KD*P crystal can yield the same output light intensity with a voltage less than half of that needed by the other crystals, such as KDP and ADP.

If maximum retardation is required, the power needed for obtaining the required maximum phase difference must be maximum. This means, from Eq. (10-6-1), that the phase angle is

$$\theta_m = \frac{2\pi n_0^3 r_{63} V_m}{\lambda_0} \tag{10-6-4}$$

When the load resistance R_ℓ of a resonant circuit is chosen much larger than the circuit series resistance R_s, the finite modulation bandwidth is given by

$$\Delta f = \frac{1}{2\pi R_\ell C} \tag{10-6-5}$$

Then the required maximum power can be expressed as

$$P_m = \frac{V_m^2}{2R_\ell} = \frac{\theta_m^2 \lambda_0^2 A \epsilon \Delta f}{4\pi^2 n_0^6 r_{63}^2 L} \tag{10-6-6}$$

where $C = \epsilon A/L$
A = crystal cross section
L = crystal length
Δf = finite modulation bandwidth in the region 10^8 to 10^9 Hz

Example 10-6-2:   Power Requirement for Maximum Retardation

A KD*P Pockels cell has the following parameters:

Relative dielectric constant	$\epsilon_r = 50$
Electrooptical constant	$r_{63} = 26.4 \times 10^{-12}$ m/V
Crystal refractive index	$n_0 = 1.52$
Wavelength of light source	$\lambda_0 = 0.54$ μm
Crystal cross section	$A = 3 \times 10^{-4}$ m²

Crystal length $L = 40$ mm
Maximum retardation angle $\theta_m = 10°$
Finite modulation bandwidth $\Delta f = 0.5$ GHz

Calculate the required maximum power.

Solution   From Eq. (10-6-6) the required maximum power is

$$P_m = \frac{(\pi/18)^2(0.54 \times 10^{-6})^2(3 \times 10^{-4})(8.854 \times 10^{-12} \times 50)(0.5 \times 10^9)}{4(\pi)^2(1.52)^6(26.4 \times 10^{-12})^2(40 \times 10^{-3})}$$

$$= 42.81 \text{ W}$$

Transverse mode: Kerr effect and Kerr cell.   When a modulating voltage is applied normal to the optic axis of a crystal, the crystal will exhibit the Kerr effect. This type of device is called a *transverse electrooptic modulator* (TEOM).

Kerr effect. The Kerr effect, named after John Kerr, who discovered the effect in 1875 when he studied light refraction in a glass by an electric field, is a quadratic electrooptic effect exhibited in certain crystals or liquids when a voltage is applied normal to the optic axis of the crystal or the direction of the light propagation. The effect is not linear but varies as the square of the applied voltage.

Kerr cell. A Kerr cell is a transverse electrooptic modulator commonly made with a liquid such as nitrobenzene or a crystal such as barium titanate. Table 10-6-2 lists the Kerr constants for several commonly used liquids at $\lambda_0 = 5893$ Å and $T = 20°$C.

TABLE 10-6-2   VALUES OF KERR CONSTANTS AT $\lambda_0 = 5893$ Å AND $T = 20°$C

Material	Kerr constant K (m/V^2)
Benzene (C_6H_6)	0.67×10^{-14}
Carbon disulfide (CS_2)	3.56×10^{-14}
Nitrotoluene ($C_5H_7NO_2$)	1.37×10^{-12}
Nitrobenzene ($C_6H_5NO_2$)	2.44×10^{-12}
Water (H_2O)	5.10×10^{-14}

When a Kerr effect medium is inserted between two crossed polarizers and a modulating voltage is applied normal to the direction of light propagation, the device works as a Kerr cell electrooptic modulator. Figure 10-6-2 shows a Kerr cell. When the modulating voltage is off, no light is transmitted by the output polarizer. When the modulating voltage is on, the liquid becomes doubly refracting and the light is transmitted. The change in phase of the two polarized waves in the Kerr cell is given by

$$\phi = \frac{2\pi K V^2 L}{d^2} \qquad \text{rad} \qquad (10\text{-}6\text{-}7)$$

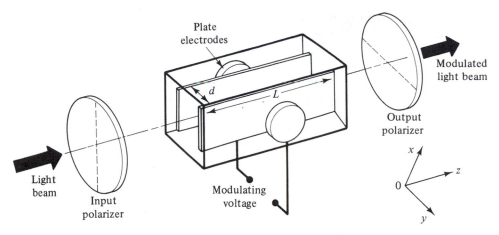

Figure 10-6-2   Kerr cell.

where K = Kerr constant of medium meters per volt squared
V = applied voltage, volts
L = length of electrodes in the optic axis, meters
d = separation of electrodes, meters

A Kerr cell requires between 5 and 10 times the voltage that a Pockels cell would need to obtain the same optical effect. For this reason, as well as the fact that the liquids used in Kerr cells are poisonous and explosive, Pockels cells have replaced Kerr cells in most laser modulators.

Example 10-6-3:   Kerr Effect Modulation

A Kerr cell is constructed with carbon disulfide (OS_2) and has the following parameters:

Wavelength of LED (GaAsP) light $\qquad \lambda_0 = 0.589 \ \mu m$
Kerr constant of carbon disulfide $\qquad K = 3.56 \times 10^{-14} \ m/V^2$
Length of electrodes $\qquad L = 1 \ cm$
Separation of electrodes $\qquad d = 2 \ mm$
Applied voltage $\qquad V = 37.5 \ kV$

Determine the changing phase in radians and in degrees.

Solution   From Eq. (10-6-7) the changing phase is

$$\Phi = \frac{2\pi K V^2 L}{d^2} = \frac{6.2832 \times 3.56 \times 10^{-14} \times (37.5 \times 10^3)^2 \times 10^{-2}}{(2 \times 10^{-3})^2}$$

$$= 0.79 \ \text{rad} = 45°$$

10-6-2 Magnetooptic Modulator

When a modulating magnetic field is applied parallel to the optic axis of a magnetic crystal and a constant magnetic field is normal to the axis, the crystal will exhibit the Faraday effect. This type of device is called a *magnetooptic modulator.*

Faraday effect. The Faraday effect, named after Michael Faraday, who discovered the effect in 1845 when he studied light refraction in a glass by a strong magnetic field, is a magnetooptic rotation exhibited in a crystal by a magnetic field. When a magnetic crystal is subjected to a magnetic field, the light passing through the crystal will be rotated. The amount of rotation is found by experiment to be proportional to the magnetic flux density B, the distance L for the light to travel through the crystal or medium, and the Verdet constant V of the crystal or medium. Therefore, the rotation is given by

$$\psi = VBL \qquad \textit{minutes of arc} \qquad (10\text{-}6\text{-}8)$$

where V = Verdet constant, which is defined as the rotation in minutes of arc per meter per tesla

B = magnetic flux density, webers per square meter (tesla)

L = length of the medium, meters

A variety of gases, liquids, and solids are used for the Faraday effect modulators. Their Verdet constants V at the wavelength of 5893 Å are tabulated in Table 10-6-3.

TABLE 10-6-3 VERDET CONSTANTS AT $\lambda = 5893$ Å

Material	T (°C)	Verdet constant $V (\times 10^4)$
Acetone	15	1.109
Carbon disulfide (CS$_2$)	20	4.230
Ethyl alcohol	25	1.112
Glass (phosphate crown)	18	1.610
Glass (light flint)	18	3.170
Phosphorus (P)	33	13.260
Quartz (perpendicular to axis)	20	1.660
Salt (NaCl)	16	3.585
Water (H$_2$O)	20	1.310
YIG (Y$_3$Fe$_5$O$_{12}$)	—	—

Faraday effect modulator. A magnetooptic modulator is a device that uses Faraday rotation, as shown in Fig. 10-6-3. The magnetic crystal yttrium–iron–garnet (YIG) is placed between two crossed polarizers. A constant magnetic field is applied normal to the crystal and the modulating magnetic field, parallel to the crystal. The resultant component of the two magnetic fields along the crystal axis causes Faraday rotation. As the current flowing through the coil is changed, the magnetic field in the optic axis is also changed. As a result, the Faraday rotation along the optic axis is also changed. Consequently, a linear maximum modulation is obtained at the output polarizer if the Faraday effect causes a rotation of 45° with respect to the input polarizer. The magnetooptic modulator is usually used for large-bandwidth frequency modulation of light.

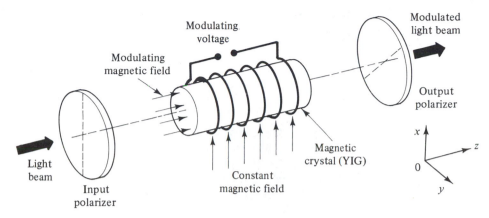

Figure 10-6-3   Magnetooptic modulator.

Example 10-6-4:   Faraday Effect Modulator

The Verdet constant V for the magnetic crystal YIG is not available. However, the product of VB measured at room temperature is 1750° per centimeter at a wavelength of 0.588 μm under a saturation magnetic intensity of 0.0199 Wb/m^2 or 199 gausses. Determine the length L of the magnetic crystal YIG for a Faraday rotation of 45°.

Solution   The crystal length is

$$L = \frac{5 \times 360 \times 45}{1750} = 1.05 \text{ cm}$$

10-6-3 Traveling-Wave Electrooptic Modulator

When an electrooptic material slab is inserted between a parallel-plate transmission line, the device will exhibit optic phase modulation. This type of device, called a *traveling-wave electrooptic modulator* is shown in Fig. 10-6-4.

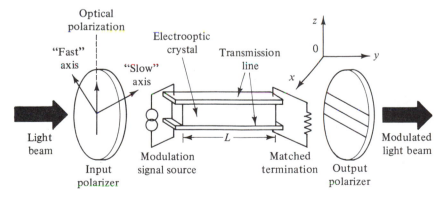

Figure 10-6-4   Traveling-wave electrooptic modulator. (From [12] *Introduction to Optical Electronics* 2/e by Ammon Yariv. Copyright © 1976 by Holt, Rinehart and Winston. Reprinted by permission of CBS College Publishing.)

The electrooptic crystal is placed between the parallel-plate transmission line. The modulating voltage is applied to the input end of the line. Two linearly polarized light waves from the input polarizer travel through the crystal slab. The time for the light to travel the crystal is

$$\tau = \frac{L}{v} = \frac{Ln}{c} = \frac{L\sqrt{\epsilon_r}}{c} \qquad \text{seconds} \qquad (10\text{-}6\text{-}9)$$

where v = velocity of light in crystal, meters per second
   L = length of crystal slab, meters
   n = refractive index of crystal
   c = velocity of light in vacuum
   ϵ_r = relative dielectric constant of crystal

The phase of the output light from the output polarizer is related to the input light at the input polarizer by

$$\Phi = \omega\tau = \frac{\omega L}{v} \qquad \text{rad} \qquad (10\text{-}6\text{-}10)$$

The phase modulation can be accomplished by varying the velocity of light in the crystal. The change or modulation of the phase angle $\Delta\Phi$ is related to the changing velocity of the light wave in the crystal by

$$\Delta\Phi = \frac{\Phi}{v}\Delta v = \frac{\omega L}{v^2}\Delta v \qquad (10\text{-}6\text{-}11)$$

The velocity of light in an electrooptic crystal is a function of the applied electric field. The changes in the refractive index in the major axes x, y, and z of the index ellipsoid are given by

$$n_x = n_o - \frac{n_o^3}{2}r_{63}E_z \qquad (10\text{-}6\text{-}12)$$

$$n_y = n_o + \frac{n_o^3}{2}r_{63}E_z \qquad (10\text{-}6\text{-}13)$$

$$n_z = n_e \qquad (10\text{-}6\text{-}14)$$

where n_o = refractive index of the ordinary wave
   r_{63} = electrooptic constant of crystal, meters per volt
   n_e = refractive index of the extraordinary wave
   E_z = applied electric field in the z direction

The changes in the velocity of light propagating parallel to the x and y axes of the crystal are given by

$$\Delta v_x = -\frac{1}{2}cn_o^2r_{63}E_z \qquad (10\text{-}6\text{-}15)$$

and

$$\Delta v_y = + \frac{1}{2} c n_o^2 r_{63} E_z \tag{10-6-16}$$

Substitution of Eq. (10-6-15) or (10-6-16) in Eq. (10-6-11) and replacing the velocity v by c/n_0 yields the changing phase angle $\Delta \Phi$ as

$$\Delta \Phi = \frac{n_o^4 \omega r_{63} L E_z}{2 c} = \frac{\pi n_o^4 r_{63} L E_z}{\lambda_0} \tag{10-6-17}$$

where λ_0 is the wavelength in vacuum in meters. This equation shows that the phase change $\Delta \Phi$ is directly proportional to the applied electric field E_z and to the length L of the path traveled by the light when the electric field can be adjusted such that maximum light intensity is transmitted. The traveling-wave electrooptic modulator is often used for large-bandwidth, low-power modulation of light.

Example 10-6-5:   Traveling-Wave Electrooptic Modulator

A traveling-wave electrooptic modulator is constructed with a KDP crystal and has the following parameters:

Refractive index of KDP	$n_0 = 1.51$
Electrooptic constart of KDP	$r_{63} = 10.60 \times 10^{-12} \text{m/V}$
Length of KDP	$L = 1 \text{ cm}$
Wavelength of Nd^{3+}:YAG laser	$\lambda_0 = 1.064 \text{ μm}$
Electric field	$E_z = 482.7 \text{ kV/m}$

Determine the changing phase in radians and degrees.

Solution   From Eq. (10-6-17) the changing phase angle is

$$\Delta \Phi = \frac{3.1416 \times (1.51)^4 \times 10.6 \times 10^{-12} \times 10^{-2} \times 482.7 \times 10^3}{1.064 \times 10^{-6}}$$

$$= 0.79 \text{ rad}$$

$$= 45°$$

REFERENCES

1. Giallorenzi, T. G., Optical communications research and technology: fiber optics. *Proc. IEEE,* Vol. 66, No. 7, July 1978.

2. Elion, G. R., and Herbert A. Elion, *Fiber Optics in Communications systems.* New York: Marcel Dekker, Inc., 1978, p. 35.

3. Jacobs, Ira, and S. E. Miller, Optical transmission of voice and data. *IEEE Spectrum,* Vol. 14, No. 2, February 1977, p. 39.

4. Schelkunoff, S. A., *Electromagnetic Waves.* New York: D. Van Nostrand, 1964, p. 247.

5. Midwinter, John E., *Optical Fibers for Transmission.* New York: John Wiley & Sons, Inc., 1979.

6. Potter, R. J., et al., *J. Opt. soc. Am.,* Vol. 53, 1963, p. 256.

7. Gloge, D., and E. A. J. Marcatili, Multimode theory of graded-core fibers. *Bell System Technical Journal,* Vol. 52, No. 9, November 1973.

8. Burrus, C. A., and B. I. Miller, Small area, double heterostructure AlGaAs electroluminescent diode source for optical-fiber transmission lines. *Opt. Common.* Vol. 4, 1971, p. 307.

9. Cherin, Allen H., *An Introduction to Optical Fibers.* New York: McGraw-Hill Book Company, 1983, p. 3.

10. Wilson, J., and J. F. B. Hawkes, *Optoelectronics: An Introduction.* Englewood Cliffs, N.J.: Prentice-Hall, Inc., 1983, p. 104.

11. Liao, Samuel Y., *Pockels-Cell Laser Modulation Subsystem for A-6E TRAM Aircraft.* El Segundo, Calif.: Hughes Aircraft Company, 1980.

12. Yariv, Ammon, *Introduction to Optical Elecgtronics,* 3rd ed. New York: Holt, Rinehart and Winston, 1976.

SELECTED READINGS

Allan, W. B., *Fiber Optics: Theory and Practice.* London: Plenum Press, 1973.

Howes, M. J., and D. V. Morgan, eds., *Optical Fibre Communications.* New York: John Wiley & Sons, Inc., 1980.

IEEE Journal of Quantum Electronics, Vol. QE-18, No. 4, April 1982. Special issue on optical guided wave technology.

IEEE Proc., Vol. 68, No. 6, June 1980. Special issue on optical communications.

IEEE Proc., Vol. 61, No. 12, December 1973. Special issue on optical fibers.

IEEE Proc., Vol. 68, No. 10, October 1980. Special issue on optical-fiber communications.

IEEE Proc., Vol. 66, No. 7, July 1978. Special issue on optical-fiber light paths.

IEEE Trans. Electron Devices. Vol. ED-29, No. 9, September 1982. Special issue on optical fibers.

IEEE Trans. Microwave Theory and Techniques, Vol. MTT-30, No. 4, April 1982. Special issue on optical guided wave technology.

Kao, Charles K., *Optical Fiber Systems: Technology, Design, and Applications.* New York: McGraw-Hill Book Company, 1982.

PROBLEMS

10-1 Optical Fibers

10-1-1. The absorption loss of an optical fiber is described by Eq. (10-1-1), where ℓ is the light-path length. Verify that

$$\ell = n_1 / (n_1^2 - \sin^2 \theta_{max})^{1/2}.$$

10-2 Operational Mechanisms

10-2-1. The numerical aperture of an optical fiber is defined as $NA = \sin\theta$, where θ is the acceptance angle. Verify Eq. (10-2-35).

10-2-2. A certain optical fiber has $n_1 = 1.30$ and $n_2 = 1.02$. Calculate the meridional light-gathering power.

10-2-3. The wave equations in the core region ($r < a$) of an optical fiber are represented by Bessel functions as shown in Eqs. (10-2-3) and (10-2-4). From these two equations and Maxwell's equations, derive Eqs. (10-2-7) and (10-2-8).

10-2-4. The wave equations in the cladding region ($r > a$) of an optical fiber are represented by Hankel functions as shown in Eqs. (10-2-5) and (10-2-6). From these two equations and Maxwell's equations, derive Eqs. (10-2-9) and (10-2-10).

10-2-5. For cutoff wavelengths, the Hankel function H_n must vanish at the interface between the core and cladding ($r = a$). Derive the free-space cutoff wavelength equation (10-2-19).

10-2-6. The cladding refractive index n_2 of an optical fiber is 1.10 and the core radius a is 30λ. Determine the wave component $K_v(u)$ in the cladding region for a distance r of 31λ.

10-2-7. Derive the equation of maximum acceptance angle as shown in Eq. (10-2-31) from Fig. 10-2-3.

10-2-8. A multimode step-index fiber has the following parameters:

Carrier wavelength	$\lambda = 0.85 \ \mu m$
Carrier frequency	$f = 3.53 \times 10^{14} \ Hz$
Core radius	$a = 30 \ \mu m$
Core refractive index	$n_1 = 1.48$
Cladding radius	$b = 60 \ \mu m$
Cladding refractive index	$n_2 = 1.20$

Compute:
(a) The phase constant in the core region.
(b) The phase constant in the cladding region.
(c) The phase constant in free space.
(d) The transverse propagation constant in the core region.
(e) The transverse propagation constant in the cladding region.
(f) The value of the Bessel function $J_1(ka)$.
(g) The value of the Hankel function $H_1(\chi b)$.

10-3 Step-Index Fibers

10-3-1. A step-index fiber operates in a single mode EH_{01}. Its radius a is 5 μm and its refractive indexes are $n_1 = 1.15$ and $n_2 = 1.08$. Determine the cutoff wavelength and frequency.

10-3-2. A multimode step-index fiber has a core radius a of 40 μm and a cladding radius b of 50 μm. the refractive index n_1 of the core is 1.40 and the refractive differential between n_1 and n_2 is 10%.
(a) Calculate the light-gathering power.
(b) Compute the cutoff wavelength.
(c) Determine the number of modes.

10-3-3. Describe the characteristics and applications of the monomode and multimode step-index fibers.

10-3-4. A multimode step-index fiber has the following parameters:

Core radius	$a = 40 \ \mu m$
Core refractive index	$n_1 = 1.52$
Cladding refractive index	$n_2 = 1.48$
Wavelength of light source LED	$\lambda = 0.87 \ \mu m$

(a) Calculate the numerical aperture of the fiber.
(b) Compute the maximum acceptance angle of the fiber.
(c) Estimate the maximum number of modes.
(d) Determine the light-gathering power.

10-3-5. A multimode step-index fiber has the following parameters:

Core refractive index	$n_1 = 1.50$
Numerical aperture	$NA = 0.37$
V number	$X_{np} = 120$
Wavelength of light source Nd^{3+}:YAG	$\lambda = 1.06 \ \mu m$

(a) Calculate the cladding refractive index n_2.
(b) Compute the core radius a.
(c) Determine the cladding radius b.

10-4 Graded-Index Fibers

10-4-1. A graded-index fiber has $n_1 = 1.50$ and $n_2 = 1.05$. Determine the refractive index at $r = a/2$ for $\alpha = 4$.

10-4-2. The graded-index fiber can support multiple modes. Its core radius a is 30 μm and its refractive indexes are $n_1 = 1.50$, $n_2 = 1.05$. Determine the number of modes.

10-4-3. Describe the characteristics and applications of multimode graded-index fibers.

10-4-4. Derive from Eq. (10-4-3) the equation for the total number of modes for a graded-index fiber as shown in Eq. (10-4-5).

10-4-5. A graded-index fiber has the following parameters:

Core radius	$a = 30 \ \mu m$
Core refractive index	$n_1 = 1.50$
Deviation index	$\Delta = 0.02$
Power-law coefficient	$\alpha = 2$
Wavelength of light source	$\lambda_0 = 1.06 \ \mu m$

(a) Calculate the number of modes for the graded-index fiber.
(b) Compute the graded refractive index.
(c) Determine the number of modes for a step-index fiber.

10-6 Light Modulators

10-6-1. When a voltage is applied to a Pockels-cell laser modulator, the light beam will be rotated a certain number of degrees depending on the magnitude of the voltage and the nature of the dielectric crystal, as shown in Eq. (10-6-1). Derive the equation.

10-6-2. The output light intensity from a Pockels-cell laser modulator is expressed by Eq. (10-6-3). Derive the equation.

10-6-3. The changing phase angle for a traveling-wave electrooptic modulator is shown in Eq. (10-6-17). Verify the equation.

10-6-4. The crystal KD*P is often used for the Pockels-cell laser modulator and it has a refractive index n of 1.52. Its electrooptic constant r_{63} is 26.4×10^{-12} m/V. The half-wave voltage is 3.72 kV.
   (a) Determine the wavelength and the frequency of the light.
   (b) Identify the type of light.
   (c) Find the applied voltage for a phase change of 90°.

10-6-5. The crystal KDP in a Pockels-cell laser modulator has a refractive index n of 1.51 and an electrooptic constant r_{63} of 10.60×10^{-12} m/V.
   (a) Compute the half-wave voltage at a wavelength of 0.5461 μm.
   (b) Identify the type of light.
   (c) Calculate the output light intensity in terms of the input light intensity for $V = V_{\lambda/2}$.

10-6-6. A Kerr cell uses liquid nitrobenzene with a Kerr constant K of 2.44×10^{-12} m/V². Its dimensions are $L = 10$ cm and $d = 2$ cm. Calculate the applied voltage for a phase change of 90°.

10-6-7. A magnetooptic laser modulator uses a flint glass as the refractor. The flint glass has a Verdet constant V of 3.17×10^4 minutes of arc per meter per tesla. The magnetic flux density is 0.2 telsa and the cell length L is 10 cm. Determine the light rotation at a wavelength of 5893 Å. (*note:* 1 tesla = 1 weber per square meter = 10^4 gauss.)

10-6-8. A maximum retardation requires a maximum power as shown in Eq. (10-6-6). Start from Eqs. (10-6-4) and (10-6-5) and $\Delta f = 1/(2\pi R_r C)$, and derive Eq. (10-6-6).

10-6-9. A KD*P Pockels cell has the following parameters:

Relative dielectric constant	$\epsilon_r = 50$
Electrooptic constant	$r_{63} = 26.4 \times 10^{-12}$ m/V
Wavelength of light source	$\lambda_0 = 0.54$ μm
Crystal cross section	$A = 2 \times 10^{-4}$ m²
Crystal length	$L = 50$ mm
Crystal refractive index	$n_0 = 1.52$
Maximum retardation angle	$\theta_m = 45°$
Finite modulation bandwidth	$\Delta f = 0.2$ GHz

Calculate the required maximum power in watts.

Chapter 11

Dielectric Planar Waveguides and Film Coating Design

11-0 INTRODUCTION

A planar waveguide consists of two parallel conducting plates and it is different from a rectangular waveguide because its top and bottom plates have no connection. This type of waveguide is similar to a two-conductor line or a hollow tube, so it can support the TE and TM nodes in addition to the TEM modes. Also, in recent monolithic microwave integrated circuit (MMIC) technology, a coplanar waveguide has been preferred for monolithic circuits.

The planar waveguide has a cutoff frequency similar to that of the hollow waveguide. In general, the planar waveguide can generate, propagate, combine, detect, deflect, divide, modulate, demodulate, and switch optical guided waves. Planar waveguides are commonly used in microwave integrated circuits for signal transmission. In this chapter we discuss several widely used dielectric planar waveguides, including the infinite parallel-plate waveguide, dielectric-slab waveguide, coplanar waveguide, thin film-on-conductor waveguide, and thin film-on-dielectric waveguide, all shown in Fig. 11-0-1.

11-1 PARALLEL-PLATE WAVEGUIDE

A parallel-plate waveguide is composed of two perfecrt conducting plates separated by a dielectric medium with constitutive parameters of μ_d and ϵ_d as shown in Fig. 11-1-1. The two plates are assumed to be infinite in the x and z directions.

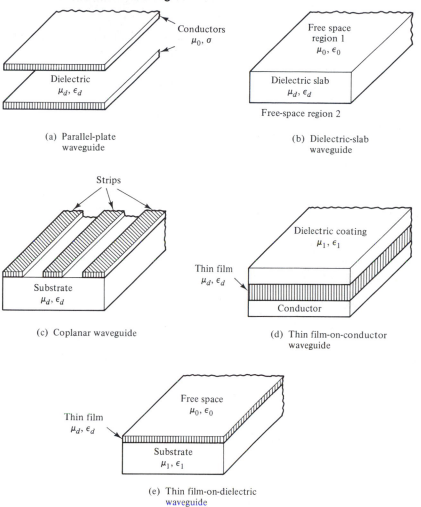

(a) Parallel-plate
waveguide

(b) Dielectric-slab
waveguide

(c) Coplanar waveguide

(d) Thin film-on-conductor
waveguide

(e) Thin film-on-dielectric
waveguide

Figure 11-0-1   Schematic diagrams of commonly used dielectric planar waveguides.

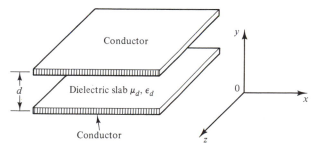

Figure 11-1-1   Schematic diagram of a parallel-plate waveguide.

In Section 6-4, the characteristics of a TEM wave propagating along a finite parallel stripline were described. The field behaviors for the TEM mode have similar characteristics to those in an unbounded dielectric medium. Besides the TEM mode, the TE and TM modes can be supported in an infinite parallel-plate waveguide. This type of waveguide is commonly used in microwave integrated circuits. The characteristics of the TE and TM modes in an infinite parallel-plate waveguide are described next.

11-1-1 TM Waves along the Plates

An infinite parallel-plate waveguide can support TM and TE waves. Let us suppose that TM waves propagate in the positive z direction; that there are no field variations in the x direction, as the width in the x direction is infinite; and that the fringing field effects outside the edges of the x direction are negligible for $x \gg d$. The plates are assumed to be infinite in extent in the x and z directions. For the TM mode ($H_z = 0$), the electric field component is expressed as

$$E_z(y,z) = E_{z0}(y)e^{-\gamma z} \tag{11-1-1}$$

The TM wave equation is

$$\nabla^2 E_z - \gamma^2 E_z = 0 \tag{11-1-2}$$

or

$$\frac{\partial^2 E_z}{\partial y^2} + \frac{\partial^2 E_z}{\partial z^2} - \gamma^2 E_z = 0 \tag{11-1-3}$$

where $\gamma^2 = -\omega^2 \mu_d \epsilon_d$. The general solution of Eq. (11-1-1) is

$$E_z(y,z) = E_{z0} \sin\left(\frac{n\pi y}{d}\right)e^{-\gamma_z z} \tag{11-1-4}$$

where $\gamma_z = [(n\pi/d)^2 - \omega^2\mu_d\epsilon_d]^{1/2}$ is the wave propagation constant
$\quad n = 0, 1, 2, 3, \ldots$
When $\gamma_z = 0$, the cutoff frequency is

$$f_c = \frac{n}{2d\sqrt{\mu_d\epsilon_d}} = \frac{nc}{2d\sqrt{\epsilon_{nd}}} \qquad \text{for } \mu_d = \mu_0 \tag{11-1-5}$$

and the cutoff wavelength is

$$\lambda_c = \frac{2d}{n} \tag{11-1-6}$$

For propagation, $(n\pi/d)^2 < \omega^2\mu_d\epsilon_d$, the propagation constant is imaginary. That is,

$$\gamma_z = j\beta_z = j\sqrt{\omega^2\mu_d\epsilon_d - \left(\frac{n\pi}{d}\right)^2} = j\omega\sqrt{\mu_d\epsilon_d}\left[1 - \left(\frac{f_c}{f}\right)^2\right]^{1/2} \tag{11-1-7}$$

Finally, the field equations for TM modes in an infinite parallel-plate waveguide are

$$E_z(y, z) = E_{z0} \sin\left(\frac{n\pi y}{d}\right)e^{-j\beta_z z} \tag{11-1-8a}$$

$$E_y = \frac{-j\beta_z}{k}E_{z0} \cos\left(\frac{n\pi y}{d}\right)e^{-j\beta_z z} \tag{11-1-8b}$$

$$H_x = \frac{j\omega\epsilon}{k}E_{z0} \cos\left(\frac{n\pi y}{d}\right)e^{-j\beta_z z} \tag{11-1-8c}$$

where

$$k = \left[\omega^2\mu\epsilon - \left(\frac{n\pi}{d}\right)^2\right]^{1/2}$$

11-1-2 TE Waves along the Plates

It is assumed that there are no field variations in the x direction. For the TE mode ($E_z = 0$), the magnetic field component is given by

$$H_z(y, z) = H_{z0}(y)e^{-\gamma z} \tag{11-1-9}$$

The wave equation is

$$\nabla^2 H_z - \gamma^2 H_z = 0 \tag{11-1-10}$$

or

$$\frac{\partial^2 H_z}{\partial y^2} + \frac{\partial^2 H_z}{\partial z^2} - \gamma^2 H_z = 0 \tag{11-1-11}$$

Similarly, the general solution is

$$H_z(y, z) = H_{z0} \cos\left(\frac{n\pi y}{d}\right)e^{-j\beta_z z} \tag{11-1-12}$$

where

$$\beta_z = \omega\sqrt{\mu_d\epsilon_d}\left[1 - \left(\frac{f_c}{f}\right)^2\right]^{1/2} \tag{11-1-13}$$

$$n = 1, 2, 3, \dots$$

The cutoff frequency is

$$f_c = \frac{n}{2d\sqrt{\mu_d\epsilon_d}} \tag{11-1-14}$$

The cutoff wavelength is

$$\lambda_c = \frac{2d}{n} \tag{11-1-15}$$

The phase velocity of the wave in the waveguide is

$$v_p = \frac{\omega}{\beta_z} = \frac{1}{\sqrt{\mu_d \epsilon_d}} \left[1 - \left(\frac{f_c}{f} \right)^2 \right]^{-1/2} \tag{11-1-16}$$

The final field equations for TE modes in a parallel-plate waveguide are

$$E_x(y, z) = \frac{j\omega\mu}{k} H_{z0} \sin\left(\frac{n\pi y}{d} \right) e^{-j\beta_{zz}} \tag{11-1-17}$$

$$H_y(y, z) = \frac{j\beta_z}{k} H_{z0} \sin\left(\frac{n\pi y}{d} \right) e^{-j\beta_{zz}} \tag{11-1-18}$$

$$H_z(y, z) = H_{z0} \cos\left(\frac{n\pi y}{d} \right) e^{-j\beta_{zz}} \tag{11-1-19}$$

11-1-3 Attenuations and Intrinsic Impedance

The intrinsic wave propagation constant in a medium is given by

$$\gamma = \sqrt{j\omega\mu(\sigma + j\omega\epsilon)} = j\omega\sqrt{\mu\epsilon} \left(1 + \frac{\sigma}{j\omega\epsilon} \right)^{1/2} \tag{11-1-20}$$

The propagation constant due to losses in the dielectric at frequencies above the cutoff frequency f_c is then expressed as

$$\begin{aligned}
\gamma_z &= \sqrt{\left(\frac{n\pi}{d} \right)^2 - \omega^2 \mu_d \left(\epsilon_d + \frac{\sigma_d}{j\omega\epsilon_d} \right)} \\
&= \left[\left(\frac{n\pi}{d} \right)^2 - \omega^2 \mu_d \epsilon_d + \frac{j\omega\sigma_d\mu_d}{\epsilon_d} \right]^{1/2} \\
&= \frac{\sigma}{2} \sqrt{\frac{\mu_d}{\epsilon_d}} \left[1 - \left(\frac{f_c}{f} \right)^2 \right]^{-1/2} + j\omega\sqrt{\mu_d\epsilon_d} \left[1 - \left(\frac{f_c}{f} \right)^2 \right]^{1/2}
\end{aligned} \tag{11-1-21}$$

where $f_c = n/(2d\sqrt{\mu_d\epsilon_d})$ is defined in Eq. (11-1-5).

Finally, the attenuation constant in the dielectric for the TM mode is

$$\alpha_d = \frac{\sigma}{2} \sqrt{\frac{\mu_d}{\epsilon_d}} \left[1 - \left(\frac{f_c}{f} \right)^2 \right]^{-1/2} \qquad \text{Np/m} \tag{11-1-22}$$

For the TE mode, the attenuation constant is the same:

$$\alpha_d = \frac{\sigma\eta}{2} \left[1 - \left(\frac{f_c}{f} \right)^2 \right]^{-1/2} \qquad \text{Np/m} \tag{11-1-23}$$

When $n = 0$, $E_z = 0$ and only the transverse components H_x and E_y exist; hence the TEM mode, described earlier, is the dominant mode in a parallel-plate waveguide and its cutoff frequency is zero.

The intrinsic impedance of one conducting plate is

$$\eta = \sqrt{\frac{j\omega\mu}{\sigma + j\omega\epsilon}} = \sqrt{\frac{j\omega\mu}{\sigma}} \qquad \text{for } \sigma \gg \omega\epsilon$$

$$= \sqrt{\frac{\omega\mu}{\sigma}}\underline{/45°} = (1 + j)\sqrt{\frac{\omega\mu}{2\sigma}} \tag{11-1-24}$$

$$= (1 + j)\frac{1}{\sigma\delta} = (1 + j)R_s$$

where $R_s = \sqrt{\omega\mu/(2\sigma)}$ is known as the *skin effect* or surface resistance.

Example 11-1-1:   Characteristics of an Infinite Parallel-Plate Waveguide

An infinite parallel-plate waveguide has the following parameters:

Operating frequency	$f = $ 10 GHz
Separation distance	$d = $ 8 mm
Conductivity of gold plate	$\sigma_{au} = $ 4.1×10^7 mhos/m
Relative dielectric constant of	$\epsilon_{nd} = $ 13.10
GaAs insulator	
Conductivity of GaAs substrate	$\sigma_d = $ 1.17×10^{-2} mhos/m

Calculate (a) the cutoff frequency for the TM mode, (b) the phase velocity for the TM wave, (c) the intrinsic impedance of the plate, and, (d) the attenuation constant for the TM mode.

Solution   (a) From Eq. (11-1-5) the cutoff frequency is

$$f_c = \frac{1 \times 3 \times 10^8}{2 \times 8 \times 10^{-3}\sqrt{13.1}} = 5.18 \text{ GHz}$$

(b) From Eq. (11-1-16) the phase velocity is

$$v_p = \frac{3 \times 10^8}{\sqrt{13.1}}\left[1 - \left(\frac{5.18}{10}\right)^2\right]^{-1/2}$$

$$= 0.97 \times 10^8 \text{ m/s}$$

(c) From Eq. (11-1-24) the intrinsic impedance is

$$\eta = (1 + j)\left(\frac{2\pi \times 10^9 \times 4\pi \times 10^{-7}}{2 \times 4.1 \times 10^7}\right)^{1/2}$$

$$= 0.031 + j0.031$$

(d) From Eq. (11-1-22) the attenuation is

$$\alpha_d = \frac{1.17 \times 10^{-2}}{2} \times \frac{377}{\sqrt{13.1}}\left[1 - \left(\frac{5.18}{10}\right)^2\right]^{-1/2}$$

$$= 0.713 \text{ Np/m} = 6.20 \text{ dB/m}$$

11-2 DIELECTRIC-SLAB WAVEGUIDE

Thin dielectric-slab waveguides are commonly used in optical integrated circuits for signal transmission. Figure 11-2-1 shows the schematic diagram of a dielectric-slab waveguide. The thin dielectric slab, of thickness d, permittivity ϵ_d, and permeability μ_d, is situated in two free-space regions (μ_0, ϵ_0).

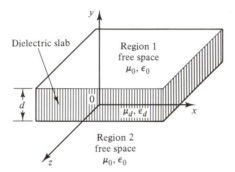

Figure 11-2-1 Schematic diagram of a dielectric-slab waveguide.

It is assumed that there is no field dependence on the x coordinate, the dielectric is lossless, and the waves propagate in the positive z direction. As described for wave propagation in a rectangular waveguide, wave propagation in a dielectric-slab waveguide can also be analyzed as TE and TM modes.

11-2-1 TE Waves along Dielectric-Slab Waveguide

For TE mode ($E_z = 0$) and no x dependence, the wave equation in a dielectric-slab waveguide is

$$\frac{\partial^2 H_z(y)}{\partial y^2} + k_y^2 H_z(y) = 0 \qquad (11\text{-}2\text{-}1)$$

where $k_y^2 = \omega^2 \mu_d \epsilon_d - \beta^2$ is the characteristic equation in the dielectric slab

$k_y = \dfrac{n\pi}{d}$ is wave number in the y coordinate

The solution of Eq. (11-2-1) can be written as

$$H_z(y) = H_0 \sin(k_y y) + H_e \cos(k_y y) \qquad \text{for } |y| \le \frac{d}{2} \qquad (11\text{-}2\text{-}2)$$

where $\sin(k_y y)$ = odd function (unsymmetric waves) of the coordinate y

$\cos(k_y y)$ = even function (symmetric waves) of the coordinate y

$k_y = \dfrac{n\pi}{d}$ is wave number in the y coordinate

In free-space region 1 ($y > d/2$) and region 2 ($y < -d/2$) the waves must decay exponentially as

$$H_z(y) = H_1 e^{-\alpha(y-d/2)} \qquad \text{for } y \geq \frac{d}{2} \tag{11-2-3}$$

$$H_z(y) = H_2 e^{-\alpha(y + d/2)} \qquad \text{for } y \leq -\frac{d}{2} \tag{11-2-4}$$

where $\alpha = \sqrt{\beta^2 - \omega^2 \mu_0 \epsilon_0}$ is the characteristic equation in free-space regions 1 and 2.

Odd TE modes.   The odd TE modes are described by the function

$$H_z(y) = H_0 \sin(k_y y) \tag{11-2-5}$$

In the dielectric slab, $|y| \leq d/2$, the field equations can be written after the boundary conditions are applied as

$$E_x(y) = \frac{-j\omega\mu_d}{k_y} H_0 \cos(k_y y) \tag{11-2-6}$$

$$H_y(y) = \frac{-j\beta}{k_y} H_0 \cos(k_y y) \tag{11-2-7}$$

$$H_z(y) = H_0 \sin(k_y y) \tag{11-2-8}$$

In region 1 ($y \geq d/2$) the field equations can be expressed after the boundary conditions are applied as

$$E_x(y) = \frac{-j\omega\mu_0}{\alpha} H_0 \sin\left(k_y \frac{d}{2}\right) e^{-\alpha(y-d/2)} \tag{11-2-9}$$

$$H_y(y) = \frac{-j\beta}{\alpha} H_0 \sin\left(k_y \frac{d}{2}\right) e^{-\alpha(y-d/2)} \tag{11-2-10}$$

$$H_z(y) = H_0 \sin\left(k_y \frac{d}{2}\right) e^{-\alpha(y-d/2)} \tag{11-2-11}$$

In region 2 ($y \leq -d/2$) the field equations can be described after the boundary conditions are applied as

$$E_x(y) = \frac{-j\omega\mu_0}{\alpha} H_0 \sin\left(k_y \frac{d}{2}\right) e^{\alpha(y+d/2)} \tag{11-2-12}$$

$$H_y(y) = \frac{-j\beta}{\alpha} H_0 \sin\left(k_y \frac{d}{2}\right) e^{\alpha(y + d/2)} \tag{11-2-13}$$

$$H_z(y) = -H_0 |\sin|\left(k_y \frac{d}{2}\right) e^{\alpha(y+d/2)} \tag{11-2-14}$$

The characteristic equation between α and k_y can be obtained by equation Eqs. (11-2-6) and (11-2-9) at $y = d/2$ and is given by

$$\frac{\alpha}{k_y} = \frac{\mu_0}{\mu_d}\tan\left(k_y\frac{d}{2}\right) \qquad \text{for odd TE modes} \qquad (11\text{-}2\text{-}15)$$

Surface Impedance. The surface impedance of the dielectric-slab waveguide looking down from the top surface of the dielectric slab is expressed as

$$Z_s = \frac{E_x}{H_z}\bigg|_{-y} = -j\frac{\omega\mu_0}{\alpha} \qquad \text{for odd TE modes} \qquad (11\text{-}2\text{-}16)$$

The surface impedance can also be defined from the bottom surface of the dielectric slab looking up as

$$Z_s = \frac{-E_x}{H_z}\bigg|_{+y} = -j\frac{\omega\mu_0}{\alpha} \qquad \text{for odd TE modes} \qquad (11\text{-}2\text{-}17)$$

These indicate that an odd TE surface wave can be supported by a capacitive surface.

Cutoff Frequency. By adding α and k_y together as shown in Eqs. (11-2-1) and (11-2-4), we have

$$\alpha^2 + k_y^2 = \omega^2(\mu_d\epsilon_d - \mu_0\epsilon_0) \qquad (11\text{-}2\text{-}18)$$

Then the attenuation constant can be expressed as

$$\alpha = \left[\omega^2(\mu_d\epsilon_d - \mu_0\epsilon_0) - k_y^2\right]^{1/2} \qquad (11\text{-}2\text{-}19)$$

When the phase constant β approaches the value of $\omega\sqrt{\mu_0\epsilon_0}$ as shown in Eq. (11-2-4), the attenuation α becomes zero. This means that the waves are no longer bound to the dielectric-slab region. In other words, the waves are cut off. Therefore, the cutoff condition is

$$\alpha = 0 = \frac{\mu_0}{\mu_d}k_c \tan\left(k_y\frac{d}{2}\right)$$

$$= \tan\left(\frac{\omega_c d}{2}\sqrt{\mu_d\epsilon_d - \mu_0\epsilon_0}\right) \qquad (11\text{-}2\text{-}20)$$

where $k_c = k_y = \omega_c\sqrt{\mu_d\epsilon_d - \mu_0\epsilon_0}$ has been replaced

$$\pi f_c d\sqrt{\mu_d\epsilon_d - \mu_0\epsilon_0} = (n-1)\pi$$
$$n = 1, 2, 3, \ldots$$

Therefore, the cutoff frequency for the odd TE modes is

$$f_{c0} = \frac{n-1}{d\sqrt{\mu_d\epsilon_d - \mu_0\epsilon_0}} \qquad \text{odd TE modes} \qquad (11\text{-}2\text{-}21)$$

Let $n = 1$, $f_c = 0$. This means that the lowest-order odd TE mode can propagate along a dielectric-slab waveguide for any thickness d. As the frequency increases the TE waves will be confined in the dielectric slab as the attenuation constant α increases.

Example 11-2-1:   Characteristics of Odd TE Waves in Dielectric-Slab Waveguide

An odd TE wave propagates through a dielectric-slab waveguide and the guide has the following parameters:

Relative dielectric constant of	ϵ_{rd}	$= 10$
alumina slab		
Slab thickness	d	$= 15$ mm
Frequency	f	$= 10$ GHz
Integer	n	$= 2$

Compute (a) the phase constant β, (b) the attenuation constant α, (c) the surface impedance Z_s, and (d) the cutoff frequency f_{co}.

Solution   (a) The phase constant is

$$\beta = \sqrt{\omega^2 \mu_d \epsilon_d - \left(\frac{n\pi}{d}\right)^2}$$

$$= \left[(2\pi \times 10^{10})^2 \times 4\pi \times 10^{-7} \times 8.854 \times 10^{-12} \times 10 - \left(\frac{2\pi}{15 \times 10^{-3}}\right)^2\right]^{1/2}$$

$$= [43.93 \times 10^4 - 17.55 \times 10^4]^{1/2}$$

$$= 5.14 \times 10^2 \text{ rad/m}$$

(b) The attenuation constant is

$$\alpha = \left[(2\pi \times 10^{10})^2 \times 4\pi \times 10^{-7} \times 8.854 \times 10^{-12} (10 - 1) - \left(\frac{2\pi}{15 \times 10^{-3}}\right)^2\right]^{1/2}$$

$$= [39.54 \times 10^4 - 17.55 \times 10^4]^{1/2}$$

$$= 469 \text{ Np/m}$$

(c) The surface impedance is

$$Z_s = -j\frac{2\pi \times 10^{10} \times 4\pi \times 10^{-7}}{469} = -j168.35\Omega$$

(d) The cutoff frequency is

$$f_{co} = \frac{n - 1}{d\sqrt{\mu_o \epsilon_0 (\epsilon_{nd} - 1)}} = \frac{(2 - 1) \times 3 \times 10^8}{15 \times 10^{-3}\sqrt{10 - 1}} = 6.67 \text{ GHz}$$

Even TE modes.   The even TE modes are described by the function

$$H_z(y) = H_e \cos(k_y y) \tag{11-2-22}$$

In the dielectric slab, $|y| \leq d/2$, the field equations can be written after the boundary conditions are applied as

$$E_x(y) = \frac{j\omega\mu_d}{k_y} H_e \sin(k_y y) \tag{11-2-23}$$

$$H_y(y) = \frac{j\beta}{k_y} H_e \sin(k_y y) \tag{11-2-24}$$

$$H_z(y) = H_e \cos(k_y y) \tag{11-2-25}$$

In region 1 ($y \geq d/2$) the field equations can be expressed after the boundary conditions are applied as

$$E_x(y) = \frac{-j\omega\mu_0}{\alpha} H_e \cos\left(k_y \frac{d}{2}\right) e^{-\alpha(y-d/2)} \tag{11-2-26}$$

$$H_y(y) = \frac{-j\beta}{\alpha} H_e \cos\left(k_y \frac{d}{2}\right) e^{-\alpha(y-d/2)} \tag{11-2-27}$$

$$H_z(y) = H_e \cos\left(k_y \frac{d}{2}\right) e^{-\alpha(y-d/2)} \tag{11-2-28}$$

In region 2 ($y \leq d/2$) the field equations can be described after the boundary conditions are applied as

$$E_x(y) = \frac{j\omega\mu_0}{\alpha} H_e \cos\left(k_y \frac{d}{2}\right) e^{\alpha(y+d/2)} \tag{11-2-29}$$

$$H_y(y) = \frac{j\beta}{\alpha} H_e \cos\left(k_y \frac{d}{2}\right) e^{\alpha(y+d/2)} \tag{11-2-30}$$

$$H_z(y) = H_e \cos\left(k_y \frac{d}{2}\right) e^{\alpha(y+d/2)} \tag{11-2-31}$$

The characteristic equation between α and k_y for the even TE modes at $y = d/2$ is

$$\frac{\alpha}{k_y} = -\frac{\mu_0}{\mu_d} \cot\left(k_y \frac{d}{2}\right) \qquad \text{for even TE modes} \tag{11-2-32}$$

Surface Impedance.   The surface impedance of the dielectric-slab waveguide either looking down from the top or looking up from the bottom surface is

$$Z_s = \frac{E_x}{H_z}\bigg|_{-y} = \frac{\frac{-j\omega\mu_0}{\alpha}}{1} = -j\frac{\omega\mu_0}{\alpha} \qquad \text{for even TE modes} \tag{11-2-33}$$

or

$$Z_s = \left.\frac{-E_x}{H_z}\right|_{+y} = \frac{-\dfrac{+j\omega\mu_0}{\alpha}}{1} = -j\frac{\omega\mu_0}{\alpha} \qquad \text{for even TE modes} \qquad (11\text{-}2\text{-}34)$$

These indicate that an even TE surface wave can be supported by a capacitive surface.

Cutoff Frequency.   The cutoff frequency can be found by using the same method as before and it is

$$\alpha = 0 = -\frac{\mu_0}{\mu_d}k_c \cot\left(k_y\frac{d}{2}\right) \qquad\qquad\qquad (11\text{-}2\text{-}35)$$

$$= -\frac{\mu_0}{\mu_d}k_c \cot\left(\frac{\omega_c d}{2}\sqrt{\mu_d\epsilon_d - \mu_0\epsilon_0}\right)$$

where

$$\pi f_c d\sqrt{\mu_d\epsilon_d - \mu_0\epsilon_0} = \left(n - \frac{1}{2}\right)\pi$$

$$n = 1, 2, 3, \ldots$$

Therefore, the cutoff frequency for the even TE modes is

$$f_{ce} = \frac{n - 0.5}{d\sqrt{\mu_d\epsilon_d - \mu_0\epsilon_0}} \qquad \text{even TE modes} \qquad (11\text{-}2\text{-}36)$$

11-2-2 TM Waves along Dielectric-Slab Waveguide

The analysis of the TM waves along a dielectric-slab waveguide is similar to that for TE waves, described earlier. For TM modes ($H_z = 0$) and no x dependence, the wave equation in a dielectric-slab waveguide is

$$\frac{d^2E_z(y)}{dy^2} + k_y^2E_z(y) = 0 \qquad\qquad (11\text{-}2\text{-}37)$$

where $k_y^2 = \gamma^2 + \omega^2\mu_d\epsilon_d$ is the characteristic equation in the dielectric slab. The solution of Eq. (11-2-37) can be written as

$$E_z(y) = E_o \sin(k_y y) + E_e \cos(k_y y) \qquad\qquad (11\text{-}2\text{-}38)$$

In free-space region 1 ($y > d/2$) and region 2 ($y < d/2$) the waves must decay exponentially as

$$E_z(y) = E_1 e^{-\alpha(y - d/2)} \qquad \text{for } y \geq d/2 \qquad (11\text{-}2\text{-}39)$$

$$E_z(y) = E_2 e^{-\alpha(y + d/2)} \qquad \text{for } y \leq d/2 \qquad (11\text{-}2\text{-}40)$$

where $\alpha = \sqrt{\beta^2 - \omega^2\mu_0\epsilon_0}$ is the characteristic equation in free-space regions 1 and 2.

Odd TM modes.   The odd TM modes are described by the function

$$E_z(y) = E_o \sin (k_y y) \tag{11-2-41}$$

In the dielectric slab for $|y| \leq d/2$ the field equations can be written after the boundary conditions are applied as

$$E_y(y) = \frac{-j\beta}{k_y} E_o \cos (k_y y) \tag{11-2-42}$$

$$E_z(y) = E_o \sin (k_y y) \tag{11-2-43}$$

$$H_x(y) = \frac{j\omega \epsilon_d}{k_y} E_o \cos (k_y y) \tag{11-2-44}$$

In region 1 ($y \geq d/2$) the field equations can be written after the boundary conditions are applied as

$$E_y(y) = \frac{-j\beta}{\alpha} E_o \sin \left(k_y \frac{d}{2}\right) e^{-\alpha(y-d/2)} \tag{11-2-45}$$

$$E_z(y) = E_o \sin \left(k_y \frac{d}{2}\right) e^{-\alpha(y-d/2)} \tag{11-2-46}$$

$$H_x(y) = \frac{j\omega \epsilon_0}{\alpha} E_o \sin \left(k_y \frac{d}{2}\right) e^{-\alpha(y-d/2)} \tag{11-2-47}$$

In region 2 ($y \leq -d/2$) the field equations can be expressed after the boundary conditions are applied as

$$E_y(y) = \frac{-j\beta}{\alpha} E_o \sin \left(k_y \frac{d}{2}\right) e^{\alpha(y+d/2)} \tag{11-2-48}$$

$$E_z(y) = -E_o \sin \left(k_y \frac{d}{2}\right) e^{\alpha(y+d/2)} \tag{11-2-49}$$

$$H_x(y) = \frac{j\omega \epsilon_0}{\alpha} E_o \sin \left(k_y \frac{d}{2}\right) e^{\alpha(y+d/2)} \tag{11-2-50}$$

The characteristic equation between α and k_y can be obtained from Eqs. (11-2-44) and (11-2-47) as

$$\frac{\alpha}{k_y} = \frac{\epsilon_0}{\epsilon_d} \tan \left(k_y \frac{d}{2}\right) \tag{11-2-51}$$

The cutoff frequency is given by

$$f_{co} = \frac{n - 1}{d\sqrt{\mu_d \epsilon_d} - \mu_0 \epsilon_0} \tag{11-2-52}$$

The surface impedance is expressed by

$$Z_s = \frac{E_z(y)}{H_x(y)}\bigg|_y = \frac{-1}{j\omega\epsilon_0/\alpha} = j\frac{\alpha}{\omega\epsilon_0} \tag{11-2-53}$$

This indicates that an odd TM surface wave can be supported by an inductive surface.

Even TM modes.   The even TM modes are described by the field equation

$$E_z(y) = E_e \cos(k_y y) \tag{11-2-54}$$

In the dielectric slab ($|y| \leq d/2$) the field equations can be expressed after the boundary conditions are applied as

$$E_y(y) = \frac{j\beta}{k_y} E_e \sin(k_y y) \tag{11-2-55}$$

$$E_z(y) = E_e \cos(k_y y) \tag{11-2-56}$$

$$H_x(y) = \frac{-j\omega\epsilon_d}{k_y} E_e \sin(k_y y) \tag{11-2-57}$$

In region 1 ($y \geq d/2$) the field equations can be written after the boundary conditions are applied as

$$E_y(y) = \frac{-j\beta}{\alpha} E_e \cos\left(k_y\frac{d}{2}\right) e^{-\alpha(y-d/2)} \tag{11-2-58}$$

$$E_z(y) = E_e \cos\left(k_y\frac{d}{2}\right) e^{-\alpha(y-d/2)} \tag{11-2-59}$$

$$H_x(y) = \frac{j\omega\epsilon_0}{\alpha} E_e \cos\left(k_y\frac{d}{2}\right) e^{-\alpha(y-d/2)} \tag{11-2-60}$$

In region 2 ($y \leq -d/2$) the field equations can be expressed after the boundary conditions are applied as

$$E_y(y) = \frac{j\beta}{\alpha} E_e \cos\left(k_y\frac{d}{2}\right) e^{\alpha(y+d/2)} \tag{11-2-61}$$

$$E_z(y) = E_e \cos\left(k_y\frac{d}{2}\right) e^{\alpha(y+d/2)} \tag{11-2-62}$$

$$H_x(y) = \frac{-j\omega\epsilon_0}{\alpha} E_e \cos\left(k_y\frac{d}{2}\right) e^{\alpha(y+d/2)} \tag{11-2-63}$$

The characteristic equation between α and k_y can be obtained from Eqs. (11-2-57) and (11-2-60) as

$$\frac{\alpha}{k_y} = -\frac{\epsilon_0}{\epsilon_d} \cot\left(k_y\frac{d}{2}\right) \tag{11-2-64}$$

Then the cutoff frequency is

$$f_{ce} = \frac{n - 0.5}{d\sqrt{\mu_d \epsilon_d - \mu_0 \epsilon_0}} \tag{11-2-65}$$

The surface impedance is expressed by

$$Z_s = \frac{-E_z(y)}{H_x(y)}\bigg|_{-y} = \frac{-1}{j\omega\epsilon_0/\alpha} = j\frac{\omega\epsilon_0}{\alpha} \tag{11-2-66}$$

This indicates that an odd TM surface wave can be supported by an inductive surface.

Example 11-2-2:   Characteristics of Even TM Waves in Dielectric-Slab Waveguides

An even TM wave propagates through a dielectric-slab waveguide and the guide has the following parameters:

Relative dielectric constant of the beryllia slab	$\epsilon_{rd} = 6$
Slab thickness	$d = 15$ mm
Frequency	$f = 9$ GHz
Integer	$n = 1$

Compute (a) the phase constant β, (b) the attenuation constant α, (c) the surface impedance Z_s, and (d) the cutoff frequency f_{ce}.

Solution   (a) The phase constant is

$$\beta = [\omega^2 \mu_d \epsilon_d - k_y^2]^{1/2}$$

$$= \left[(2\pi \times 9 \times 10^9)^2 \times 4\pi \times 10^{-7} \times 8.854 \times 10^{-12} \times 6 - \left(\frac{1\pi}{15 \times 10^{-3}}\right)^2\right]^{1/2}$$

$$= [21.36 \times 10^4 - 4.387 \times 10^4]^{1/2}$$

$$= 412 \text{ rad/m}$$

(b) The attenuation constant is

$$\alpha = [\beta^2 - \omega^2 \mu_0 \epsilon_0]^{1/2}$$

$$= [(412)^2 - (2\pi \times 9 \times 10^9)^2 \times 4\pi \times 10^{-7} \times 8.854 \times 10^{-12}]^{1/2}$$

$$= 367 \text{ Np/m}$$

(c) The surface impedance is

$$Z_s = j\frac{\omega\epsilon_0}{\alpha} = j\frac{2\pi \times 9 \times 10^9 \times 8.854 \times 10^{-12}}{367}$$

$$= j1.36 \times 10^{-3} \ \Omega$$

(d) From Eq. (11-2-65) the cutoff frequency is

$$f_{ce} = \frac{n - 0.5}{d\sqrt{\mu_d \epsilon_d - \mu_0 \epsilon_0}} = \frac{n - 0.5}{d\sqrt{\mu_0 \epsilon_0 (\epsilon_{nd} - 1)}} = \frac{(1 - 0.5) \times 3 \times 10^8}{15 \times 10^{-3}\sqrt{6 - 1}}$$

$$= 4.46 \text{ GHz}$$

11-3 COPLANAR WAVEGUIDE

A coplanar waveguide (CPW) consists of a center strip of thin metallic film deposited on the surface of a dielectric substrate with two ground electrodes parallel to and separate from the center strip on the same surface, as shown in Fig. 11-3-1.

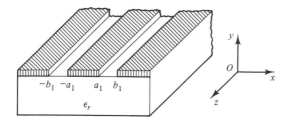

Figure 11-3-1 Coplanar waveguide.

The coplanar waveguide permits easy connections of external elements such as active devices as well as capacitors. It is also ideal for connecting various elements in monolithic microwave integrated circuits (MMMICs) built on one semi-insulating substrate. The future solid-state airborne phased-array radar system requires a large number of identical circuits and the coplanar waveguide is a suitable component for the system.

11-3-1 Characteristic Impedance

Because of the high dielectric constant of the substrate, most of the RF energy is stored in the dielectric and the loading effect of the grounded cover is negligible if it is more than two slot widths away from the center strip. The characteristic impedance Z_0 of a coplanar waveguide fabricated on a dielectric half-plane with relative dielectric constant ϵ_r can be calculated as a function of the ratio a_1/b_1, where $2a_1$ is the width of the center strip and $2b_1$ is the distance between two ground electrodes. A zero-order quasi-static approximation is employed. The dielectric half-plane Z_1 may be transformed to the interior of a rectangle in the Z-plane as shown in Fig. 11-3-2. The transformation equation can be expressed [1] as

$$\frac{dZ}{dZ_1} = \frac{A}{\sqrt{(Z_1^2 - a_1^2)(Z_1^2 - b_1^2)}} \tag{11-3-1}$$

where A = a constant
$2a_1$ = center strip width
$2b_1$ = distance between two ground electrodes

The ratio a/b of the rectangle in the Z plane is shown in Fig. 11-3-2(b) may be evaluated by multiplying both sides of Eq. (11-3-1) by dZ_1 and carrying out the integration [1]

$$a + jb = \int_0^{b_1} \frac{A\,dz_1}{\sqrt{(Z_1^2 - a_1^2)(Z_1^2 - b_1^2)}} \tag{11-3-2}$$

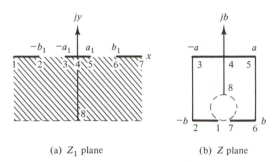

Figure 11-3-2 Transformation of mapping for a CPW.

(a) Z_1 plane

(b) Z plane

This equation is one form of an elliptic integral. The ratio a/b can be conveniently expressed in terms of tabulated complete elliptic integrals [1]

$$\frac{a}{b} = \frac{K(k)}{K'(k)} \tag{11-3-3}$$

where $k = a_1/b_1$ is the modulus of the first kind integral
$K(k) = $ complete elliptic integral of the first kind of modulus k [2]
$k' = (1 - k^2)^{1/2}$ is the complementary modulus
$k'(k) = K(k')$ is the complete integral of the first kind of complementary modulus k'

By assuming a semi-infinite dielectric, in parallel with a half-space of air, the equivalent static capacitance per unit lentgh for a pure TEM mode propagating in the line [1] is

$$C = \epsilon_0(\epsilon_r + 1)\frac{2a}{b} = 2\epsilon_0(\epsilon_r + 1)\frac{K(k)}{K'(k)} \qquad F/m \tag{11-3-4}$$

where ϵ_n is the relative dielectric constant of the substrate. The phase velocity of the propagating wave is approximately related to the effective dielectric constant [1] as

$$v_p = \left(\frac{2}{\epsilon_r + 1}\right)^{1/2} c \tag{11-3-5}$$

where $c = 3 \times 10^8$ m/s is the velocity of light in vacuum. Then the characteristic impedance of a coplanar waveguide [1] is

$$Z_0 = \frac{1}{Cv_p} \qquad ohms \tag{11-3-6}$$

Figure 11-3-3 shows the curves or the characteristic impedance Z_0 of the coplanar waveguide as a function of the ratio a_1/b_1 with the relative dielectric constant ϵ_r of the substrate as a parameter.

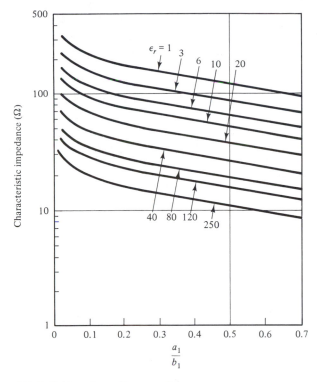

Characteristic impedance (Ω)

$\frac{a_1}{b_1}$

Figure 11-3-3 Characteristic impedance Z_0 of CPW as a function of the ratio a_1/b_1 with the relative dielectric constant ϵ_r as a parameter. (After Cheng P. Wen [1]. Copyright © 1969 IEEE. Reprinted by permission of the IEEE, Inc.)

11-3-2 Design Example

The coplanar waveguide can be designed by following the procedures listed below.

1. Specify the desired characteristic impedance.
2. Choose the proper substrate.
3. Estimate the value of a_1/b_1 from Fig. 11-3-3.
4. Compute the ratio of a/b from the complete elliptic integrals $K(k)$ and $K'(k)$ in Jahnke's tables of functions ([2, p. 78] or see Appendix D).
5. Calculate the static capacitance.
6. Find the phase velocity.
7. Verify the characteristic impedance.

Design Example 11-3-2: Coplanar Waveguide with a 50-Ohm Characteristic Impedance

A coplanar waveguide is to be designed for a 50-Ω characteristic impedance. The parameters chosen are as follows:

Relative dielectric constant $\epsilon_n = 6$
of substrate beryllia
Ratio of center strip width over $a_1/b_1 = 0.71$
the width from the center line
of the center strip to the first
edge of the ground strip

Compute (a) the capacitance of the CPW, (b) the phase velocity of the propagating wave, and (c) the characteristic impedance of the CPW.

Solution: (a) Choose $k = a_1/b_1 = 0.71$. From Jahnke's complete elliptic integral table [2, p. 78] we have

$$K(k) = K(0.71) = 1.8541$$

Then

$$k' = (1 - k^2)^{1/2} = 0.71$$

$$K'(k) = K(k') = K(0.71) = 1.8541$$

$$\frac{a}{b} = \frac{K(k)}{K'(k)} = \frac{1.8541}{1.8541} = 1.0$$

From Eq. (11-3-4) the capacitance is

$$C = 2 \times 8.854 \times 10^{-12}(6 + 1) \times 1.0$$
$$= 123.96 \text{ pF/m}$$

(b) The phase velocity is

$$v_p = \left(\frac{2}{6 + 1}\right)^{1/2} \times 3 \times 10^8 = 1.6 \times 10^8 \text{ m/s}$$

(c) The characteristic impedance is

$$Z_0 = \frac{1}{123.96 \times 10^{-12} \times 1.6 \times 10^8} = 50.61 \text{ }\Omega$$

11-3-3 Radiation Losses

When the distance between the ground electrodes approaches one wavelength of the signal wave, radiation is a problem. Because the coplanar waveguide is, in essence, an "edge-coupled" structure with a high concentration of charge and current near the strip edges, the losses tend to be higher than for microstrip line. Another loss is ohmic loss due to the surface resistance of the conducting strip.

11-3-4 Applications

In microwave integrated circuits (MICs), wire bonds have always been a serious factor in reliability and reproducibility. However, in monolithic microwave integrated circuits (MMICs) all active and passive circuit elements or components and interconnections are formed into the surface of a semi-insulating substrate. The coplanar waveguide structure is suitable to the MMIC design, so it is used widely in the batch processing of hundreds of circuits per wafer of substrate.

11-4 THIN FILM-ON-CONDUCTOR WAVEGUIDE

A thin film-on-conductor waveguide is made of a thin film of lossless dielectric (μ_d, ϵ_d) of thickness d on a conducting ground plane as shown in Fig. 11-4-1. The thin film is protected by a lossless dielectric coating (μ_1, ϵ_1). Both the widths and lengths of the coating and ground plane are assumed to extend to infinity. This type of dielectric thin-film waveguide is commonly used in microwave integrated circuits (MICs). It is assumed that the wave is propagating in the positive z direction and there is no field dependence on the x coordinate. The wave propagation in a dielectric waveguide can be described as TE and TM modes.

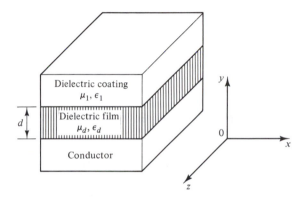

Figure 11-4-1 Thin film-on-conductor waveguide.

11-4-1 TM Modes in Dielectric Thin Film

For TM modes $(H_z = 0)$ and no dependence on x, the wave equation and its solution are

$$\frac{d^2 E_z(y)}{dy^2} + k_y^2 E_z(y) = 0 \tag{11-4-1}$$

and

$$E_z(y) = E_o \sin(k_y y) + E_e \cos(k_y y) \tag{11-4-2}$$

The boundary condition at $y = 0$ requires the vanishing of the tangential E field and it is

$$E_z(y)\bigg|_{y=0} = 0\bigg|_{y=0} + E_e\bigg|_{y=0} = 0 \rightarrow E_e = 0$$

Then the wave equation is an odd function only and it is

$$E_z(y) = E_0 \sin(k_y y) \tag{11-4-3}$$

In the thin-film region ($0 \leq y \leq d$) the field equations are

$$E_e(y) = -\frac{j\beta}{k_y}E_o \cos{(k_y y)} \tag{11-4-4}$$

$$E_z(y) = E_o \sin{(k_y y)} \tag{11-4-5}$$

$$H_x(y) = \frac{j\omega\epsilon_d}{k_y}E_o \cos{(k_y y)} \tag{11-4-6}$$

In the dielectric coating region 1 ($y \geq d$) the field equations are

$$E_y(y) = -\frac{j\beta}{\alpha}E_o \sin{(k_y y)}e^{-\alpha(y-d)} \tag{11-4-7}$$

$$E_z(y) = E_o \sin{(k_y y)}e^{-\alpha(y-d)} \tag{11-4-8}$$

$$H_x(y) = \frac{j\omega\epsilon_1}{\alpha}E_o \sin{(k_y y)}e^{-\alpha(y-d)} \tag{11-4-9}$$

where $\alpha = \sqrt{\beta^2 - \omega^2\mu_1\epsilon_1}$. The characteristic equation between α and k_y at $y = d$ is

$$\frac{j\omega\epsilon_d}{k_y}E_o \cos{(k_y d)} = \frac{j\omega\epsilon_1}{\alpha}E_o\sin{(k_y d)} \tag{11-4-10}$$

$$\frac{\alpha}{k_y} = \frac{\epsilon_1}{\epsilon_d} \tan{(k_y d)} \tag{11-4-11}$$

The cutoff frequency is

$$f_{co} = \frac{n-1}{d\sqrt{\mu_d\epsilon_d - \mu_1\epsilon_1}} \tag{11-4-12}$$

where $n = 1, 2, 3, \ldots$.

11-4-2 TE Modes in Dielectric Thin Film

For TE modes ($E_z = 0$) and no dependence on x, the wave equation and its solution are

$$\frac{d^2H_z(y)}{dy^2} + k_y^2 H_z(y) = 0 \tag{11-4-13}$$

and

$$H_z(y) = H_o \sin{(k_y y)} + H_e \cos{(k_y y)} \tag{11-4-14}$$

The boundary condition at $y = 0$ requires the continuity of the tangential H field to be

$$H_z(y)\bigg|_{y=0} = 0 + H_e$$

Then the wave equation is an even function only and it is

$$H_z(y) = H_e \cos(k_y y) \tag{11-4-15}$$

In the thin-film region $(0 \leq y \leq d)$ the field equations are

$$E_x(y) = \frac{j\omega\mu_d}{k_y} H_e \sin(k_y y) \tag{11-4-16}$$

$$H_y(y) = \frac{j\beta}{k_y} H_e \sin(k_y y) \tag{11-4-17}$$

$$H_z(y) = H_e \cos(k_y y) \tag{11-4-18}$$

In the dielectric coating region 1 $(y \geq d)$ the field equations are

$$E_x(y) = \frac{-j\omega\mu_1}{\alpha} H_e \cos(k_y y) e^{-\alpha(y-d)} \tag{11-4-19}$$

$$H_y(y) = \frac{-j\beta}{\alpha} H_e \cos(k_y y) e^{-\alpha(y-d)} \tag{11-4-20}$$

$$H(y)_z = H_e \cos(k_y y) \tag{11-4-21}$$

where $\alpha = \sqrt{\omega^2(\mu_d\epsilon_d - \mu_1\epsilon_1) - k_y^2}$. The characteristic equation between α and k_y are $y = d$ is

$$\frac{j\omega\mu_d}{k_y} H_e \sin(k_y d) = \frac{-j\omega\mu_1}{\alpha} H_e \cos(k_y d) \tag{11-4-22}$$

$$\frac{\alpha}{k_y} = -\frac{\mu_1}{\mu_d} \cot(k_y d) \tag{11-4-23}$$

For the cutoff frequency

$$\alpha = 0 = \cot(\omega_c d \sqrt{\mu_d\epsilon_d - \mu_1\epsilon_1}) \tag{11-4-24}$$

Then the cutoff frequency is

$$f_{ce} = \frac{n + 0.5}{2d\sqrt{\mu_d\epsilon_d - \mu_1\epsilon_1}} \tag{11-4-25}$$

Example 11-4-1:   Cutoff Frequency of Thin Film-on-Conductor Waveguide

A thin film-on-conductor waveguide has the following parameters:

Thin-film relative dielectric constant $\epsilon_{dr} = 1.50$
Film-coating relative dielectric constant $\epsilon_{ir} = 1.498$
Permeability $\mu_d = \mu_1 = \mu_0$
Thickness of thin film $d = 1\ \mu m$
Integer $n = 1$

Calculate the cutoff frequency.

Solution: From Eq. (11-4-25) we obtain the cutoff frequency as

$$f_{ce} = \frac{1 + 0.5}{2 \times 10^{-6} \times \sqrt{4\pi \times 10^{-7} \times 8.854 \times 10^{-12}\ (1.5 - 1.498)}}$$
$$= 5.12 \times 10^{15}\ Hz$$

11-5 THIN FILM-ON-DIELECTRIC WAVEGUIDE

A thin-film waveguide can easily be fabricated on a dielectric substrate in microwave integrated circuits (MICs). Figure 11-5-1 shows a dielectric thin film of thickness d with permeability μ_d and permittivity ϵ_d situated on a dielectric substrate of μ_1 and ϵ_1. The dimensions of both the x and z directions are assumed to extend to infinity. The wave propagations of either TE modes or TM modes are similar to those in a dielectric-slab waveguide, discussed in Section 11-2.

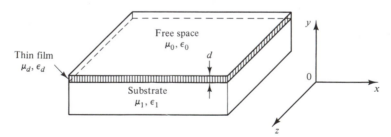

Figure 11-5-1   Thin film-on-dielectric waveguide.

11-5-1 TE Waves along Thin Film

The wave equation for TE modes along the thin film is

$$\frac{d^2 H_z(y)}{dy^2} + k_y^2 H_z(y) = 0 \qquad (11\text{-}5\text{-}1)$$

where $k_y^2 = \omega^2 \mu_d \epsilon_d - \beta^2$. In the free-space region ($y \geq d$) and the substrate region ($y \leq 0$) the wave must decay exponentially as

$$H_z(y) = H_0 e^{-\alpha(y-d)} \qquad \text{for } y \geq d \qquad\qquad (11\text{-}5\text{-}2)$$

$$H_z(y) = H_1 e^{+\alpha(y-0)} \qquad \text{for } y \leq 0 \qquad\qquad (11\text{-}5\text{-}3)$$

where $\alpha^2 = \beta^2 - \omega^2 \mu_0 \epsilon_0$ for the free-space region
$\quad\quad\;\; \alpha_d^2 = \beta^2 - \omega^2 \mu_1 \epsilon_1$ for the substrate region

11-5-2 TM Waves along Thin Film

The wave equation for TM modes along the thin film is

$$\frac{d^2 E_z(y)}{dy^2} + k_y^2 E_z(y) = 0 \qquad\qquad (11\text{-}5\text{-}4)$$

where $k_y^2 = \omega^2 \mu_d \epsilon_d - \beta^2$. In the free-space region ($y \geq d$) and the substrate region ($y \leq 0$), the waves must decay exponentially as

$$E_z(y) = E_0 e^{-\alpha(y-d)} \qquad \text{for } y \geq d \qquad\qquad (11\text{-}5\text{-}5)$$

$$E_z(y) = E_1 e^{+\alpha_d(y-0)} \qquad \text{for } y \leq 0 \qquad\qquad (11\text{-}5\text{-}6)$$

where $\alpha^2 = \beta^2 - \omega^2 \mu_0 \epsilon_0$ for the free space region
$\quad\quad\;\; \alpha_d^2 = \beta^2 - \omega^2 \mu_1 \epsilon_1$ for the substrate region

Example 11-5-1: Attenuation Constants of a Dielectric Thin Film-on-Dielectric Waveguide

A dielectric thin film-on-dielectric waveguide has the following parameters:

Operating frequency	j =	18 GHz
Relative dielectric constant of thin film	ϵ_{rd} =	10
Relative dielectric constant of substrate	ϵ_{r1} =	2.1
Film Thickness	d =	3 mm
Integer	n =	1

Compute (a) the attenuation constants for TE and TM modes in the free-space region, and (b) the attenuation constants for TE amd TM modes in the substrate region.

Solution: (a) The attenuation constant in free space for the TE and TM modes is

$$\alpha = \sqrt{\beta^2 - \omega^2 \mu_0 \epsilon_0} = \sqrt{\omega^2 \mu_0 \epsilon_0 (\epsilon_{nd} - 1) - k_y^2}$$

$$= \left[(2\pi \times 18 \times 10^9)^2 \times 4\pi \times 10^{-7} \times 8.854 \right.$$

$$\left. \times 10^{-12} (10 - 1) - \left(\frac{\pi}{3 \times 10^{-3}} \right)^2 \right]^{1/2}$$

$$= [(1.28 \times 10^6 - 1.10 \times 10^6)]^{1/2}$$

$$= 4.24 \times 10^2 \text{ Np/m}$$

(b) The attenuation constant in the substrate region for the TE and TM modes is

$$
\begin{aligned}
\alpha_d &= \Bigg[(2\pi \times 1.8 \times 10^{10})^2 \times 4\pi \times 10^{-7} \times 8.854 \\
&\quad \times 10^{-12}(10 - 2.1) - \left(\frac{\pi}{3 \times 10^{-3}} \right) \Bigg]^{1/2} \\
&= [(1.124 \times 10^6 - 1.10 \times 10^6)]^{1/2} \\
&= 4.90 \times 10^2 \text{ Np/m}
\end{aligned}
$$

11-6 GOLD-FILM-COATING DESIGN

In certain engineering applications, it is often desirable to use a metallic-film-coated glass to attenuate optimum electromagnetic radiation at microwave frequencies and also to transmit as much light intensity as possible at visible-light frequencies. The coated metallic film is normally required to have the properties of a high melting point, high electrical conductivity, high adhesion to glass, high resistance to oxidation, and insensitivity to light and water; and the capability to dissipate some power for deicing, defogging, or maintaining a certain level of temperature. Metallic film coatings on a plastic substrate are being used in such applications as windshields on airplanes or automobiles, in hospital medical equipment, and on dome windows of space vehicles or military devices.

11-6-1 Surface Resistance of Gold Film

Very thin metallic films have a much higher resistivity than that of bulk metal because of the scattering of electrons from the surface of the film. If the film thickness is very large compared to the electron mean free path, the resistivity is expected to be nearly the same as that of a bulk metal. When the film thickness is of the order of the electron mean free path, the role of surface scattering becomes dominant. Fuchs [3] and Sondheimer [4] considered the general form of the solution of the Boltzmann equation for the case of a conducting film and found the film conductivity σ_f in terms of the bulk conductivity σ, the film thickness t, and the electron mean free path p as given by

$$
\sigma_f = \frac{3t\sigma}{4p} \left[\ell n\!\left(\frac{p}{t} \right) + 0.4228 \right] \qquad \text{for } t \ll p \tag{11-6-1}
$$

The surface resistance of conducting films is normally quoted in units of ohms per square. This is because in the equation for resistance

$$
R = \frac{\text{specific resistivity} \times \text{length}}{\text{thickness} \times \text{width}} = \frac{\rho \ell}{tw} \tag{11-6-2}
$$

When the units of length ℓ and width w are chosen to have equal magnitude (i.e., resulting in a square), the resistance R in ohms per square is independent of the dimensions of the square and equals

$$R_s = \frac{\rho_f}{t} = \frac{1}{t\sigma_f} \quad \text{ohms/square} \tag{11-6-3}$$

From the Fuchs–Sondheimer theory, the surface resistance of a metallic film is decreased as the thickness of the film is increased.

At room temperature, the properties of bulk gold are:

Conductivity	$\sigma = 4.10 \times 10^7$ mhos/m
Resistivity	$\rho = 2.44 \times 10^{-8}$ $\Omega - $ m
Electron mean free path	$p = 570$ Å

It is assumed that the thickness t of the gold film is various from 10 to 100 Å. The surface resistances of the gold film are computed by using Eqs. (11-6-1) and (11-6-3) and are tabulated in Table 11-6-1.

TABLE 11-6-1 SURFACE RESISTANCE OF GOLD FILM

Thickness t (Å)	Conductivity σ_f (mhos/m^{-1} $\times$ 10^7)	Resistivity ρ_f (Ω $-$m $\times$ 10^{-7})	Surface resistance R_s (Ω/square)
100	1.17	0.86	8.60
90	1.11	0.90	10.00
80	1.03	0.97	12.13
70	0.96	1.04	14.86
60	0.85	1.17	19.50
50	0.77	1.30	26.00
40	0.68	1.48	37.00
30	0.54	1.86	62.00
20	0.42	2.41	120.00
10	0.22	4.48	448.00

Figure 11-6-1 shows the surface resistances of the gold film in ohms per square against the thicknesses of the gold film from 10 to 100 Å. From the Fuchs–Sondheimer theory, gold films have a typical surface resistance at about 10 to 30 Ω per square for a thickness of about 90 to 45 Å. The surface resistance is decreased as the thickness of the gold film is increased.

11-6-2 Optical Properties of Plastic Substrates

The optical properties of materials are usually characterized by two constants, the refractive index n and the extinction index k. The *refractive index* is defined as the ratio of the phase velocities of light in a medium and in a vacuum. The *extinction index* is related to the exponential decay of the wave as it passes through a medium.

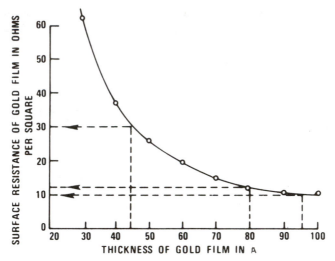

Figure 11-6-1 Surface resistance of gold film versus thickness of gold film.

Most optical plastics are suitable as substrate materials for a dome window and for application of a metallic film. Table 11-6-2 lists the values of the refractive index n of several commonly used nonabsorbing plastic substrate materials [5].

TABLE 11-6-2 SUBSTRATE MATERIALS

Substrate material	Refractive index n
Corning Vycor	1.458
Crystal quartz	1.540
Fussed silica	1.458
Plexiglass	1.490
Polycyclohexyl methacrylate	1.504
Polyester glass	1.500
Polymethyl methacrylate	1.491
Zinc crown glass	1.508

The complex refractive index of an optical material is usually given by

$$N = n - jk \tag{11-6-4}$$

where n = refractive index
$\quad\ k$ = extinction index

The measured values of the refractive index n and of the extinction index k of thin metallic-film coatings deposited in a vacuum [5] are tabulated in Table 11-6-3.

11-6-3 Optical Attenuation of Gold-Film Coating on Plastic Substrate

A conductor of high conductivity and low permeability has low intrinsic impedance. When a radio wave propagates from a medium of high intrinsic impedance into a

TABLE 11-6-3 REFRACTIVE INDEX *n* AND EXTINCTION INDEX *k* OF THIN METALLIC FILMS

Wavelength (Å)	Cooper film		Gold film		Silver film	
	n	*k*	*n*	*k*	*n*	*k*
2000			1.427	1.215	1.13	1.23
2200					1.32	1.29
2300					1.38	1.31
2400					1.37	1.33
2500					1.39	1.34
2600					1.45	1.35
2700					1.51	1.33
2800					1.57	1.27
2900					1.60	1.17
3000					1.67	0.96
3100					1.54	0.54
3200					1.07	0.32
3300					0.30	0.55
3400					0.16	1.14
3500					0.12	1.35
3600					0.09	1.52
3700					0.06	1.70
3800						
4000						
4500	0.870	2.200	1.400	1.880		
4920						
5000	0.880	2.420	0.840	1.840		
5460						
5500	0.756	2.462	0.331	2.324		
6000	0.186	2.980	0.200	2.897		
6500	0.142	3.570	0.142	3.374		
7000	0.150	4.049	0.131	3.842		
7500	0.157	4.463	0.140	4.266		
8000	0.170	4.840	0.149	4.654		
8500	0.182	5.222	0.157	4.993		
9000	0.190	5.569	0.166	5.335		
9500	0.197	5.900	0.174	5.691		
10,000	0.197	6.272	0.179	5.044		

Source: Adapted from *The American Institute of Physics Handbook* by the American Institute of Physics. Copyright
© 1972 by McGraw-Hill, Inc. Used with permission of McGraw-Hill Book Company.

medium of low intrinsic impedance, the reflection coefficient is high. From electromagnetic planewave theory in the far field, high attenuation occurs in a medium made of material of high conductivity and low permeability. Good conductors, such as gold, silver, and copper, have high conductivity and are often used as the material for attenuating electromagnetic energy. Microwave radiation attenuation by a metallic-film coating on substrate consists of three parts [7] as given by

$$\text{attenuation} = A + R + C \quad \text{dB} \qquad (11\text{-}6\text{-}5)$$

where A = absorption of penetration loss in decibles inside the metallic-film coating while the substrate is assumed to be nonabsorbing plastic glass

R = reflection loss in decibels from the multiple boundaries of a metallic-film coating on substrate

C = correction term in decibels required to account for multiple internal reflections when the absorption loss A is much less than 10 dB for electrically thin film

Figure 11-6-2 shows the absorption and reflection of a metallic-film coating on a plastic substrate.

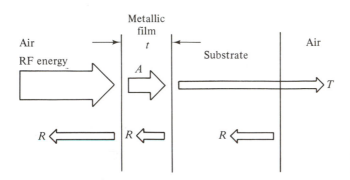

Figure 11-6-2   Absorption and reflection of film coating on plastic substrate.

Absorption loss A.   From transmission-line theory, the propagation constant γ for a uniform plane wave in a good conducting material is given by

$$\gamma = \alpha + j\beta = (1 + j)\sqrt{\pi f \mu \sigma_f} \qquad \text{for } \sigma_f \gg \omega\epsilon \qquad (11\text{-}6\text{-}6)$$

If the plastic substrate is assumed to be nonabsorbing material, the absorption loss A of the metallic-film coating on a substrate is related only to the thickness t of the coated film and the attenuation α as shown:

$$A = 20 \log_{10} e^{\alpha t} = 20(\alpha t) \log_{10} e = (20)(0.4343)(\alpha t)$$
$$= 8.686 t \sqrt{\pi f \mu \sigma_f} \quad \cdot \quad \text{dB} \qquad (11\text{-}6\text{-}7)$$

where t = thickness of the film coating, meters
μ = permeability of the film, henrys per meter
f = frequency, hertz
σ_f = conductivity of the coated film, mhos per meter

Since the thickness of the coated film is very thin, for example, 100 Å (Å = 10^{-10} m) at most, the absorption loss A is very small and can be ignored.

Reflection loss R.   The reflection loss R due to the multiple boundaries of the substrate glass coated with a metallic film can be analyzed by means of energy-transmission theory and it is expressed as

$$R = -20 \log \frac{2|\eta_f|}{|\eta_a + \eta_f|} - 20 \log \frac{2|\eta_g|}{|\eta_f + \eta_g|} - 20 \log \frac{2|\eta_a|}{|\eta_g + \eta_a|}$$

$$= 20 \log \frac{|\eta_a + \eta_f| \, |\eta_f + \eta_g| \, |\eta_g + \eta_a|}{8|\eta_f| \, |\eta_g| \, |\eta_a|} \quad \text{dB}$$

(11-6-8)

where η_f = intrinsic impedance of the coated metallic film
  η_g = intrinsic impedance of the glass substrate
  η_a = intrinsic impedance of air or free space = 377 Ω
The intrinsic impedance of a metallic film is given by

$$|\eta_f| = \left| (1 + j)\sqrt{\frac{\mu\omega}{2\sigma_f}} \right| = \sqrt{\frac{\mu\omega}{\sigma_f}}$$

(11-6-9)

The intrinsic impedance of a glass substrate is expressed by

$$\eta_g = \frac{\eta_a}{\sqrt{\epsilon_r}} = \frac{377}{\sqrt{3.78}} = 194 \ \Omega \qquad \text{for } \sigma_g \ll \omega\epsilon_g$$

(11-6-10)

where σ_g = about 10^{-12} mho/m is the conductivity of the glass substrate
  ϵ_g = 4.77×10^{-11} F/m is the permittivity of the glass substrate
  ϵ_r = 3.78 is the relative permittivity of the glass substrate
Substitution of the values of the intrinsic impedances η_f, η_g, and η_a in Eq. (11-6-8) yields the reflection loss as

$$R \simeq 20 \log \left(28.33 \sqrt{\frac{\sigma_f}{\eta_f}} \right) = 88 + 10 \log \left(\frac{\sigma_f}{f} \right) \quad \text{dB}$$

(11-6-11)

(c) **Correction term C.** For very electrically thin film, the value of the absorption loss A is much less than 10 dB and the correction term is given by [7]

$$C = 20 \log |1 - K 10^{-A/10}(\cos \theta - j \sin \theta)|$$

(11-6-12)

where $K = \left(\dfrac{\eta_f - \eta_a}{\eta_f + \eta_a} \right)^2 \simeq 1$ for $\eta_a \gg \eta_f$
  $\theta = 3.59t\sqrt{f\mu\sigma_f}$
Over the frequency range from 100 MHz to 40 GHz, the angle θ is much smaller than 1°, so that $\cos \theta \simeq 1$ and $\sin \theta \simeq \theta$. Thus the correction term of Eq. (11-6-12) can be simplified to

$$C = 20 \log (3.59t\sqrt{f\mu\sigma_f}) \simeq -48 + 20 \log (t\sqrt{f\sigma_f}) \quad \text{dB}$$

(11-6-12)

Finally, the total microwave radiation attenuation by a metallic-film coating on a glass substrate defined in Eq. (11-6-5) in the far field becomes

$$\text{attenuation} = 40 - 20 \log (R_s) \quad \text{dB}$$

(11-6-14)

It is interesting to note that the microwave radiation attenuation due to the coated metallic film on a glass substrate in the far field is independent of frequency and is related only to the surface resistance of the coated metallic film.

11-6-4 Light Transmission of Gold-Film Coating on Plastic Substrate

The complex refractive index of an optical material is given by [5] as

$$N = n - jk \qquad (11\text{-}6\text{-}15)$$

It is assumed that light in air is normally incident on a thin absorbing film N_1 of the thickness t_1 and that it is transmitted through an absorbing substrate of complex refractive index N_2 and then emerges into air. The incidence and the emergence media are dielectric of refractive index n_0. The reflection loss between the substrate and the air is small and, for convenience, is taken as zero. Figure 11-6-3 shows light transmittance, reflection, and absorption through a thin absorbing metallic film and a plastic substrate.

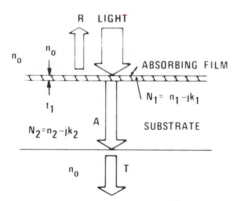

Figure 11-6-3 Light transmittance, reflection, and absorption through thin metallic film coated on plastic substrate.

From the multireflection and transmission theory, the reflection loss is expressed by

$$R = \frac{a_1 e^\alpha + a_2 e^{-\alpha} + a_3 \cos v + a_4 \sin v}{b_1 e^\alpha + b_2 e^{-\alpha} + b_3 \cos v + b_4 \sin v} \qquad (11\text{-}6\text{-}16)$$

where $a_1 = [(n_0 - n_1)^2 + k_1^2][(n_1 + n_2)^2 + (k_1 + k_2)^2]$

$a_2 = [(n_0 + n_1)^2 + k_1^2][(n_1 - n_2)^2 + (k_1 - k_2)^2]$

$a_3 = 2\{[n_0^2 - (n_1^2 + k_1^2)][(n_1^2 + k_1^2) - (n_2^2 + k_2^2)] + 4n_0 k_1(n_1 k_2 - n_2 k_1)\}$

$a_4 = 4\{[n_0^2 - (n_1^2 + k_1^2)](n_1 k i_2 - n_2 k_1) - n_0 k_1[(n_1^2 + k_1^2) - [n_2^2 + k_2^2)]\}$

$\alpha = \dfrac{4\pi k_1 t_1}{\lambda_0}$

$\lambda_0 = c / f$ is the wavelength in a vacuum

$c = 3 \times 10^8$ m/s is the velocity of light in vacuum; f is the frequency in hertz

$$\nu = \frac{4\pi n_1 t_1}{\lambda_0}$$

$$b_1 = [(n_0 + n_1)^2 + k_1^2][(n_1 + n_2)^2 + (k_1 + k_2)^2]$$

$$b_2 = [(n_0 - n_1)^2 + k_1^2][(n_1 - n_2)^2 + (k_1 - k_2)^2]$$

$$b_3 = 2\{[n_0^2 - (n_1^2 + k_1^2)][(n_1^2 + k_1^2) - (n_2^2 + k_2^2)] - 4n_0 k_1(n_1 k_2 - n_2 k_1)\}$$

$$b_4 = 4\{[n_0^2 - (n_1^2 + k_1^2)](n_1 k_2 - n_2 k_1) + n_0 k_1[(n_1^2 + k_1^2) - (n_2^2 + k_2^2)]\}$$

The transmittance T is given by Ref. 5 as

$$T = \frac{16 n_0 n_2(n_1^2 + k_1^2)}{b_1 e^\alpha + b_2 e^{-\alpha} + b_3 \cos \nu + b_4 \sin \nu} \tag{11-6-17}$$

The absorption loss A is given by

$$A = 1 - R - T \tag{11-6-18}$$

The total attenuation loss L is given by

$$L = A + R \tag{11-6-19}$$

However, when the concave surface of a plastic dome is uniformly coated with an electromagnetic interference shield of metallic film, the light is normally incident on the plastic substrate N_2, transmits through the thin metallic film N_1, and then emerges into the air n_0. From the electromagnetic theory of luminous transmission in transparent media, the light transmittance is the same whether the light is normally incident on the substrate medium N_2 or on the absorbing film N_1. Hence the total attenuation loss is the same in both cases.

Example 11-6-1:   Calculation of Gold-Film Coating

A gold film of 80 Å is coated on a plastic substrate with a refractive index of 1.50. Calculate (a) the gold-film surface resistance in ohms per square, (b) the microwave attenuation in decibels, (c) the light transmittance T at $\lambda = 0.6$ μm, and (d) the light reflection loss R at $\lambda = 0.6$ μm.

Solution: (a) Gold-film surface resistance

1. From Eq. (11-6-1), the gold-film conductivity is

$$\sigma_f = \frac{3 \times 80 \times 10^{-10} \times 4.1 \times 10^7}{4 \times 570 \times 10^{-10}}\left[\ell n\left(\frac{570}{80}\right) + 0.4228\right]$$

$$= 1.03 \times 10^7 \text{ mhos/m}$$

2. The gold-film resistivity is

$$\rho_f = \frac{1}{\sigma_f} = \frac{1}{1.03 \times 10^7} = 0.97 \times 10^{-7} \ \Omega$$

3. From Eq. (11-6-3), the gold-film surface resistance is

$$R_s = \frac{\rho_f}{t} = \frac{0.97 \times 10^{-7}}{80 \times 10^{-10}} = 12.12 \ \Omega/\text{square}$$

(b) From Eq. (11-6-14), the microwave attenuation is

$$\text{micro-attenuation} = 40 - 20 \log (12.12) = 18 \text{ dB}$$

(c) The light transmittance T is computed with a program from Eq. (11-6-17) to
be

$$T = 80\%$$

(d) The light reflection loss R is computed with a program from Eq. (11-6-16) as

$$R = 20\%$$

11-6-5 Design Example

Gold–film coatings on plastic glasses have many engineering applications [8]. For
example, a gold film is coated on the concave surface of a plastic-glass dome so that
an optimum amount of microwave radiation may be attenuated by the gold film while,
at the same time, sufficient light intensity may be transmitted through the gold film.

Design Procedure.   The following design procedures are suggested:

Step 1. Choose a specific film coating on a specific substrate for a particular application.
Here a gold-film coating on plastic is to be designed for the dome window of
a missile.

Step 2. Calculate the surface resistance of the designed gold-film coating. The surface
resistance of a gold film is tabulated in Table 11-6-1 and the relationships
between the film resistivity and film thickness are plotted in Fig. 11-6-1.

Step 3. Estimate the radiation attenuation of the gold-film coating on plastic. Substitution of the values of the surface resistances for gold films in Eq. (11-6-
14) yields the microwave radiation attenuation in decibels by the gold-film
coating on a plastic glass. Figure 11-6-4 shows graphically the microwave
radiation attenuation versus the surface resistance of the gold-film coating.
For a coated gold film which has a surface resistance of 12 Ω per square, the
microwave radiation attenuation is about 19 dB. The data agree with Hawthorne's conclusion [9].

Step 4. Compute the light transmittance of the gold-film coating on plastic. For the
visible-light region, the values of the refractive index n and of the extinction
index k of a gold-film coating on a plastic glass deposited in vacuum are
taken from Table 11-6-3. The refractive index n_0 of air or vacuum is unity.
The refractive index n_2 of the nonabsorbing plastic glass is taken as 1.50. The
light transmittance T and the light reflection loss R of the gold-film coating
on a plastic glass are computed using Eqs. (11-6-17) and (11-6-16), respectively.
From the values of T and R, the absorption loss A and the total attenuation
L are calculated. The results are graphically presented in Fig. 11-6-5. It can
be seen that for a light transmittance of 80%, the thickness of the gold-film

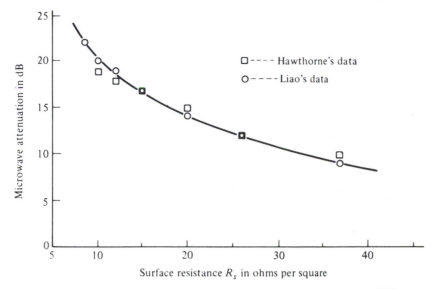

Figure 11-6-4   Microwave radiation attenuation versus surface resistance of gold film.

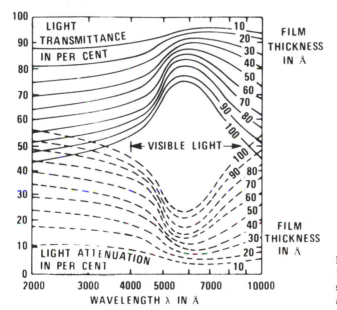

Figure 11-6-5   Light transmittance T and light attenuation loss L versus wavelength λ with film thickness t as parameter for gold film.

coating is about 80 Å. However, when the absorption loss in the substrate material is considered, the light transmittance may be a little less than 80%.

Step 5. Evaluate the optimum condition. The surface resistance of a metallic film is decreased as the thickness of the film coating is increased. However, the luminous transmittance is decreased as the surface resistance of the metallic film is decreased. This relationship for the visible-light region is shown in Fig.

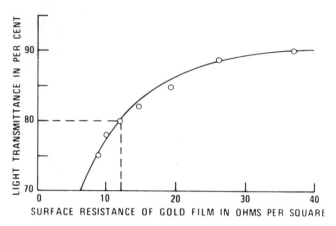

- LIGHT TRANSMITTANCE DECREASES, AS SURFACE RESISTANCE DECREASES
- FOR 80 PER CENT OF LIGHT TRANSMITTANCE, SURFACE RESISTANCE IS ABOUT 12 OHMS PER SQUARE

Figure 11-6-6   Light transmittance versus surface resistance of gold film.

11-6-6. Figure 11-6-7 illustrates the relationship of light transmittance versus wavelength for a given surface resistance of the gold film. If a power dissipation of 5 W per square is allowed for deicing and defogging or keeping warm by the gold-film coating on a plastic substrate and if the effective area of the

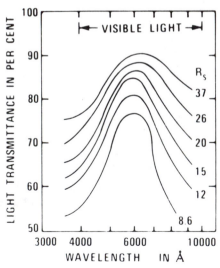

- LIGHT TRANSMITTANCE INCREASES AS SURFACE RESISTANCE INCREASES
- LIGHT TRANSMITTANCE OF 80 PER CENT OCCURS AT 12 OHMS PER SQUARE

Figure 11-6-7   Light transmittance versus wavelength with surface resistance R_s as parameter for gold film.

coated film is 13 in.^{2}, the surface resistance of the coated film must be 12 Ω per square. The power dissipation can be expresed as

$$P = \frac{V^2}{R_s} = \frac{(28)^2}{12 \times 13} = 5 \text{ W}$$

in which the voltage applied to the film-coating terminations is 28 V. The optimum condition occurs at 18 dB of microwave radiation attenuation and 90% of light transmittance.

Example 11-6-2:   Gold-Film Coating on Plastic

A gold-film coating on a plastic is to be designed for a missile dome window so that the microwave radiation can be attenuated and the light intensity will be transmitted to a certain level. The specifications are:

Minimum microwave attenuation	attenuation $=$ 18 dB
Light transmittance	$T =$ 80%
Allowable power dissipation	$P =$ 5 W/square
Missile dome window	$A =$ 14 in.^{2}
Refractive index of plastic substrate	$n =$ 1.5

Determine (a) the surface resistance in ohms per square, (b) the thickness of the gold-film coating, and (c) the microwave attenuation in decibels. (d) Calculate the power dissipation in watts per square.

Solution   (a) From the graph in Fig. 11-6-6, the surface resistance for 80% of light transmittance is

$$R_s = 12 \ \Omega/\text{square}$$

(b) From Table 11-6-1, the thickness of the gold-film coating is

$$t = 80 \text{ Å}$$

(c) From Fig. 11-6-4, the microwave attenuation is

$$\text{attenuation} = 18 \text{ dB}$$

(d) The power dissipation is

$$P = \frac{V^2}{R_s} = \frac{(28)^2}{12 \times 14} = 4.67 \text{ W/square}$$

where 28 V is the dc power supply in the missile.

REFERENCES

1. Wen, Cheng P., Coplanar waveguide: A surface strip transmission line suitable for non-reciprocal gyromagnetic devices applications. *IEEE Trans. Microwave Theory and Techniques*, Vol MTT-17, No. 12, December 1969, pp. 1087–1090.

2. Jahnke, E., and F. Emde, *Tables of Functions with Formulae and Curves*, 4th ed. New York: Dover Publications, Inc., 1945.

3. Fuchs, K., The conductivity of thin metallic films according to the electron theory of metals, *Proc. Camb. Phil. Soc.*, Vol. 30, 1938, p. 100.

4. Sondheimer, E. H., The Mean-Free Path of Electrons in Metals. *Advances in Physics*, Vol. 1, 1952, p. 1.

5. *American Institute of Physics Handbook*, New York: McGraw-Hill Book Company, 1972, pp. 6-12, 6-119–6-121, and 6-138.

6. Schulz, Richard B., et al., Shielding theory and practice. *Proc. 9th Tri-Service Conference Electromagnetic Compatibility*, October 1963.

7. Vasaka, C. S. Problems in Shielding Electrical and Electronic Equipments. U.S. Naval Air Development Center. Rept. NO. NACD-EL-N5507, Johnsville, Pa. June 1955.

8. Liao, Samuel Y., Design of a gold film on a glass substrate for maximum light transmittance and RF shielding effectiveness. *IEEE Electromagnetic Compatibility Symposium Records*, San Antonio, Texas, 1975; Light transmittance and microwave attenuation of a gold-film coating on a plastic substrate. *IEEE Trans. Microwave Theory and Techniques*, Vol. MTT-23, No. 10, 1975; Light transmittance and RF shielding effectiveness of a gold film on a glass substrate. *IEEE Trans. Electromagnetic Compatibility*, Vol. EMC-17, No. 4, 1975.

9. Hawthorne, E. I., Electromagnetic shielding with transparent coated glass. *Proc. IRE*, Vol. 42, March 1954, pp. 548-553.

PROBLEMS

11-1 Parallel-Plate Waveguide

11-1-1. An infinite gold parallel-plate waveguide has the following parameters:

Separation distance	$d = 10$ mm
Gold conductivity	$\sigma_{au} = 4.1 \times 10^7$ mhos/m
Relative dielectric constant of alumina	$\epsilon_{rd} = 10$
Operating frequency	$f = 10$ GHz
Integer	$n = 1$
Conductivity of GaAs substrate	$\sigma_d = 1.17 \times 10^{-2}$ mho/m

Compute:
(a) The cutoff frequency for the TE wave.
(b) The phase velocity for the TE wave.
(c) The attenuation constant for the TE wave.
(d) The intrinsic impedance for the plate.

11-1-2. An infinite gold parallel-plate waveguide has the following parameters:

Separation distance	$d = 12$ mm
Gold conductivity	$\sigma_{au} = 4.1 \times 10^7$ mhos/m
Relative dielectric constant of beryllia	$\epsilon_{rd} = 6$
Operating frequency	$f = 9$ GHz
Integer	$n = 1$
Conductivity of GaAs substrate	$\sigma_d = 1.17 \times 10^{-2}$ mho/m

Calculate:
(a) The cutoff frequency for the TM mode.
(b) The phase velocity of the TM mode.
(c) The intrinsic impedance of the plate.
(d) The attenuation constant of the TM mode.

11-2 Dielectric-Slab Waveguide

11-2-1. An even TE wave propagates through a dielectric-slab waveguide and the guide has the following parameters:

Relative dielectric constant of GaAs insulator	$\epsilon_{rd} = 13.1$
Slab thickness	$d = 12$ mm
Operating frequency	$f = 10$ GHz
Integer	$n = 1$

Determine:
(a) The phase constant β.
(b) The attenuation constant α.
(c) The surface impedance Z_s.
(d) The cutoff frequency f_{ce}.

11-2-2. An odd TM wave propagates through a dielectric-slab waveguide and the guide has the following parameters:

Relative dielectric constant of alumina	$\epsilon_{rd} = 10$
Slab thickness	$d = 10$ mm
Operating frequency	$f = 11$ GHz
Integer	$n = 2$

Compute:
(a) The phase constant β.
(b) The attenuation constant α.
(c) The surface impedance Z_s.
(d) The cutoff frequency f_{co}.

11-3 Coplanar Waveguide

11-3-1. A coplanar waveguide has the following parameters:

Relative dielectric constant of alumina	$\epsilon_{rd} = 10$
Dimension ratio	$a_1/b_1 = 0.33$

Calculate:
(a) The capacitance of the CPW.
(b) The phase velocity of the propagating wave.
(c) The characteristic impedance of the CPW.

11-3-2. A coplanar waveguide has the following parameters:

Relative dielectric constant of beryllia	$\epsilon_{rd} = 6$
Dimension ratio	$a_1.b_1 = 0.4$

Determine:

(a) The capacitance of the CPW.

(b) The phase velocity of the propagating wave.

(c) The characteristic impedance of the CPW.

11-4 Thin Film-on-Conductor Waveguide

11-4-1. A dielectric thin film-on-conductor waveguide has the following parameters:

Operating frequency	$f = 18$ GHz
Relative dielectric constant of thin film	$\epsilon_{rd} = 10$
Relative dielectric constant of film coating	$\epsilon_{r1} = 1.54$
Film thickness	$d = 6$ mm
Integer	$n = 2$

Calculate:

(a) The cutoff frequency for the TM wave.

(b) The attenuation constant for the TM wave.

11-4-2. A dielectric thin film-on-conductor waveguide has the following parameters:

Operating frequency	$f = 20$ GHz
Relative dielectric constant of thin film	$\epsilon_{rd} = 6$
Relative dielectric constant of film coating	$\epsilon_{r1} = 1.50$
Film thickness	$d = 6$ mm
Integer	$n = 1$

Compute:

(a) The cutoff frequency for the TE mode.

(b) The attenuation constant for the TE mode.

11-5 Thin Film-on-Dielectric Waveguide

11-5-1. A dielectric thin film-on-dielectric waveguide has the following parameters:

Operating frequency	$f = 18$ GHz
Relative dielectric constant of thin film	$\epsilon_{rd} = 13.1$
Relative dielectric constant of the substrate	$\epsilon_{r1} = 1.50$
Film thickness	$d = 3.75$ mm
Integer	$n = 1$

Determine:

(a) The attenuation constant for the TE and TM waves in the free-space region.

(b) The attenuation constant for the TE and TM waves in the substrate region.

11-6 Gold-Film-Coating Design

11-6-1. Bulk gold has a conductivity of 4.1×10^7 mhos/m, a resistivity of 2.44×10^{-8} Ω-m, and an electron mean free path of 570 Å. Calculate the surface conductivity, surface resistivity,and surface resistance of gold film for thicknesses of 10 to 100 Å with an increment of 10 Å for each step.

11-6-2. Silver has a conductivity of 6.170×10^7 mhos/m, a resistivity of 1.620×10^{-8} Ω-m, and an electron mean free path of 570 Å. Calculate the surface conductivity, surface resistivity, and surface resistance of silver film for thicknesses of 10 to 100 Å with an increment of 10 Å for each step.

11-6-3. Copper has a conductivity of 5.800×10^7 mhos/m, a resistivity of 1.724×10^{-8} Ω-m, and an electron mean free path of 420 Å. Calculate the surface conductivity, surface resistivity, and surface resistance of copper film for thicknesses of 10 to 100 Å with an increment of 10 Å for each step.

11-6-4. Write a FORTRAN program to compute the light transmittance and light reflection of a gold-film coating on a nonabsorbing plastic glass for thicknesses of 10 to 100 Å with an increment of 10 Å each step. The wavelengths vary from 2000 to 10,000 Å with an increment of 500 Å each step. The values of the refracive index n and the extinction index k of a gold film are listed in Table 2-1-2. (The values for 2200 to 4000 Å may be projected.) The refractive index n of the nonabsorbing plastic glass is 1.50. Use F10.5 format for numerical outputs, Hollerith format for character outputs, and data statements to read the input values.

11-6-5. Write a FORTRAN program to compute the light transmittance and light reflection of a silver-film coating on a nonabsorbing plastic glass for the wavelengths from 2000 to 3700 Å with an increment of 100 Å each step. The other requirements are the same as specified in Problem 11-6-4.

11-6-6. Write a FORTRAN program to compute the light transmittance and light reflection of a copper-film coating on a nonabsorbing plastic glass for the wavelengths 4500 to 10,000 Å with an increment of 500 Å each step. The other requirements are the same as specified in Problem 11-6-4.

Chapter 12

Microwave Measurements and Evaluations

12-0 INTRODUCTION

The measurements and evaluations of microwave components and systems [1] are important processes of microwave amplifier and oscillator circuit analysis and design. The microwave components should be tested and evaluated individually according to their specifications and standards before they are incorporated into a microwave electronic system. After the components are assembled, the microwave system should be tested and evaluated again.

12-1 MEASURING UNITS

Many electronic components and systems, such as radar or missles, require a field intensity in microvolts per centimeter (μV/cm). However, many of the field intensity meters at the receiver station read voltages in microvolts (μV) or in decibels above 1 μV(dBμV). Then it is necessary to convert the intensity meter reading in μV or dBμV to μV/cm. Although the receiver reads power in watts or milliwatts, it is still necessary to convert the receiving power to power density in milliwatts per square centimeter or to field intensity in volts per centimeter. A few widely used measuring units are defined below.

    1. *dB*. The decibel (dB) is a dimensionless number that expresses the ratio of two power levels. It is defined as

$$dB \equiv 10 \log_{10} \left(\frac{P_2}{P_1} \right) \qquad (12\text{-}1\text{-}1)$$

The two power levels are relative to each other. If power level P_2 is higher than power level P_1, the decibel value is positive, and vice versa. Since $P = V^2/R$, when their voltages are measured across the same or equal resistors, the number of decibels is given by

$$dB \equiv 20 \log_{10}\left(\frac{V_2}{V_1}\right) \qquad (12\text{-}1\text{-}2)$$

The voltage definition of decibels has no meaning at all unless the two voltages under consideration appear across equal impedances. Above 10 GHz the impedance of waveguides varies with frequency, and the decibel calibration is limited to power levels only. Table 12-1-1 shows the conversion of voltage and power ratios to decibels.

TABLE 12-1-1   CONVERSION OF VOLTAGE AND POWER RATIOS TO DECIBELS

Voltage ratio	Power ratio	−dB+	Voltage ratio	Power ratio	Voltage ratio	Power ratio	−dB+	Voltage ratio	Power ratio
1.000	1.000	0	1.000	1.000	0.692	0.479	3.2	1.445	2.089
0.989	0.977	0.1	1.012	1.023	0.684	0.568	3.3	1.462	2.138
0.977	0.955	0.2	1.023	1.047	0.676	0.457	3.4	1.479	2.188
0.966	0.933	0.3	1.035	1.072	0.668	0.447	3.5	1.496	2.239
0.955	0.912	0.4	1.047	1.096	0.661	0.437	3.6	1.514	2.291
0.944	0.891	0.5	1.059	1.122	0.653	0.427	3.7	1.531	2.344
0.933	0.871	0.6	1.072	1.148	0.646	0.417	3.8	1.549	2.399
0.923	0.851	0.7	1.084	1.175	0.638	0.407	3.9	1.567	2.455
0.912	0.832	0.8	1.095	1.202	0.631	0.398	4.0	1.585	2.512
0.902	0.813	0.9	1.109	1.230	0.624	0.389	4.1	1.603	2.570
0.891	0.794	1.0	1.122	1.259	0.617	0.380	4.2	1.622	2.630
0.881	0.776	1.1	1.135	1.288	0.610	0.372	4.3	1.641	2.692
0.871	0.759	1.2	1.148	1.318	0.603	0.363	4.4	1.660	2.754
0.861	0.741	1.3	1.161	1.349	0.596	0.355	4.5	1.679	2.818
0.851	0.724	1.4	1.175	1.380	0.589	0.347	4.6	1.698	2.884
0.841	0.708	1.5	1.189	1.413	0.582	0.339	4.7	1.718	2.951
0.832	0.692	1.6	1.202	1.445	0.575	0.331	4.8	1.738	3.020
0.822	0.676	1.7	1.216	1.479	0.569	0.324	4.9	1.758	3.090
0.813	0.661	1.8	1.230	1.514	0.562	0.316	5.0	1.778	3.162
0.804	0.646	1.9	1.245	1.549	0.556	0.309	5.1	1.799	3.236
0.794	0.631	2.0	1.259	1.585	0.550	0.302	5.2	1.820	3.311
0.785	0.617	2.1	1.274	1.622	0.543	0.295	5.3	1.841	3.388
0.776	0.603	2.2	1.288	1.660	0.537	0.288	5.4	1.862	3.467
0.767	0.589	2.3	1.303	1.698	0.531	0.282	5.5	1.884	3.548
0.759	0.575	2.4	1.318	1.738	0.525	0.275	5.6	1.905	3.631
0.750	0.562	2.5	1.334	1.778	0.519	0.269	5.7	1.928	3.715
0.741	0.550	2.6	1.349	1.820	0.513	0.263	5.8	1.950	3.802
0.733	0.537	2.7	1.365	1.862	0.507	0.257	5.9	1.972	3.890
0.724	0.525	2.8	1.380	1.905	0.501	0.251	6.0	1.995	3.981
0.716	0.513	2.9	1.390	1.950	0.496	0.246	6.1	2.018	4.074
0.708	0.501	3.0	1.413	1.995	0.490	0.240	6.2	2.042	4.159
0.700	0.490	3.1	1.429	2.042	0.484	0.234	6.3	2.065	4.265

TABLE 12-1-1 (continued)

Voltage ratio	Power ratio	−dB+	Voltage ratio	Power ratio	Voltage ratio	Power ratio	−dB+	Voltage ratio	Power ratio
0.479	0.229	6.4	2.089	4.365	0.275	0.0759	11.2	3.631	13.18
0.473	0.224	6.5	2.113	4.467	0.272	0.0741	11.3	3.673	13.49
0.468	0.219	6.6	2.138	4.571	0.269	0.0724	11.4	3.715	13.80
0.462	0.214	6.7	2.163	4.677	0.266	0.0708	11.5	3.758	14.13
0.457	0.209	6.8	2.188	4.786	0.263	0.0692	11.6	3.802	14.45
0.452	0.204	6.9	2.215	4.898	0.260	0.0676	11.7	3.846	14.79
0.447	0.200	7.0	2.239	5.012	0.257	0.0661	11.8	3.890	15.14
0.442	0.195	7.1	2.265	5.129	0.254	0.0646	11.9	3.936	15.49
0.437	0.191	7.2	2.291	5.248	0.251	0.0631	12.0	3.981	15.85
0.432	0.186	7.3	2.317	5.370	0.248	0.0617	12.1	4.027	16.22
0.427	0.182	7.4	2.344	5.495	0.246	0.0603	12.2	4.074	16.60
0.422	0.178	7.5	2.371	5.623	0.243	0.0589	12.3	4.121	16.98
0.417	0.174	7.6	2.399	5.754	0.240	0.0575	12.4	4.169	17.38
0.412	0.170	7.7	2.427	5.888	0.237	0.0562	12.5	4.217	17.78
0.407	0.166	7.8	2.455	6.026	0.234	0.0550	12.6	4.266	18.20
0.403	0.162	7.9	2.483	6.166	0.232	0.0537	12.7	4.315	18.62
0.398	0.159	8.0	2.512	6.310	0.229	0.0525	12.8	4.365	19.05
0.394	0.155	8.1	2.541	6.457	0.227	0.0513	12.9	4.416	19.50
0.389	0.151	8.2	2.570	6.607	0.224	0.0501	13.0	4.467	19.95
0.385	0.148	8.3	2.600	6.761	0.221	0.0490	13.1	4.519	20.42
0.380	0.145	8.4	2.630	6.918	0.219	0.0479	13.2	4.571	20.89
0.376	0.141	8.5	2.661	7.079	0.216	0.0468	13.3	4.624	21.38
0.372	0.138	8.6	2.692	7.244	0.214	0.0457	13.4	4.677	21.88
0.367	0.135	8.7	2.723	7.413	0.211	0.0447	13.5	4.732	22.39
0.363	0.132	8.8	2.754	7.586	0.209	0.0437	13.6	4.786	22.91
0.359	0.129	8.9	2.786	7.762	0.207	0.0427	13.7	4.842	23.44
0.355	0.126	9.0	2.818	7.943	0.204	0.0417	13.8	4.898	23.99
0.351	0.123	9.1	2.851	8.128	0.202	0.0407	13.9	4.955	24.55
0.347	0.120	9.2	2.884	8.318	0.200	0.0398	14.0	5.012	25.12
0.343	0.118	9.3	2.917	8.511	0.197	0.0389	14.1	5.070	25.70
0.339	0.115	9.4	2.951	8.710	0.195	0.0380	14.2	5.129	26.30
0.335	0.112	9.5	2.985	8.913	0.193	0.0372	14.3	5.188	26.92
0.331	0.110	9.6	3.020	9.120	0.191	0.0363	14.4	5.248	27.54
0.327	0.107	9.7	3.055	9.333	0.188	0.0355	14.5	5.309	28.18
0.324	0.105	9.8	3.090	9.550	0.186	0.0347	14.6	5.370	28.84
0.320	0.102	9.9	3.126	9.772	0.184	.0399	14.7	5.433	29.51
0.316	0.100	10.0	3.162	10.000	0.182	0.0331	14.8	5.495	30.20
0.313	0.0977	10.1	3.199	10.23	0.180	0.0324	14.9	5.559	30.90
0.309	0.0955	10.2	3.236	10.47	0.178	0.0316	15.0	5.623	31.62
0.306	0.0933	10.3	3.273	10.72	0.176	0.0309	15.1	5.689	32.36
0.302	0.0912	10.4	3.311	10.96	0.174	0.0302	15.2	5.754	33.11
0.299	0.0891	10.5	3.350	11.22	0.172	0.0295	15.3	5.821	33.88
0.295	0.0871	10.6	3.388	11.48	0.170	0.0288	15.4	5.888	34.67
0.292	0.0851	10.7	3.428	11.75	0.168	0.0282	15.5	5.957	35.48
0.288	0.0832	10.8	3.467	12.02	0.166	0.0275	15.6	6.026	36.31
0.283	0.0813	10.9	3.508	12.30	0.164	0.0269	15.7	6.095	37.15
0.282	0.0794	11.0	3.548	12.59	0.162	0.0263	15.8	6.166	38.02
0.279	0.0776	11.1	3.589	12.88	0.160	0.0257	15.9	6.237	38.90

TABLE 12-1-1 (continued)

Voltage ratio	Power ratio	−dB+	Voltage ratio	Power ratio	Voltage ratio	Power ratio	−dB+	Voltage ratio	Power ratio
0.159	0.0251	16.0	6.310	39.81	0.119	0.0141	18.5	8.414	70.79
0.157	0.0246	16.1	6.383	40.74	0.118	0.0138	18.6	8.511	72.44
0.155	0.0240	16.2	6.457	41.69	0.116	0.0135	18.7	8.610	74.13
0.153	0.0234	16.3	6.531	42.66	0.115	0.0132	18.8	8.710	75.86
0.151	0.0229	16.4	6.607	43.65	0.114	0.0129	18.9	8.811	77.62
0.150	0.0224	16.5	6.683	44.67	0.112	0.0126	19.0	8.913	79.43
0.148	0.0219	16.6	6.761	45.71	0.111	0.0123	19.1	9.016	81.28
0.146	0.0214	16.7	6.839	46.77	0.110	0.0120	19.2	9.120	83.18
0.145	0.0209	16.8	6.918	47.86	0.108	0.0118	19.3	9.226	85.11
0.143	0.0204	16.9	6.998	48.98	0.107	0.0115	19.4	9.333	87.10
0.141	0.0200	17.0	7.079	50.12	0.106	0.0112	19.5	9.441	89.13
0.140	0.0195	17.1	7.161	51.29	0.103	0.0110	19.6	9.550	91.20
0.138	0.0191	17.2	7.244	52.48	0.104	0.0107	19.7	9.661	93.33
0.137	0.0186	17.3	7.328	53.70	0.102	0.0105	19.8	9.772	95.50
0.135	0.0182	17.4	7.413	54.95	0.101	0.0102	19.9	9.886	97.72
0.133	0.0178	17.5	7.499	56.23	0.100	0.0100	20.0	10.000	100.00
0.132	0.0174	17.6	7.586	57.54		10^{-3}	30		10^3
0.130	0.0170	17.7	7.674	58.88	10^{-2}	10^{-4}	40	10^2	10^4
0.129	0.0166	17.8	7.762	60.26		10^{-5}	50		10^5
0.127	0.0162	17.9	7.852	61.66	10^{-3}	10^{-6}	60	10^3	10^6
0.126	0.0159	18.0	7.943	63.10		10^{-7}	70		10^7
0.125	0.0155	18.1	8.035	64.57	10^{-4}	10^{-8}	80	10^4	10^8
0.123	0.0151	18.2	8.128	66.07		10^{-9}	90		10^9
0.122	0.0148	18.3	8.222	67.61	10^{-5}	10^{-10}	100	10^5	10^{10}
0.120	0.0145	18.4	8.318	69.18		10^{-11}	110		10^{11}
					10^{-6}	10^{-12}	120	10^6	10^{12}

2. *dBW*. The decibel above 1 watt (dBW) is another useful measure for expressing power level P_2 with respect to a reference power level P_1 of 1 watt. Similarly, if the power level P_2 is lower than 1 watt, the dBW value is negative.

3. *dBm*. The decibel above 1 milliwatt (dBm) is also a useful measure of expressing power level P_2 with respect to a reference power level P_1 of 1 milliwatt (mW). Since the power level in the microwave region is quite low, the dBm unit is very useful in that frequency range. It is customary to designate *milli* by a lowercase letter m and *mega* by an uppercase letter M.

4. *dBV*. The decibel above 1 volt (dBV) is a dimensionless voltage ratio in decibels referred to a reference voltage of 1 volt.

5. *dBμV*. The decibel above 1 microvolt (dBμV) is another dimensionless voltage ratio in decibels referred to a reference voltage of 1 microvolt (μV). The field intensity meters used for the measurements in the microwave region often have a scale in dBμV, since the power levels to be measured are usually extremely low.

6. *μV/m*. Microvolts per meter (μV/m) are units of 10^{-6} volt per meter, expressing the electric field intensity.

7. *dBμV/m*. The decibel above 1 microvolt per meter (dBμV/m) is a dimensionless electric field intensity ratio in decibels relative to 1 μV/m. This unit is also often used for field intensity measurements in the microwave region.

8. *μV/m/MHz*. The microvolts per meter per megahertz (μV/m/MHz) are units of 10^{-6} volt per meter per broadband electric field intensity distribution. This is a two-dimensional distribution, in space and in frequency.

9. *dBμV/m/MHz*. The decibel above 1 microvolt per meter per megahertz (dBμV/m/MHz) is a dimensionless broadband electric field intensity distribution ratio with respect to 1 μV/m/MHz.

10. *μV/MHz*. Microvolts per megahertz per second of bandwidth (μV/MHz) are units of 10^{-6} volt-second of broadband voltage distribution in the frequency domain. The use of this unit is based on the assumption that the voltage is evenly distributed over the bandwidth of interest.

12-2 MICROWAVE AMPLIFIER TEST

After the microwave amplifier is designed and built it should be tested and evaluated according to its test procedures. In general, there are three tests required for a designed amplifier: power gain test, noise figure test, and stability test. These tests can be carried out in the ordinary microwave laboratory with an automatic network analyzer (ANA).

Power gain test.   The latest HP 8408S automatic network analyzer is a complete microwave network measurement system for a frequency range 0.5 to 18 GHz. In comparison to a magnitude measurement system, a tuned receiver provides a 60-dB measurement range. By using vector error correction and the appropriate calibration standards, the effective system directivity is better than 40 dB at the measurement test port. The power gain of a microwave amplifier can be measured by the ANA system and plotted on a Smith chart in terms of frequency for evaluations.

Noise-figure test.   The noise figure F of any linear two-port electronic network can be defined as

$$F = \frac{\text{available noise power at output}}{\text{available noise power at input}} = \frac{N_o}{kTBG} \qquad (12\text{-}2\text{-}1)$$

where N_o = available noise power at output
    kTB = available noise power of the standard noise source
        in a bandwidth B at temperature T
    T = 300°K
    k = 1.381×10^{-23} J/°K is the Boltzmann constant
    G = available power gain at the frequency band considered
The output noise N_o can be expressed as

$$N_o = N_n + kTBG \qquad (12\text{-}2\text{-}2)$$

where N_n is the noise generated by the network. Therefore, the single-frequency noise figure results in

$$F = 1 + \frac{N_n}{kTBG} \qquad (12\text{-}2\text{-}3)$$

In general, when the noise powers are expressed by their noise temperatures, Eq. (12-2-3) becomes

$$F = 1 + \frac{T_n}{T_o} \qquad (12\text{-}2\text{-}4)$$

where T_n = noise temperature of the network, degrees Kelvin
      T_o = ambient noise temperature at 300°K
By using a known input noise source the noise figure of a microwave amplifier can be measured easily by the HP 8408S ANA system.

Stability test.   The stability factor K of a microwave amplifier is defined as

$$K = \frac{1 + |\Delta|^2 - |S_{11}|^2 - |S_{22}|^2}{2\,|S_{12}S_{21}|} > 1 \qquad (12\text{-}2\text{-}5)$$

where $\Delta = S_{11}S_{22} - S_{12}S_{21}$. The HP 8408S ANA system can carry out the stability factor test and make sure that the amplifier will not oscillate at the designed frequencies.

12-3 MICROWAVE OSCILLATOR TEST

There are three tests for a designed microwave oscillator:

1. *Oscillation condition test.* The first oscillation condition is defined as

$$K = \frac{1 + |\Delta|^2 - |S_{11}|^2 - |S_{22}|^2}{2|S_{12}S_{21}|} < 1 \qquad (12\text{-}3\text{-}1)$$

   where $\Delta = S_{11}S_{22} - S_{12}S_{21}$.
2. *Oscillation frequency test.* An oscillator is usually designed for a specific frequency and an oscillation frequency test should be conducted for that frequency.
3. *Bandpass frequency test.* For a broadband oscillator a bandpass frequency test should be carried out for that purpose.

12-4 MICROWAVE ELECTRONIC SYSTEM MEASUREMENTS

Measurement of a microwave electronic system or subsystem is usually conducted in an anechoic chamber. An anechoic chamber is a physical enclosure lined by anechoic material that has a high absorption coefficient. The more electromagnetic energy is absorbed by the anechoic material, the less the energy is reflected and the more the environment looks like free space to the wave. The absorption property of the material is expressed in decibels of the reflectivity at a given frequency of operation. The reflectivity is defined as

$$\text{reflectivity} = 10 \log \left(\frac{W_{\text{ref}}}{W_{\text{inc}}}\right) \quad \text{dB} \qquad (12\text{-}4\text{-}1)$$

where W_{ref} = reflected electromagnetic energy
$\quad W_{\text{inc}}$ = incident electromagnetic energy

An absorption of 20 to 60 dB represents 1 to 0.0001% reflection from the material. An absorber must be able to dissipate the energy incident on it. The materials usually have power ratings in the range 0.1 to 3 W/in.^{2}. In MKS units, these same power ratings are equivalent from 15.5 to 465 mW/cm^2.

In the anechoic chamber, there is a quiet zone. The quiet zone is defined as a volume space within the chamber where electromagnetic waves reflected from the walls, floor, and ceiling are stated to be below a certain specified minimum level. A quiet-zone performance rating of 40 dB, for example, means that the reflected energy arriving at any point within a specific volume space will be 40 dB below the direct wave arriving at the same point. The microwave electronic system or subsystem to be tested must be placed in the quiet zone for test. Figure 12-4-1 shows a test setup in an anechoic chamber.

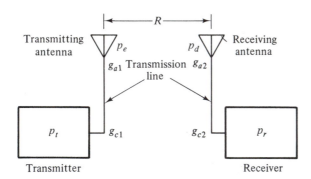

Figure 12-4-1 Test setup in an anechoic chamber.

The receiver power is equal to the power density at the antenna head times the aperture of the receiving antenna, the antenna gain, and the cable loss. The power density at the receiving antenna is equal to the transmitter power times the antenna gain and the cable loss and is then divided by the spherical surface area. Therefore, the receiving power can be written

$$P_r = \frac{p_e}{4\pi R^2}\frac{\lambda^2}{4\pi}g_{a2}g_{c2} = p_t g_{a1}g_{c1}g_{a2}g_{c2}\left(\frac{\lambda}{4\pi R}\right)^2 \quad \text{watts} \qquad (12\text{-}4\text{-}2)$$

where p_r = receiving power, watts
$\quad p_d$ = receiving power density, watts/m^2
$\quad p_t$ = transmitter power, watts
$\quad p_e$ = effective radiated power, watts
$\quad g_{a1}$ = transmitting antenna gain (numeric)
$\quad g_{c1}$ = transmitting line loss (numeric)
$\quad g_{a2}$ = receiving antenna gain (numeric)
$\quad g_{c2}$ = receiving line loss (numeric)
$\quad R$ = range between two antennas, meters
$\quad \lambda$ = wavelength, meters

The factor $\lambda^2/(4\pi)$ is the antenna aperture. It does not imply that a higher-frequency wave decreases in magnitude more rapidly than a lower-frequency wave. It is simply a result of the fact that, for a given gain, the aperture of a higher-frequency antenna is smaller than that of a lower-frequency antenna and so it intercepts a smaller amount of the power from the wave.

If the transmitter power is in dBW, the antenna gains and the cable losses in dB, the receiving power given by Eq. (12-4-2) can be expressed in dBW:

$$P_r = P_t + G_{a1} + G_{a2} - G_{c1} - G_{c2} - 20 \log_{10}\left(\frac{4\pi R}{\lambda}\right) \quad \text{dBW} \quad (12\text{-}4\text{-}3)$$

The last term is called the *free-space attenuation*. That is,

$$\text{free-space attenuation} = -20 \log_{10}\left(\frac{4\pi R}{\lambda}\right) \quad \text{dB} \quad (12\text{-}4\text{-}4)$$

The free-space attenuation can be found from the standard nomogram, which is shown in Fig. 12-4-2. For instance, if the wavelength is 0.30 m and the range is 3 m, the free-space attenuation is about 42 dB. Table 12-4-1 shows the free-space attenuation in dB for different distances in feet. However, if the distances are in meters, all values in dB should be increased by 10. It should be noted that the uppercase letters in Eq. (12-4-3) are designated for the decibel values.

12-4-1 Conversion of Transmitting Power to Electric Field Intensity

The electric field intensity at a distance R in meters from the transmitting antenna may be computed from the transmitter power. From Fig. 12-4-1 the power density at the point of R from the transmitting antenna is given by

$$p_d = \frac{p_t g_{c1} g_{a1}}{4\pi R^2} \quad \text{W/m}^2 \quad (12\text{-}4\text{-}5)$$

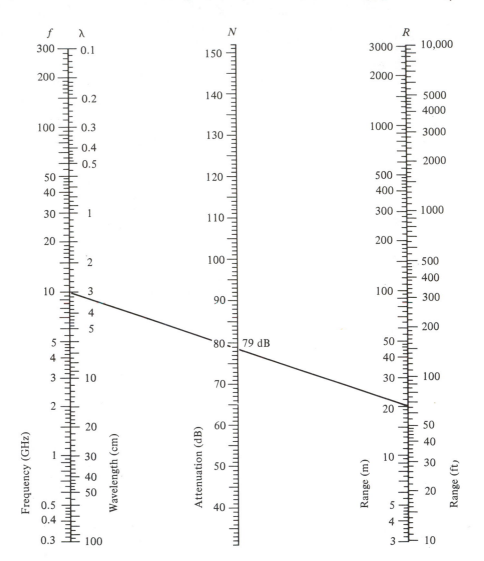

Figure 12-4-2 Nomogram of free-space attenuation.

From field theory, the average power density carried by an electromagnetic plane wave at the far field in free space is given by

$$p_d = \frac{E^2}{\eta_0} = \frac{E^2}{120\pi} \qquad \text{W/m}^2 \tag{12-4-6}$$

In Eq. (12-4-6) the electric field intensity is assumed to be in volts per meter. Therefore,

TABLE 12-4-1 FREE-SPACE ATTENUATION IN (dB)

Frequency (MH:)	Distance (ft)							
	10	50	100	500	1,000	5,000	10,000	50,000
100	22	36	42	56	62	76	82	96
200	28	42	48	62	68	82	88	102
300	32	46	52	66	72	86	92	106
400	34	48	54	68	74	88	94	108
500	36	50	56	70	76	90	96	110
600	38	52	58	72	78	92	98	112
700	39	53	59	73	79	93	99	113
800	40	54	60	74	80	94	100	114
900	41	55	61	75	81	95	101	115
1,000	42	56	62	76	82	96	102	116
1,500	46	60	66	80	86	100	106	120
2,000	48	62	68	82	88	102	108	122
2,500	50	64	70	84	90	104	110	124
3,000	52	66	72	86	92	106	112	126
3,500	53	67	73	87	93	107	113	127
4,000	54	68	74	88	94	108	114	128
4,500	55	69	75	89	95	109	115	129
5,000	56	70	76	90	96	110	116	130
5,500	57	71	77	91	97	111	117	131
6,000	58	72	78	92	98	112	118	132
7,000	59	73	79	93	99	113	119	133
8,000	60	74	80	94	100	114	120	134
9,000	61	75	81	95	101	115	121	135
10,000	62	76	82	96	102	116	122	136
15,000	66	80	86	100	106	120	126	140

Note: If the distance is in meters, all numbers in decibels should be increased by 10.

the field intensity E can be computed from Eqs. (12-4-6) and (12-4-5), and it is given by

$$E = \sqrt{120\pi p_d} = 19.4\sqrt{p_d} = \frac{5.48}{R}(p_t g_{c1} g_{a1})^{1/2} \qquad V/m \qquad (12\text{-}4\text{-}7)$$

It is often desirable to express the power density and the field intensity in terms of decibels. That is,

$$P_d = -11 \text{ dB} - 20 \log_{10}(R) + P_t(\text{dBW}) \qquad (12\text{-}4\text{-}8)$$

$$+ G_{a1}(\text{dB}) - G_{c1}(\text{dB}) \qquad \text{dBW/m}^2$$

$$E = 14.8 \text{ dB} - 20 \log_{10}(R) + P_t(\text{dBW}) \qquad (12\text{-}4\text{-}9)$$

$$+ G_{a1}(\text{dB}) - G_{c1}(\text{dB}) \qquad \text{dBV/m}$$

Subtraction of eq. (12-4-8) from (12-4-9) results in the conversion of power density in dBW/m^2 to field intensity in dBV/m, and vice versa. That is,

$$E = 25.8 \text{ dB} + P_d(\text{dBW}/\text{m}^2) \qquad \text{dBV/m} \qquad (12\text{-}4\text{-}10)$$

It should be noted that if values per square centimeter and per centimeter are desired, a term of -40 dB must be added to each of the preceding three equations. Table 12-4-2 tabulates the values of electric field intensity and power density conversions.

TABLE 12-4-2 CONVERSION OF FIELD INTENSITY AND POWER DENSITY (REFER TO FREE-SPACE RESISTANCE 377 Ω)

Field intensity			Power density		
V/m	dBV/m	dBµV/m	W/m²	dBW/m²	dBm/cm²
1×10^{-6}	-120	0	27×10^{-16}	-146	-156
2×10^{-6}	-114	6	11×10^{-15}	-140	-150
3×10^{-6}	-110	10	24×10^{-15}	-136	-146
4×10^{-6}	-108	12	42×10^{-15}	-134	-144
5×10^{-6}	-106	14	66×10^{-15}	-132	-142
7×10^{-6}	-103	17	13×10^{-14}	-129	-139
8×10^{-6}	-102	18	17×10^{-14}	-128	-138
10×10^{-6}	-100	20	27×10^{-14}	-126	-136
20×10^{-6}	-94	26	11×10^{-13}	-120	-130
30×10^{-6}	-90	30	24×10^{-13}	-116	-126
40×10^{-6}	-88	32	42×10^{-13}	-114	-124
50×10^{-6}	-86	34	66×10^{-13}	-112	-122
70×10^{-6}	-83	37	13×10^{-12}	-109	-119
80×10^{-6}	-82	38	17×10^{-12}	-108	-118
100×10^{-6}	-80	40	27×10^{-12}	-106	-116
200×10^{-6}	-74	46	11×10^{-11}	-100	-110
300×10^{-6}	-70	50	24×10^{-11}	-96	-106
400×10^{-6}	-68	52	42×10^{-11}	-94	-104
500×10^{-6}	-66	54	66×10^{-11}	-92	-102
700×10^{-6}	-63	57	13×10^{-10}	-89	-99
800×10^{-6}	-62	58	17×10^{-10}	-88	-98
1×10^{-3}	-60	60	27×10^{-10}	-86	-96
10×10^{-3}	-40	80	27×10^{-8}	-66	-76
100×10^{-3}	-20	100	27×10^{-6}	-46	-56
1	0	120	27×10^{-4}	-26	-36
10	20	140	27×10^{-2}	-6	-16
100	40	160	27	14	4
1×10^3	60	180	27×10^2	34	24
10×10^3	80	200	27×10^4	54	44

12-4-2 Conversion of Receiving Power to Electric Field Intensity

From Fig. 12-4-1 the power density at the receiving antenna is equal to the receiver power divided by the antenna gain, the cable loss, and the antenna aperture. That is,

$$p_d = \frac{4\pi p_r}{\lambda^2 g_{a2} g_{c2}} \quad \text{W/m}^2 \tag{12-4-11}$$

Substitution of Eq. (12-4-11) in (12-4-7) yields the field intensity at that point as

$$E = \frac{68.77}{\lambda} \left(\frac{p_r}{g_{a2} g_{c2}} \right)^{1/2} \quad \text{V/m} \tag{12-4-12}$$

Similarly, the power density and the field intensity can be expressed in decibels:

$$P_d = 11 \text{ dB} - 20 \log_{10}(\lambda) + P_r(\text{dBW}) - G_{a2}(\text{dB}) \tag{12-4-13}$$

$$-G_{c2}(\text{dB}) \quad \text{dBW/m} \tag{12-4-14}$$

$$E = 36.8 \text{ dB} - 20 \log_{10}(\lambda) + P_r(\text{dBW}) - G_{a2}(\text{dB})$$

$$-G_{c2}(\text{dB}) \quad \text{dBV/m}$$

where λ is the wavelength in meters.

12-4-3 Conversion of Receiving Voltage to Electric Field Intensity

Generally, the field intensity meter at the receiving position reads voltages in either μV or dBμV. A conversion of the receiving voltage to field intensity is necessary. Figure 12-4-3 shows a diagram for computing the field intensity from the receiving voltages. The input impedance of the field intensity receiving meter is normally specified by the manufacturer to be 50 Ω, since the coaxial line connected to the meter usually has a characteristic impedance of 50 Ω. This means that the input impedance of the meter perfectly matches the coaxial line. The power input to the intensity meter is given by

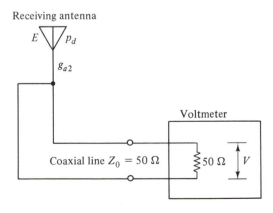

Figure 12-4-3 Field intensity in terms of receiving voltage.

$$p_r = \frac{V^2}{50} \quad \text{watts} \tag{12-4-15}$$

where V is the intensity meter reading in volts (rms).

If the input impedance is specified by another value, it is necessary to choose a coaxial line with the same value of characteristic impedance as the input impedance for impedance matching. The power density at the antenna aperture is computed from Eq. (12-4-11) as

$$p_d = \frac{0.251 V^2}{\lambda^2 g_{a2} g_{c2}} \quad \text{W/m}^2 \tag{12-4-16}$$

where λ is the wavelength in meters.

The field intensity can be computed from Eq. (12-4-7) as

$$E = \frac{9.7 V}{\lambda} \left(\frac{1}{g_{a2} g_{c2}} \right)^{1/2} \quad \text{V/m} \tag{12-4-17}$$

Similarly, the power density and the field intensity can be expressed in decibels as

$$P_d = -6 \text{ dB} - 20 \log_{10}(\lambda) + V(\text{dBV}) - G_{a2}(\text{dB}) \tag{12-4-18}$$

$$- G_{c2}(\text{dB}) \quad \text{dBW/m}^2 \tag{12-4-19}$$

$$E = 19.8 \text{ dB} - 20 \log_{10}(\lambda) + V(\text{dBV}) - G_{a2}(\text{dB})$$

$$- G_{c2}(\text{dB}) \quad \text{dBV/m}$$

To simplify the computation, the manufacturer often specifies the antenna factor for a specific antenna at a given wavelength. From Eq. (12-4-19) the antenna factor is defined as

$$\text{AF (antenna factor)} = 19.8 \text{ dB} - 20 \log_{10}(\lambda) - G_{a2}(\text{dB}) - G_{c2}(\text{dB}) \text{ dB}, \tag{12-4-20}$$

Thus the antenna factor is a quantity that is used to convert the receiver reading in dBV or dBμV to the field intensity value in dBV/m or dBμV/m for 1-m spacing. It should be noted that for a given frequency the antenna factor for wavelengths in centimeters is less than the one for wavelengths in meters by 40 dB.

Example 12-4-1   Field Intensity Computation.

The receiving antenna has a gain of 10 dB, and the coaxial line connecting the antenna to the receiving intensity meter is assumed lossless. The characteristic impedance of the coaxial line and the input impedance of the meter are both 50 Ω. The signal frequency is 3 GHz, and the meter reading is 100 μV. Determine the power density and the field intensity at the receiving antenna aperture.

Solution

1. The intensity meter reading is 40 dBμV or -80 dBV.
2. The power input to the receiver with an input impedance of 50 Ω is computed

from Eq. (12-4-15) as

$$p_r = \frac{V^2}{R} = \frac{(100 \times 10^{-6})^2}{50} = 200 \times 10^{-12} \text{ W}$$

3. The power density at the antenna aperture is determined from Eq. (12-4-11) or (12-4-16) to be

$$p_d = \frac{0.251 V^2}{\lambda^2 g_{a2} g_{c2}} = \frac{(0.251)(100 \times 10^{-6})^2}{(0.1)^2(10)} = 251 \times 10^{-10} \text{ W/m}^2$$

4. The field intensity is computed from Eq. (12-4-7) or (12-4-17) as

$$E = \sqrt{120\pi p_d} = \sqrt{(120\pi)(251 \times 10^{-10})} = 3076.7 \times 10^{-6} \text{ V/m}$$

5. The power density can be expressed in dB by use of Eq. (12-4-18).

$$
\begin{aligned}
p_d &= -6 \text{ dB} - 20 \log_{10}(0.1) - 80(\text{dBV}) - 10(\text{dB}) \\
&= -76 \text{ dBW/m}^2 \\
&= -6 \text{ dB} - 20 \log_{10}(10) - 20(\text{dB}\mu\text{V}) - 10(\text{dB}) \\
&= -56 \text{ dB}\mu\text{W/cm}^2 \\
&= -6 \text{ dB} - 20 \log_{10}(0.1) - 20(\text{dB}\mu\text{V}) - 10(\text{dB}) \\
&= -16 \text{ dB}\mu\text{W/m}^2
\end{aligned}
$$

6. The field intensity can be expressed in decibels by means of Eq. (12-4-19)

$$
\begin{aligned}
E &= 19.8 \text{ dB} - 20 \log_{10}(0.1) - 80(\text{dBV}) - 10(\text{dB}) \\
&= -50.2 \text{ dBV/m} \\
&= 19.8 \text{ dB} - 20 \log_{10}(10) + 40(\text{dB}\mu\text{V}) - 10(\text{dB}) \\
&= 29.8 \text{ dB}\mu\text{V/cm} \\
&= 19.8 \text{ dB} - 20 \log_{10}(0.1) + 40(\text{dB}\mu\text{V}) - 10(\text{dB}) \\
&= 69.8 \text{ dB}\mu\text{V/m}
\end{aligned}
$$

7. The antenna factor is computed from Eq. (12-4-20) as

$$
\begin{aligned}
\text{AF} &= 19.8 \text{ dB} - 20 \log_{10}(10) - 10(\text{dB}) = -10.2 \text{ dB for } \lambda \text{ in centimeters} \\
&= 19.8 \text{ dB} - 20 \log_{10}(0.1) - 10(\text{dB}) = 29.8 \text{ dB for } \lambda \text{ in meters}
\end{aligned}
$$

12-5 MEASUREMENT AND MICROWAVE ANALYSIS

The measured data for an electronic component or system from a HP automatic network analyzer (HP ANA8409B) in the frequency domain exhibit only the reflection of the device in terms of frequency and show no information of the connector itself. The MAMA (Measurement and Microwave Analysis) program is designed to convert the frequency-domain data into time-domain data and to reveal each reflection as it appears at distance. The program can be used to eliminate the reflections produced by connectors, fixtures, and adapters, and leave the device under test (DUT) deembedded from those unwanted reflections. In MAMA program the frequency-domain data can be converted into time-domain data by using a Construction command and the time-domain data can then be displayed at a distance scale.

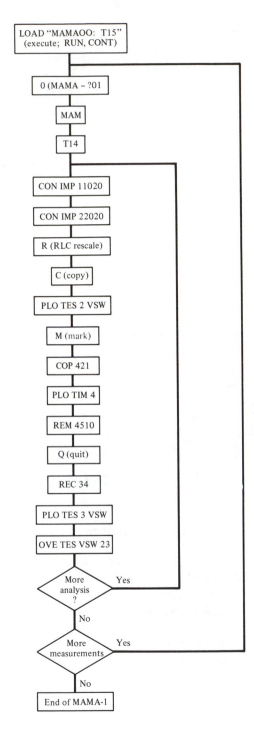

Figure 12-5-1 Flowchart of MAMA program.

However, the data in the time domain are a composite reflection of the entire connecting system. It is necessary to erase the unwanted data by using a Remove command. The resulting time-domain data can then be deconvoluted back into the frequency domain by using the Reconstruction command for display. Before applying the Remove command the Copy command must be used to copy the data to another time area from the existing time area in order to save the data in that time area for future use. Finally, the frequency-domain and time-domain data can be overlaid on the same plot for comparison by using an Overlay command, as shown in the program flow chart in Fig. 12-5-1. The MAMA computer program is supplied by the Made-It Associates, Inc., Burlington, Massachusetts.

12-6 ELECTROMAGNETIC COMPATIBILITY TESTS

Electromagnetic compatibility (EMC) is the science of containing the damaging effects of electromagnetic interference (EMI). When electromagnetic interference comes from within a system or equipment, it is called *intrasystem interference*. When electromagnetic interference develops between two systems, it is called *intersystem interference*. Both intra- and intersystem interferences may exist simutaneously. The central theme of electromagnetic compatibility is to contain, control, and/or eliminate electromagnetic interference. Electronic equipment or systems must operate in conjunction with other equipment without causing malfunction or degradation of operation of any of the equipment or systems. The area of electromagnetic compatibility consists of various subareas, such as the EMV (electromagnetic vulnerability) test and EMS (electromagnetic susceptibility) test. The definitions of various EMC measurements are described below.

EMC. Electromagnetic compatibility is defined as the ability of communications-electronics (C-E) equipment, subsystems, and systems to operate in their intended operational environment without suffering from or causing unacceptable degradation to any other system because of unintentional electromagnetic radiation or response. In other words, the equipment or system should not adversely affect the operation of, or be adversely affected by, any other equipment or system.

EMV. Electromagnetic vulnerability is defined as the operation of equipment or systems uncompromised by electromagnetic environment.

EMS. Electromagnetic susceptibility is defined as the tolerance of a system and/or subsystem to all sources of extraneous signal or unwanted electromagnetic energy.

EMI. Electromagnetic interference is defined as the impairment of the reception of a wanted electromagnetic signal caused by an electromagnetic disturbance.

EME. Electromagnetic environment is defined as the electromagnetic field or fields existing in a transmission medium.

EMP. Electromagnetic pulse is defined as the radiation detonated by nuclear element.

The accuracy of microwave measurements is a major concern of electronics engineers. Theoretically, a transmitter should only transmit electric energy, and a receiver should only receive energy. This situation hardly ever happens in the ordinary laboratory because a piece of electronics equipment can either transmit or receive electric energy. Furthermore, the walls of and the objects in the ordinary laboratory can reflect a fraction of the electric energy that they have received. Therefore, electronic equipment in the ordinary laboratory is constantly subjected to unwanted sources of energy and is constantly producing energy that adjacent equipment does not want. An ideally designed piece of equipment should not radiate any unwanted energy, nor should it be susceptible to any unwanted energy. To achieve this goal, a medium would have to enclose the equipment so that unwanted energy either leaving or attempting to enter the equipment would be effectively attenuated. Shielded rooms and anechoic chambers are the ideal enclosures for microwave measurements.

REFERENCES

1. Liao, Samuel Y., Measurements and computations of microwave electric field intensity and power density. *IEEE Trans. Instrumentation and Measurements*, Vol. IM-25, No. 1, March 1977.

PROBLEMS

12-1 Measuring Units

12-1-1. A transmitting antenna has a power gain of 10 dB and the coaxial line connecting the antenna to the transmitter is assumed lossless. The characteristic impedance of the coaxial line and the input impedance of the transmitter are both 50 Ω. The signal frequency is 2 GHz and the transmitting power is 2 kW. Determine the power density in dBm/m^2 and the electric field intensity in dBμV/m at the receiving antenna aperture for a distance of 1 m.

12-4 Microwave Measurements

12-4-1. A transmitter has an output power of 10 W at a frequency of 3 GHz. The coaxial line connected to the transmitter has a loss of 3 dB and the antenna connected to the other end of the line has a power gain of 5 dB. The test location in the quiet zone of an anechoic chamber is 3 m away from the transmitting antenna.

    (a) Determine the electric field intensity in volts per meter and power density in watts per square meter at the test site.

    (b) Convert the electric field intensity and power density into dBV/m and dBW/m^2.

12-4-2. A power density of 15.50 mW/cm^2 (0.1 W/in.^{2}) is the maximum safe power level that can be applied to the absorber of an anechoic chamber without generating enough heat for combustion. A short-wire antenna with a gain of 2 dB has an effective aperture $3\lambda^2/(8\pi)$ at 1 GHz. The performance of the absorber is 30 dB at 1 GHz. The shortest distance between the antenna and the absorber is 2 m. Determine the maximum power level from the transmitter that will not damage the absorber.

12-4-3. The receiving antenna has a power gain of 15 dB and coaxial line connecting the antenna to the receiving field intensity meter has a power loss of 3 dB. The characteristic impedance of the coaxial line and the input impedance of the meter are both 50 Ω. The signal frequency is 5 GHz and the meter reading is 500 μV.

(a) Convert the intensity meter reading 500 μV to dBμV and dBV.

(b) Calculate the power input to the receiver in watts.

(c) Determine the power density at the antenna aperture in watts per square meter.

(d) Convert the power density from watts per square meter to dBW/m^2, dBμW/m^2, and dBμW/cm^2.

(e) Compute the electric field intensity at the antenna aperture in volts per meter.

(f) Convert the field intensity from volts per meter to dBV/m and dBμV/cm.

(g) Find the antenna factor.

Bibliography

BOOKS

Besser, L., Microwave circuit design. *Electronic Engineering*, October 1980.

Carson, R. S., *High-Frequency Amplifiers*, 2nd ed. New York: John Wiley & Sons, Inc., 1982.

Chou, W. K., *Active Network and Feedback Amplifier Theory*. New York: McGraw-Hill Book Company, 1980.

Davis, W. Alan., *Microwave Semiconductor Circuit Design*. New York: Van Nostrand Reinhold Company, Inc., 1984.

Gonzalez, Guillermo, *Microwave Transistor Amplifiers: Analysis and Design*. Englewood Cliffs, N.J.: Prentice-Hall, Inc., 1984.

Ha, T. T., *Solid-State Microwave Amplifier Design*. New York: John Wiley & Sons, Inc., 1981.

Morgan, D. V., and M. J. Howes, eds., *Microwave Solid-State Devices and Applications*. Stevenage, England: Peter Peregrinus Ltd., 1980.

Vendeline, George D., *Design of Amplifiers and Oscillators by the S-parameter Method*. New York: John Wiley & Sons, Inc., 1982.

White, J., *Microwave Semiconductor Engineering*. New York: Van Nostrand Reinhold Company, Inc., 1982.

JOURNALS

Amplifiers

Kinman, Darry, et al., Symmetrical combiner analysis using S-parameters. *IEEE Trans. Microwave Theory and Techniques*, Vol. MTT-30, No. 3, March 1982, pp. 268–277.

Kuno, H. J., and D. L. English, Millimeter-wave IMPATT power amplifier combiner. *IEEE Trans. Microwave Theory and Techniques*, Vol. MTT-24, No. 11, November 1976, pp. 758–767.

Kuno, H. J., and D. L. English, Nonlinear and intermodulation characteristics of millimeter-wave IMPATT amplifiers. *IEEE Trans. on Microwave Theory and Techniques*, Vol. MTT-24, No. 11, November 1976, pp. 744–751.

Kuno, H. J., and D. L. English, Nonlinear and large-signal characteristics of millimeter-wave IMPATT amplifiers. *IEEE Trans. Microwave Theory and Techniques*, Vol. MTT-21, No. 11, November 1973, pp. 703–706.

Willing, Harry A., A two-stage IMPATT-diode amplifier. *IEEE Trans. Microwave Theory and Techniques*, Vol. MTT-21, No. 11, November 1973, pp. 694–702.

Oscillators

Basawapatna, G. R., and R. B. Stancliff, A unified approach to the design of wideband microwave solid-state oscillators. *IEEE Trans. Microwave Theory and Techniques*, Vol. MTT-27, No. 5, May 1979, pp. 379–385.

Chao, Chente, et al., Y-band (170-260 GHz) tunable CW IMPATT diode oscillators. *IEEE Trans. Microwave Theory and Techniques*, Vol. MTT-25, No. 12, December 1977, pp. 985–991.

Johnson, Kenneth M., Large signal GaAs MESFET oscillator design. *IEEE Trans. Microwave Theory and Techniques*, Vol. MTT-27, No. 3, March 1979, pp. 217–227.

Maeda, Minoru, et al., Design and performance of X-band oscillators with GaAs Schottky-gate field-effect transistors. *IEEE Trans. Microwave Theory and Techniques*, Vol. MTT-23, No. 8, August 1975, pp. 661–667.

Midford, T. A., and R. L. Bernick, Millimeter-wave CW IMPATT diodes and oscillators. *IEEE Trans. Microwave Theory and Techniques*, Vol. MTT-27, No. 5, May 1979, pp. 483–492.

Mitsui, Yasuo, et al., Design of GaAs MESFET oscillator using large-signal S-parameters. *IEEE Trans. Microwave Theory and Techniques*, Vol. MTT-25, No. 12, December 1977, pp. 981–984.

Pergal, Frank, Detail a Colpitts as a tuned one port. *Microwaves*, April 1979, pp. 110–115.

Sun, C., et al., A tunable high-power V-band Gunn oscillator. *IEEE Trans. Microwave Theory and Techniques*, Vol. MTT-27, No. 5, May 1979, pp. 512–514.

Talwar, Ashok K., A dual-diode 73-GHz Gunn oscillator. *IEEE Trans. Microwave Theory and Techniques*, Vol. MTT-27, No. 5, May 1979, pp. 510–512.

Trew, Robert J., Design theory for broadband YIG-tuned FET oscillators. *IEEE Trans. Microwave Theory and Techniques*, Vol. MTT-27, No. 1, January 1979, pp. 8–14.

Wade, Paul C., Say hello to power FET oscillators. *Microwaves*, April 1979, pp. 104–109.

Appendix A: Constants of Materials

CONDUCTIVITY σ (MHOS/M)

Conductor	σ	Insulator	σ
Silver	6.17×10^7	Quartz	10^{-17}
Copper	5.80×10^7	Polystyrene	10^{-16}
Gold	4.10×10^7	Rubber (hard)	10^{-15}
Aluminum	3.82×10^7	Mica	10^{-14}
Tungsten	1.82×10^7	Porcelain	10^{-13}
Zinc	1.67×10^7	Diamond	10^{-13}
Brass	1.50×10^7	Glass	10^{-12}
Nickel	1.45×10^7	Bakelite	10^{-9}
Iron	1.03×10^7	Marble	10^{-8}
Bronze	1.00×10^7	Soil (sandy)	10^{-5}
Solder	0.70×10^7	Sands (dry)	2×10^{-4}
Steel (stainless)	0.11×10^7	Clay	10^{-4}
Nichrome	0.10×10^7	Ground (dry)	$10^{-4} - 10^{-5}$
Graphite	7.00×10^4	Ground (wet)	$10^{-2} - 10^{-3}$
Silicon	1.20×10^3	Water (distilled)	2×10^{-4}
Water (sea)	$3-5$	Water (fresh)	10^{-3}
		Ferrite (typical)	10^{-2}

DIELECTRIC CONSTANT: RELATIVE PERMITTIVITY ϵ_r

Material	ϵ_r	Material	ϵ_r
Air	1	Porcelain (dry process)	6
Alcohol (ethyl)	25	Quartz (fused)	3.80
Bakelite	4.8	Rubber	2.5-4
Glass	4-7	Sands (dry)	4
Ground (dry)	2-5	Silica (fused)	3.8
Ground (wet)	5-30	Snow	3.3
Ice	4.2	Sodium chloride	5.9
Mica (ruby)	5.4	Soil (dry)	2.8
Nylon	4	Styrofoam	1.03
Paper	2-4	Teflon	2.1
Plexiglass	2.6-3.5	Water (distilled)	80
Polyethylene	2.25	Water (fresh)	80
Polystyrene	2.55	Water (sea)	20
Polystyrene	6	Wood (dry)	1.5-4

RELATIVE PERMEABILITY μ_r

Material	μ_r	Material	μ_r
Diamagnetic		Ferromagnetic	
Bismuth	0.99999860	Nickel	50
Paraffin	0.99999942	Cast iron	60
Wood	0.99999950	Cobalt	60
Silver	0.99999981	Machine steel	300
		Ferrite (typical)	1,000
Paramagnetic		Transformer iron	3,000
Aluminum	1.00000065	Silicon iron	4,000
Beryllinum	1.00000079	Iron (pure)	4,000
Nickel chloride	1.00004	Mumetal	20,000
Manganese sulfate	1.0001	Supermalloy	100,000

PROPERTIES OF FREE SPACE

Velocity of light in vacuum c	2.997925×10^8	meters per second
Permittivity ϵ_o	8.854×10^{-12}	farad per meter
Permeability μ_o	$4\pi \times 10^{-7}$	henry per meter
Intrinsic impedance η_o	377 or 120π	ohms

PHYSICAL CONSTANTS

Charge of electron e	1.60×10^{-19}	coulomb
Mass of electron m	9.1×10^{-31}	kilogram
Charge to mass ratio of electron e/m	1.76×10^{11}	coulombs per kilogram

Appendix B:
Characteristic Impedances
for Coupled Microstrip

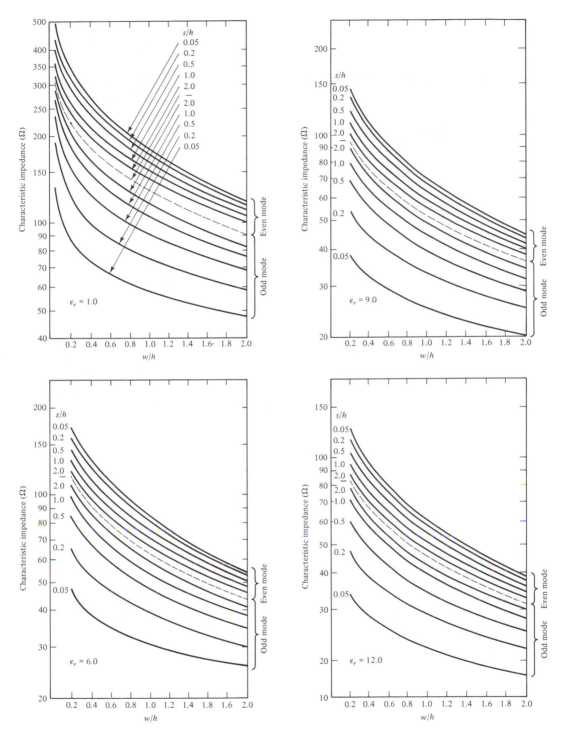

After Bryant and Weiss, in T. Saad, *Microwave Engineering Handbook*, Vol. 1, 1971, p. 40.
Reprinted with permission of Artech House, Inc.

Appendix C: Parameter Conversion Table

CONVERSIONS BETWEEN TWO-PORT PARAMETERS NORMALIZED TO $Z_0=1$ WITH $\Delta^k = K_{11}K_{22} - K_{12}K_{21}$

	S	Z	Y	H	A
S	$\begin{bmatrix} b_1 \\ b_2 \end{bmatrix} = \begin{bmatrix} S_{11} & S_{12} \\ S_{21} & S_{22} \end{bmatrix} \begin{bmatrix} a_1 \\ a_2 \end{bmatrix}$	$S_{11} = \dfrac{(Z_{11}-1)(Z_{22}+1)-Z_{12}Z_{21}}{(Z_{11}+1)(Z_{22}+1)-Z_{12}Z_{21}}$ $S_{12} = \dfrac{2Z_{12}}{(Z_{11}+1)(Z_{22}+1)-Z_{12}Z_{21}}$ $S_{21} = \dfrac{2Z_{21}}{(Z_{11}+1)(Z_{22}+1)-Z_{12}Z_{21}}$ $S_{22} = \dfrac{(Z_{11}+1)(Z_{22}-1)-Z_{12}Z_{21}}{(Z_{11}+1)(Z_{22}+1)-Z_{12}Z_{21}}$	$S_{11} = \dfrac{(1-Y_{11})(1+Y_{22})+Y_{12}Y_{21}}{(1+Y_{11})(1+Y_{22})-Y_{12}Y_{21}}$ $S_{12} = \dfrac{-2Y_{12}}{(1+Y_{11})(1+Y_{22})-Y_{12}Y_{21}}$ $S_{21} = \dfrac{-2Y_{21}}{(1+Y_{11})(1+Y_{22})-Y_{12}Y_{21}}$ $S_{22} = \dfrac{(1+Y_{11})(1-Y_{22})+Y_{12}Y_{21}}{(1+Y_{11})(1+Y_{22})-Y_{12}Y_{21}}$	$S_{11} = \dfrac{(h_{11}-1)(h_{22}+1)-h_{12}h_{21}}{(h_{11}+1)(h_{22}+1)-h_{12}h_{21}}$ $S_{12} = \dfrac{2h_{12}}{(h_{11}+1)(h_{22}+1)-h_{12}h_{21}}$ $S_{21} = \dfrac{-2h_{21}}{(h_{11}+1)(h_{22}+1)-h_{12}h_{21}}$ $S_{22} = \dfrac{(1+h_{11})(1-h_{22})+h_{12}h_{21}}{(h_{11}+1)(h_{22}+1)-h_{12}h_{21}}$	$\dfrac{A+B-C-D}{A+B+C+D}\quad\dfrac{2(AD-BC)}{A+B+C+D}$ $\dfrac{2}{A+B+C+D}\quad\dfrac{-A+B-C+D}{A+B+C+D}$
Z	$Z_{11} = \dfrac{(1+S_{11})(1-S_{22})+S_{12}S_{21}}{(1-S_{11})(1-S_{22})-S_{12}S_{21}}$ $Z_{12} = \dfrac{2S_{12}}{(1-S_{11})(1-S_{22})-S_{12}S_{21}}$ $Z_{21} = \dfrac{2S_{21}}{(1-S_{11})(1-S_{22})-S_{12}S_{21}}$ $Z_{22} = \dfrac{(1-S_{11})(1+S_{22})+S_{12}S_{21}}{(1-S_{11})(1-S_{22})-S_{12}S_{21}}$	$\begin{bmatrix} V_1 \\ V_2 \end{bmatrix} = \begin{bmatrix} Z_{11} & Z_{12} \\ Z_{21} & Z_{22} \end{bmatrix} \begin{bmatrix} I_1 \\ I_2 \end{bmatrix}$ $\dfrac{Z_{12}}{\Delta^Z}$ $\dfrac{-Z_{21}}{\Delta^Z}\qquad\dfrac{Z_{11}}{\Delta^Z}$	$\dfrac{Y_{22}}{\Delta^Y}\qquad\dfrac{-Y_{12}}{\Delta^Y}$ $\dfrac{-Y_{21}}{\Delta^Y}\qquad\dfrac{Y_{11}}{\Delta^Y}$	$\dfrac{\Delta^h}{h_{22}}\qquad\dfrac{h_{12}}{h_{22}}$ $\dfrac{-h_{21}}{h_{22}}\qquad\dfrac{1}{h_{22}}$	$\dfrac{A}{C}\qquad\dfrac{\Delta^A}{C}$ $\dfrac{1}{C}\qquad\dfrac{D}{C}$
Y	$Y_{11} = \dfrac{(1-S_{11})(1+S_{22})+S_{12}S_{21}}{(1+S_{11})(1+S_{22})-S_{12}S_{21}}$ $Y_{12} = \dfrac{-2S_{12}}{(1+S_{11})(1+S_{22})-S_{12}S_{21}}$ $Y_{21} = \dfrac{-2S_{21}}{(1+S_{11})(1+S_{22})-S_{12}S_{21}}$ $Y_{22} = \dfrac{(1+S_{11})(1-S_{22})+S_{12}S_{21}}{(1+S_{11})(1+S_{22})-S_{12}S_{21}}$	$\dfrac{Z_{22}}{\Delta^Z}\qquad\dfrac{-Z_{12}}{\Delta^Z}$ $\dfrac{-Z_{21}}{\Delta^Z}\qquad\dfrac{Z_{11}}{\Delta^Z}$	$\begin{bmatrix} I_1 \\ I_2 \end{bmatrix} = \begin{bmatrix} Y_{11} & Y_{12} \\ Y_{21} & Y_{22} \end{bmatrix} \begin{bmatrix} V_1 \\ V_2 \end{bmatrix}$	$\dfrac{1}{h_{11}}\qquad\dfrac{-h_{12}}{h_{11}}$ $\dfrac{h_{21}}{h_{11}}\qquad\dfrac{\Delta^h}{h_{11}}$	$\dfrac{D}{B}\qquad\dfrac{-\Delta^A}{B}$ $\dfrac{-1}{B}\qquad\dfrac{A}{B}$

CONVERSIONS BETWEEN TWO-PORT PARAMETERS NORMALIZED TO $Z_0 = 1$ WITH $\Delta^k = K_{11}K_{22} - K_{12}K_{21}$

	S	Z	Y	H	A
H	$h_{11} = \dfrac{(1+S_{11})(1+S_{22})-S_{12}S_{21}}{(1-S_{11})(1+S_{22})+S_{12}S_{21}}$ $h_{12} = \dfrac{2S_{12}}{(1-S_{11})(1+S_{22})+S_{12}S_{21}}$ $h_{21} = \dfrac{-2S_{21}}{(1-S_{11})(1+S_{22})+S_{12}S_{21}}$ $h_{22} = \dfrac{(1-S_{22})(1-S_{11})-S_{12}S_{21}}{(1-S_{11})(1+S_{22})+S_{12}S_{21}}$	$\dfrac{\Delta^z}{Z_{22}} \quad \dfrac{Z_{12}}{Z_{22}}$ $\dfrac{-Z_{21}}{Z_{22}} \quad \dfrac{1}{Z_{22}}$	$\dfrac{1}{Y_{11}} \quad \dfrac{-Y_{12}}{Y_{11}}$ $\dfrac{Y_{21}}{Y_{11}} \quad \dfrac{\Delta^y}{Y_{11}}$	$\begin{bmatrix} V_1 \\ I_2 \end{bmatrix} = \begin{bmatrix} h_{11} & h_{12} \\ h_{21} & h_{22} \end{bmatrix}\begin{bmatrix} I_1 \\ V_2 \end{bmatrix}$	$\dfrac{B}{D} \quad \dfrac{\Delta^A}{D}$ $\dfrac{-1}{D} \quad \dfrac{C}{D}$
A	$A = \dfrac{(1+S_{11})(1-S_{22})+S_{12}S_{21}}{2S_{21}}$ $B = \dfrac{(1+S_{11})(1+S_{22})-S_{12}S_{21}}{2S_{21}}$ $C = \dfrac{(1-S_{11})(1-S_{22})-S_{12}S_{21}}{2S_{21}}$ $D = \dfrac{(1-S_{11})(1+S_{22})+S_{12}S_{21}}{2S_{21}}$	$\dfrac{Z_{11}}{Z_{21}} \quad \dfrac{\Delta^z}{Z_{21}}$ $\dfrac{1}{Z_{21}} \quad \dfrac{Z_{22}}{Z_{21}}$	$\dfrac{-Y_{22}}{Y_{21}} \quad \dfrac{-1}{Y_{21}}$ $\dfrac{-\Delta^y}{Y_{21}} \quad \dfrac{-Y_{11}}{Y_{21}}$	$\dfrac{-\Delta^k}{h_{21}} \quad \dfrac{-h_{11}}{h_{21}}$ $\dfrac{-h_{22}}{h_{21}} \quad \dfrac{-1}{h_{21}}$	$\begin{bmatrix} V_1 \\ I_1 \end{bmatrix} = \begin{bmatrix} A & B \\ C & D \end{bmatrix}\begin{bmatrix} V_2 \\ -I_2 \end{bmatrix}$

Appendix D:
Elliptic Integral Tables

$$K$$

k^2		0	1	2	3	4	5	6	7	8	9	d	k'^2
0.0	1.5	708	747	787	828	869	910	952	994	*037	*080	41	0.9
0.1	1.6	124	169	214	260	306	353	400	448	497	546	47	0.8
0.2		596	647	698	751	804	857	912	967	*024	*081	53	0.7
0.3	1.7	139	198	258	319	381	444	*139	573	·639	706	63	0.6
0.4		775	845	916	989	*063	*139	*216	*295	*375	*457	76	0.5
0.5	1.8	541	626	714	804	895	989	*085	*184	*285	*398	94	0.4
0.6	1.9	496	605	718	834	953	*0076	*0203	*0334	*0469	*0609	123	0.3
0.7	2.	0754	0904	1059	1221	1390	1565	1748	1940	2140	2351	175	0.2
0.8		2572	2805	3052	3314	3593	3890	4209	4553	4926	5333	397	0.1
0.9		5781	6278	6836	7471	8208	9083	*0161	*1559	*3541	*6956	875	0.0
k'^2		10	9	8	7	6	5	4	3	2	1	d	k^2

$$K'$$
$$h = K - \ell\mathrm{n}\left(\frac{4}{k'}\right)$$

k^2		0	1	2	3	4	5	6	7	8	9	d	k'^2
0.7	0.0	871	851	832	812	791	771	750	728	707	684	20	0.2
0.8		662	639	615	591	567	542	516	489	462	434	24	0.1
0.9		405	375	344	321	278	242	204	163	118	068	36	0.0
k'^2		10	9	8	7	6	5	4	3	2	1	d	k^2

$$h' = K' - \ell\mathrm{n}\left(\frac{4}{k}\right)$$

Appendix E: Commercial LASER and LED Sources

Maker	Device	Drive current (A)	Emission wavelength λ_0 (μm)	Spectral width (nm)	Power (mW)
		Lasers			
AEG	CQX-20	0.40	0.82	2.5	5.0
ITT	T901-L	0.35	0.84	2.0	7.5
Laser D Labs	LCW10	0.20	0.85	2.0	14.0
RCA	C30130	0.25	0.82	2.0	6.0
		LEDs			
ASEA	1A83	0.10	0.94	40.0	10.0
ITT	T851-S	0.20	0.84	38.0	1.5
Laser D. Labs	IRE150	0.10	0.82	35.0	1.5
MERET	TL-36C	0.30	0.91	38.0	12.0
Monsanto	ME60	0.05	0.90	40.0	0.4
Philips	CQY58	0.05	0.88	45.0	0.5
RCA	SG1009	0.10	0.94	45.0	3.5
TI	TIXL472	0.05	0.91	23.0	1.0

Appendix F:
Television-Channel
Frequencies

Channel number	Frequency range (MHz)	Video carrier (MHz)	Audio carrier (MHz)
VHF			
2	54–60	55.25	59.75
3	60–66	61.25	65.75
4	66–72	67.25	71.75
5	76–82	77.25	81.75
6	82–88	83.25	87.75
7	174–180	175.25	179.75
8	180–186	181.25	185.75
9	186–192	187.25	191.75
10	192–198	193.25	197.75
11	198–204	199.25	203.75
12	204–210	205.25	209.75
13	210–216	211.25	215.75
UHF			
14	470–476	471.25	475.75
15	476–482	477.25	481.75
16	482–488	483.25	487.75
17	488–494	489.25	493.75
18	494–500	495.25	499.75
19	500–506	501.25	505.75
20	506–512	507.25	511.75
21	512–518	513.25	517.75
22	518–524	519.25	523.75
23	524–530	525.25	529.75
24	530–536	531.25	535.75
25	536–542	537.25	541.75
26	542–548	543.25	547.75
27	548–554	549.25	553.75
28	554–560	555.25	559.75
29	560–566	561.25	565.75
30	566–572	567.25	571.75
31	572–578	573.25	577.75
32	578–584	579.25	583.75
33	584–590	585.25	589.75
34	590–596	591.25	595.75

Channel number	Frequency range (MHz)	Video carrier (MHz)	Audio carrier (MHz)
35	596–602	597.25	601.75
36	602–608	603.25	607.75
37	608–614	609.25	613.75
38	614–620	615.25	619.75
39	620–626	621.25	625.75
40	626–632	627.25	631.75
41	632–638	633.25	637.75
42	638–644	639.25	643.75
43	644–650	645.25	649.75
44	650–656	651.25	655.75
45	656–662	657.25	661.75
46	662–668	663.25	667.75
47	668–674	669.25	673.75
48	674–680	675.25	679.75
49	680–686	681.25	685.75
50	686–692	687.25	691.75
51	692–698	693.25	697.75
52	698–704	699.25	703.75
53	704–710	705.25	709.75
54	710–716	711.25	715.75
55	716–722	717.25	721.75
56	722–728	723.25	727.75
57	728–734	729.25	733.75
58	734–740	735.25	739.75
59	740–746	741.25	745.75
60	746–752	747.25	751.75
61	752–758	753.25	757.75
62	758–764	759.25	763.75
63	764–770	765.25	769.75
64	770–776	771.25	775.75
65	776–782	777.25	781.75
66	782–788	783.25	787.75
67	788–794	789.25	793.75
68	794–800	795.25	799.75
69	800–806	801.25	805.75
70	806–812	807.25	811.75
71	812–818	813.25	817.75
72	818–824	819.25	823.75
73	824–830	825.25	829.75
74	830–836	831.25	835.75
75	836–842	837.25	841.75
76	842–848	843.25	847.75
77	848–854	849.25	853.75
78	854–860	855.25	859.75
79	860–866	861.25	865.75
80	866–872	867.25	871.75
81	872–878	873.25	877.75
82	878–884	879.25	883.75
83	884–890	885.25	889.75

Notes: Each channel has a 6-MHz bandwidth. Video carrier has the lower frequency plus 1.25 MHz. Audio carrier has the upper frequency minus 0.25 MHz.

Appendix G: First-Order Bessel Function Values

x	$J_1(x)$	x	$J_1(x)$	x	$J_1(x)$	x	$J_1(x)$	x	$J_1(x)$
0.00	0.000	0.92	0.413	1.86	0.582	2.86	0.389	3.84	−0.003
0.02	+0.010	0.94	0.420	1.88	0.5815	2.88	0.3825	3.86	0.011
0.04	0.020	0.96	0.427	1.90	0.581	2.90	0.375	3.88	0.019
0.06	0.030	0.98	0.4335	1.92	0.5805	2.92	0.368	3.90	0.027
0.08	0.040	1.00	0.440	1.94	0.580	2.94	0.361	3.92	0.035
0.10	0.050	1.02	0.4465	1.96	0.579	2.96	0.354	3.94	0.043
0.12	0.060	1.04	0.453	1.98	0.578	2.98	0.3465	3.96	0.051
0.14	0.070	1.06	0.459	2.00	0.577	3.00	0.339	3.98	0.058
0.16	0.080	1.08	0.465	2.02	0.575	3.02	0.3315	4.00	0.066
0.18	0.090	1.10	0.471	2.04	0.574	3.04	0.324	4.10	0.103
0.20	0.0995	1.12	0.477	2.06	0.572	3.06	0.316	4.20	0.139
0.22	0.109	1.14	0.482	2.08	0.570	3.08	0.309	4.30	0.172
0.24	0.119	1.16	0.488	2.10	0.568	3.10	0.301	4.40	0.203
0.26	0.129	1.18	0.493	2.12	0.566	3.12	0.293	4.50	0.231
0.28	0.139	1.20	0.498	2.14	0.564	3.14	0.285	4.60	0.2565
0.30	0.148	1.22	0.503	2.16	0.561	3.16	0.277	4.70	0.279
0.32	0.158	1.24	0.508	2.18	0.559	3.18	0.269	4.80	0.2985
0.34	0.1675	1.26	0.513	2.20	0.556	3.20	0.261	4.90	0.315
0.36	0.177	1.28	0.5175	2.22	0.553	3.22	0.253	5.00	0.3275
0.38	0.187	1.30	0.522	2.24	0.550	3.24	0.245	5.05	0.334
0.40	0.196	1.32	0.526	2.26	0.547	3.26	0.237	5.10	0.337
0.42	0.205	1.34	0.5305	2.28	0.543	3.28	0.229	5.16	0.341
0.44	0.215	1.36	0.534	2.30	0.540	3.30	0.221	5.20	0.343
0.46	0.224	1.38	0.538	2.32	0.536	3.32	0.212	5.26	0.345
0.48	0.233	1.40	0.542	2.34	0.532	3.34	0.204	5.30	0.346
0.50	0.242	1.42	0.5455	2.36	0.5285	3.36	0.196	5.32	0.346
0.52	0.251	1.44	0.549	2.38	0.524	3.38	0.1865	5.34	0.346
0.54	0.260	1.46	0.552	2.40	0.520	3.40	0.179	5.36	0.346
0.56	0.269	1.48	0.555	2.42	0.516	3.42	0.171	5.38	0.346
0.58	0.278	1.50	0.558	2.44	0.511	3.44	0.1625	5.40	0.345
0.60	0.287	1.52	0.561	2.46	0.507	3.46	0.154	5.47	0.343
0.62	0.295	1.54	0.563	2.48	0.502	3.48	0.146	5.50	0.341
0.64	0.304	1.56	0.566	2.50	0.497	3.50	0.137	5.56	0.3375
0.66	0.312	1.58	0.568	2.52	0.492	3.52	0.129	5.60	0.334
0.68	0.321	1.60	0.570	2.54	0.487	3.54	0.121	5.66	0.3285
0.70	0.329	1.62	0.572	2.56	0.482	3.56	0.112	5.70	0.324
0.72	0.337	1.64	0.5735	2.58	0.476	3.58	0.104	5.80	0.311
0.74	0.345	1.66	0.575	2.60	0.471	3.60	0.0955	5.90	0.295
0.76	0.353	1.68	0.5765	2.62	0.465	3.62	0.087	6.00	0.277

APPENDIX (CONTINUED)

x	$J_1(x)$	x	$J_1(x)$	x	$J_1(x)$	x	$J_1(x)$	x	$J_1(x)$
0.78	0.361	1.70	0.578	2.64	0.4595	3.64	0.079	6.10	0.256
0.80	0.369	1.72	0.579	2.66	0.454	3.66	0.070	6.20	0.233
0.82	0.3765	1.74	0.580	2.68	0.448	3.68	0.062	6.30	0.208
0.84	0.384	1.76	0.5805	2.70	0.442	3.70	0.054	6.40	0.182
0.86	0.3915	1.78	0.581	2.72	0.435	3.72	0.0455	6.60	0.125
0.88	0.399	1.80	0.5815	2.74	0.429	3.74	0.037	6.70	0.095
0.90	0.406	1.82	0.582	2.76	0.423	3.76	0.029	6.80	0.065
		1.84	0.582	2.78	0.416	3.78	0.021	6.90	0.035
				2.80	0.410	3.80	0.013	7.00	0.005
				2.82	0.403	3.82	0.005	7.01	0.000
				2.84	0.396	3.83	0.000		

Appendix H: Characteristics of Transmission Lines at Radio Frequency

Class of cable	Type	Inner conductor	Nominal overall diameter (in.)	Nominal capacitance (pF/ft)	Maximum operating voltage (V rms)
50 Ω	RG-8A/U	7/0.0296	0.405	29.5	4000
	RG-55A/U	0.035	0.216	28.5	1900
	RG-58C/U	19/0.0071	0.195	28.5	1900
	RG-122/U	27/0.005	0.160	29.3	1900
	RG-174/U	7/0.0063	0.100	30.0	1000
	RG-196A/U	7/0.0040	0.080	29.0	1000
High attenuation	RG-126/U	7/0.0203	0.275	29.0	3000
	RG-21A/U	0.053	0.332	29.0	2700
High delay	RG-65A/U	0.008	0.405	44.0	1000
Twin conductor	RG-86/U	7/0.0285	0.30 × 0.65	7.8	—
	RG-130/U	7/0.0285	0.625	17.0	8000

Appendix I:

Hankel Functions

The Hankel functions of the first and second kinds are the combinations of the Bessel functions as

$$H_v^{(1)}(x) = J_v(x) + jN_v(x) \quad \text{(incoming wave)} \tag{MBF-1}$$

$$H_v^{(2)}(x) = J_v(x) - jN_v(x) \quad \text{(outgoing wave)} \tag{MBF-2}$$

For large argument, they become

$$H_v^{(1)}(x) \xrightarrow[x \to \infty]{} \sqrt{\frac{2}{j\pi x}} \, j^{-v} e^{jx} \tag{MBF-3}$$

$$H_v^{(2)}(x) \xrightarrow[x \to \infty]{} \sqrt{\frac{2j}{\pi x}} \, j^v e^{-jx} \tag{MBF-4}$$

When $x = ju$ is imaginary, the modified Bessel functions are

$$I_v(u) = j^v J_v(-ju) \tag{MBF-5}$$

$$K_v(u) = \frac{\pi}{2}(-j)^{v+1} H_v^{(2)}(-ju) \tag{MBF-6}$$

For large argument, I_v and K_v become

$$I_v(u) \xrightarrow[u \to \infty]{} \frac{e^u}{\sqrt{2\pi u}} \tag{MBF-7}$$

$$K_v(u) \xrightarrow[u \to \infty]{} \sqrt{\frac{\pi}{2u}} \, e^{-u} \tag{MBF-8}$$

Figure AI-1 shows a few types.

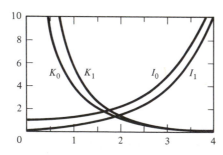

Figure AI-1   Modified Bessel functions.

Index

A band, 2
Absorption loss, 434–435
Admittance:
    of Smith chart, 40
    of transmission line, 25
Amplifier:
    balanced, 166–169
    broadband, 255–266
    feedback, 268–270
    high-gain, 139–45
    high-power, 244–250
    large-signal, 236–239 (*see also*
        Large-signal amplifier)
    low-noise, 145–149, 250–255
    narrowband, 149–155
    reflection, 308–315
    small-signal, 120–136 (*see also*
        Small-signal amplifier)
    stripline-type, 211
Amplifier stability, 96–101
Available power gain, 92–94

B band, 2
Balanced amplifier, 166–169
Bessel function, 292, 477–478
Biasing circuit (*see* dc-biasing
    circuit)
Broadband amplifier, 255–266

C band, 2
Cavity:
    circular, 305–307
    quality factor Q of, 307
    rectangular, 304–305
Characteristic impedance:
    of coplanar stripline, 217–218
    of microstrip line, 200–202
    of parallel stripline, 218–219
    of shielded stripline, 220–222
    of slot stripline, 222–225
    of transmission line, 14
Chip characterization, 169–171
Circular waveguide, 290–302
Circulator, four-port, 313
Coaxial connectors, 34
Coaxial line, 31
Coefficient:
    reflection, 15–17

transmission, 17–19
Compressed Smith chart, 52–53
Constant-gain circles:
    bilateral case, 106
    figure of merit, 104–105
    noise-figure, 115–119
    operating power-gain, 107–114
    unilateral case, 102–104

D band, 2
dc-biasing circuit:
    for GaAs MESFETs, 126–130
    for silicon transistor, 131–135
Decibel (dB), 446–450
Delta factor, 90–91
Double-stub matching, 67–72
Dynamic range, 245–246

E band, 2
Extinction index, 436

F band, 2
Faraday effect, 397–398
Feedback amplifier, 268–270
Frequency bands, 2

G band, 2
Gain compression point, 245–246
Gold-film-coating design, 430–441

H band, 2
H parameters, 78, 471
High-gain amplifier, 139–145

I band, 2
IEEE frequency bands, 3
Impedance matching:
    with double stub, 67–72
    with single stub, 65–67
Index:
    extinction, k, 436
    refractive, n, 436
Index of refraction, 436

J band, 2

K band, 2
Kerr-cell light modulator, 396–397

Ku band, 2

L band, 2
Lange coupler, 161
Large-signal amplifier, 236–239
    broadband, 255–268
    feedback, 268–270
    high-power, 244–250
    low-noise, 250–255
Large-signal measurements,
    237–238
Laser modulators (*see* Light
    modulator)
Light modulator:
    Kerr-cell, 396–397
    Magnetooptic, 397–399
    Pockels-cell, 392–396
    traveling-wave electrooptic,
        399–401
    Low-noise amplifier, 145–149,
        250–255

M band, 2
Magnetooptic modulator, 397–399
MAMA, 459–460
Mason's signal-flow rules, 87–88
Measurement and microwave
    analysis, 459–460
Microstrip line:
    Microstrip line:
        characteristic impedance,
            200–202
        dielectric loss, 203
        ohmic loss, 204
        quality factor, 208–209
        radiation loss, 207–208
        realization, 209–211
Microwave measurement:
    electric field, 451–452
    free-space attenuation, 453
    units, 446–450

Narrowband amplifier, 149–155
Negative-resistance chart, 52–53
Noise-figure circles, 115–119
Normalized impedance, 24
Numerical aperture, 377–379